Flags of the Holy Roman Empire 1545

Wappen des Heyligen Römischen Reichs Teutscher Nation

Cyriacus Jacob
Jakob Kallenberg, Engraver

Introduction and Arranged by
Marion McNealy

Published by Nadel und Faden Press LLC www.nadelundfadenpress.com

Design, Layout: Tina Vadász-Hain, www.expresso-grafik.de

Publisher's Cataloging-in-Publication Data
provided by Five Rainbows Cataloging Services

Names: McNealy, Marion, author. | Kallenberg, Jakob, active 1535-1565, engraver.
Title: Flags of the Holy Roman Empire 1545 : wappen des Heyligen Römischen Reichs Teutscher nation / Marion McNealy ; Jakob Kallenberg, engraver.
Description: Kennewick, WA : Nadel und Faden Press, 2023.
Identifiers: LCCN 2023910588 (print) | ISBN 978-0-9985977-7-5 (paperback) |
ISBN 979-8-9852835-0-1 (ebook)
Subjects: LCSH: Flags--Germany--History. | Holy Roman Empire--History--1517-1648. Heraldry. | Wood-engraving, Renaissance. | German mercenaries. | BISAC: HISTORY / Europe / Germany. | HISTORY / Europe / Renaissance. | HISTORY / Modern / 16th Century. | REFERENCE / Genealogy & Heraldry.
Classification: LCC JC347.G33 M36 2023 (print) | LCC JC347.G33 (ebook) | DDC 929.92/0943--dc23.

Table of Contents

Acknowledgments

This book is dedicated to the many Landsknecht re-enactors who have inspired me through their creative use of the woodcuts in Landsknecht Woodcuts which I published in 2013. Keep it slashy peeps!

For Anne, who once again just told me "Just finish the book already Marion! It's not doing anyone any good just sitting on your computer." For the members of my Patreon who have been incredibly patient waiting for this to be published.

Deep thanks as always to F, S & HG, for seeing me through this project, and never letting me quit, even when I wanted to give up and cancel it.

Marion McNealy, July 2023

Preface

Earlier this week, Facebook memories reminded me that it had been ten years since I unwrapped the big package that contained the large folio containing the woodcuts which would later become the book Landsknecht Woodcuts. It does not seem like ten years ago, but as I look at the thick stack of books I've published since then, there's a lot of writing and hard work that has happened in the last ten years.

For ten years, Landsknecht Woodcuts has inspired countless people around the world to up their wardrobe game and given tailors and seamstresses fits with trying to re-create various garments. I've sat back and wondered in amazement at the creativity inspired by a simple little book I put out when I knew so little about book making and publishing. I am humbled by every new photo I see of a design based on the pages of that book.

Every one of my books has a story behind it, and this one came about through a desire to inspire the current and future generations of Landsknecht re-enactors and enthusiasts to continue their efforts to create fun, authentically crazy clothing by giving them a handy source for the mid 1540s.

This book was started in mid-2020 and due to various life, family, and world issues has taken a while to come to completion. Or maybe it just was waiting for a ten-year memory to pop up?

I've only got one more thing to say:
Keep being amazing! Go forth and create great stuff!

Much love,
Marion McNealy, July 2023

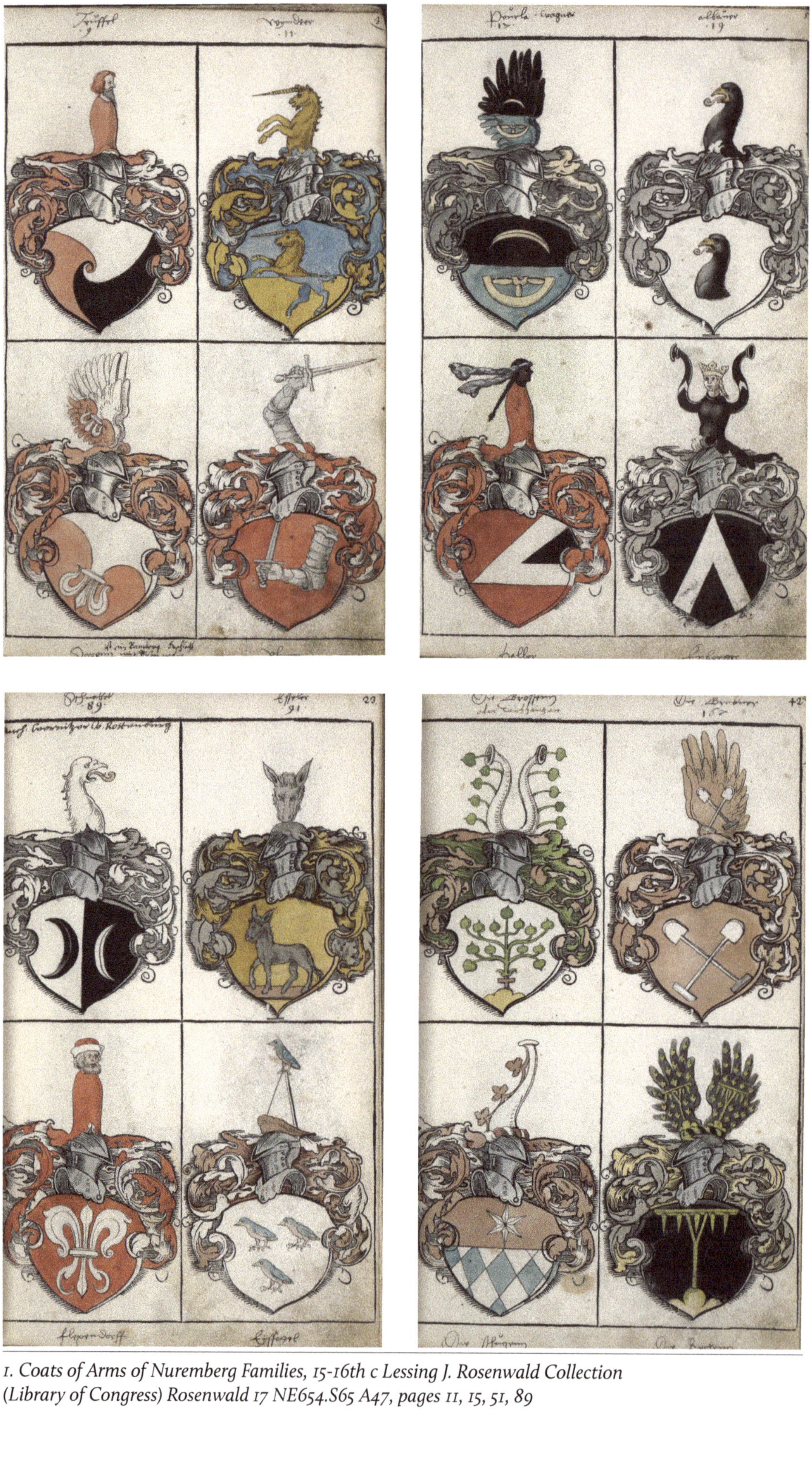

1. Coats of Arms of Nuremberg Families, 15-16th c Lessing J. Rosenwald Collection (Library of Congress) Rosenwald 17 NE654.S65 A47, pages 11, 15, 51, 89

Introduction

In early modern 16th century Germany, heraldry was an important part of everyday life as it used easily identifiable pictures to identify who and what belonged to whom. It was used by towns and families alike to mark property, to identify members of the family and staff, and to serve as a patriotic identity rally point. However, the ordinary person did not always find it easy to keep track of which flag belonged to which town or family. During this era, many heraldry books and resources were created and published to assist in identification (fig.1).

This guide, *Wappen des Heyligen Römischen Reichs Teutscher Nation* (Heraldic Flags of the Holy Roman Empire of the German Nations) was published in Frankfurt by Cyriacus Jacob in 1545 as a handy guide to the flags of the Holy Roman Empire. The manuscript which is reproduced in this book is from the Library of Congress (NE654.S65 A47). The original book is sized 8 in by 11 in (20 cm x 29 cm). It is comprised of three parts: an introduction, 143 woodcuts and a listing of the order of precedence of the members of the Imperial Diet, along with a description of the colors and symbols of their heraldic flags. It provides the modern reader with an amazing wealth of full color woodcuts depicting the flags, the Landsknecht military clothing and the armor of the time. Amazingly, the woodcuts were painted very close to the publication date, resulting in a fantastic reference source for the wide range of colors and styles in male military clothing fashions for the 1540s.

The introductory pages are mostly comprised of a short and choppy excerpt of an earlier book written by Jacob Koebel, which was an account of the founding of the Holy Roman Empire from before the birth of Christ as "collected by older wise men". This account is not reflected in any of the histories elsewhere, and the needs of the modern reader are better met by a more thoroughly researched and recently written account. After this section, a brief explanation is included of the Holy Roman Empire symbols, regalia, procedures and roles at Elector meetings. However, given the issues with the history part, this section is questionable in its accuracy. The introductory text does not fit with the following flags section. It may have been a replacement for an introductory text which was not delivered, or as a last minute addition to add more substance to the book. Therefore the introduction has not been translated.

Another curiosity about the introduction is the lack of a stated purpose for its publication. Usually, the author or publisher would state explicitly in the introductory pages why the book had been written, and for what intended audience. For the modern reader, this would be the same as a descriptive blurb on the book seller's website, back of the book, or on the inside cover flap. This omission leads to an intriguing possibility that this book may have been intended for a wide audience as a type of souvenir or patriotic memento for tourists or travelers to an Imperial Diet in 1545 or 1546.

Due to the rising tensions between Catholic and Protestants in the late 1530s, there had been several Imperial Diet meetings held in the 1540s, usually once a year, but in some years two were held. These meetings brought a large number of delegates into a city, who may have wanted to bring home a souvenir to commemorate their trip, or use it as a guide to the heraldry of the processions and meetings which were part of the Diet.

However, Charles V the Holy Roman Emperor was very busy in 1545. Between securing his Italian holdings after the truce at Crépy in 1544, negotiating a truce with the Ottoman Sultan Suleiman in late 1545, and making arrangements to deal militarily with the Protestant Schmalkaldic League in 1546, he had little time to hold a Diet. Due to the Schmalkaldic War in 1546-1547, there was a long four-year break between Imperial Diets. The next Diet was held in Augsburg in May 1548, amid a very tense atmosphere caused by Charles V's insistence that Protestants return to the Catholic faith. This issue would not be settled for a long time in Germany.

Woodcuts

Following the introduction are 143 woodcuts, each depicting a flag representing the heraldic device of a particular town, noble family, or church leader, and labeled with the name. The intention was to follow the order of precedence of the electoral council of the Holy Roman Empire Imperial Diet, or council, but many plates are out of alignment with the order of precedence published in the book. The publisher Cyriacus Jacob appears to have rearranged the woodcuts into a different order than sent by the artist IK as the plate numbers are out of order in the finished book. The artist mark of IK, visible on the woodcuts, is not unique for the time, as there were several other artists marking their work with the same initials. However, the current consensus among art historians is that the best candidate appears to be Jakob Kallenberg, a Swiss painter and woodcut master from Bern, who specialized in minor paintings, coats of arms, flags, and fire buckets. The widely held opinion is that he most likely produced these woodcuts in Bern, and not in Frankfurt, as, at Easter-time in 1545, he had become a member of the Grand Council of Bern. This may account for some of the glaringly mislabeled plates, as a Swiss citizen working from afar would not have necessarily known the proper hierarchy of the Holy Roman Empire councils.

By comparing the woodcuts to the order of precedence at the end of the book, several mislabeled plates and missing cities become apparent. The most obvious error is the mislabeling of the flag for Bavaria as belonging to Deckendorff, a small town in Bavaria, and its placement in the back of the book, not in the front where it belongs with the other Duchies.

In this new edition, when errors have been discovered either in placement or naming, the woodcut has been moved to the correct place, or a correction in naming has been noted in the caption.

Colors

Although the woodcuts were originally printed in black and white, they were later painted using the color descriptions found in the list at the end of the original manuscript and through other reference sources. While several other painted editions of *Wappen des Heyligen Römischen Reichs Teutscher Nation* are found in libraries around the world—notably at the Bavarian State Libraries (BSB call number VD16 K 1623)—this copy, located at the Library of Congress' (LOC), appears to have been painted close to the date of publication. Comparing the clothing colors of the woodcuts with paintings completed within the same time frame, it becomes clear that the artist who painted the woodcuts in this book knew what the garments would look like in real life.

The clearest example of this is that the ledergollar (leather vests/jackets) worn by many of the flag bearers are the right color for buff leather in the LOC manuscript. Comparing the ledergollar in this manuscript (fig.6, 7) to other artwork (fig.2, 3, 4, 5) created around the same time, they are all depicted as a light or medium brown, which is similar to several extant leather garments and accessories from the 16th centuries.

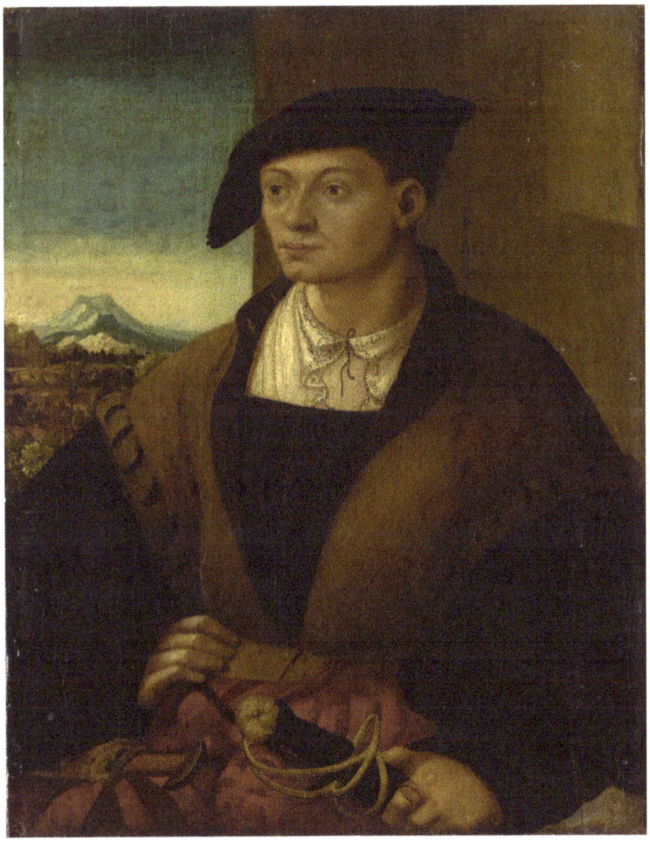

2. Portrait of a young man, about 1520
Swiss Master, 16th c.Kunstmuseum Basel Inv. 556

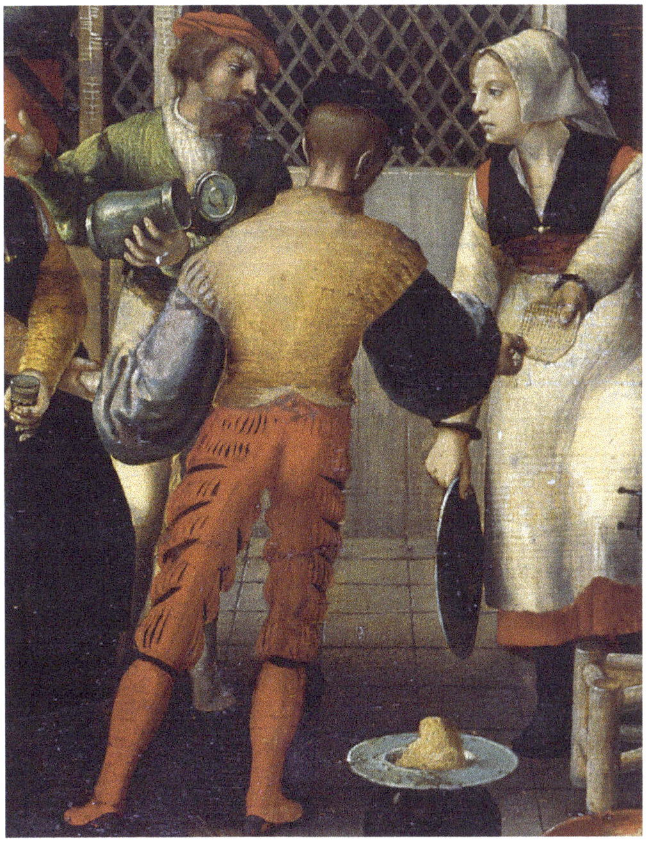

3. Detail from The Brothel Scene, ca. 1540 – 1550
The Brunsiwck Monogrammist Städel Museum, Frankfurt am Main Inventory Number 249

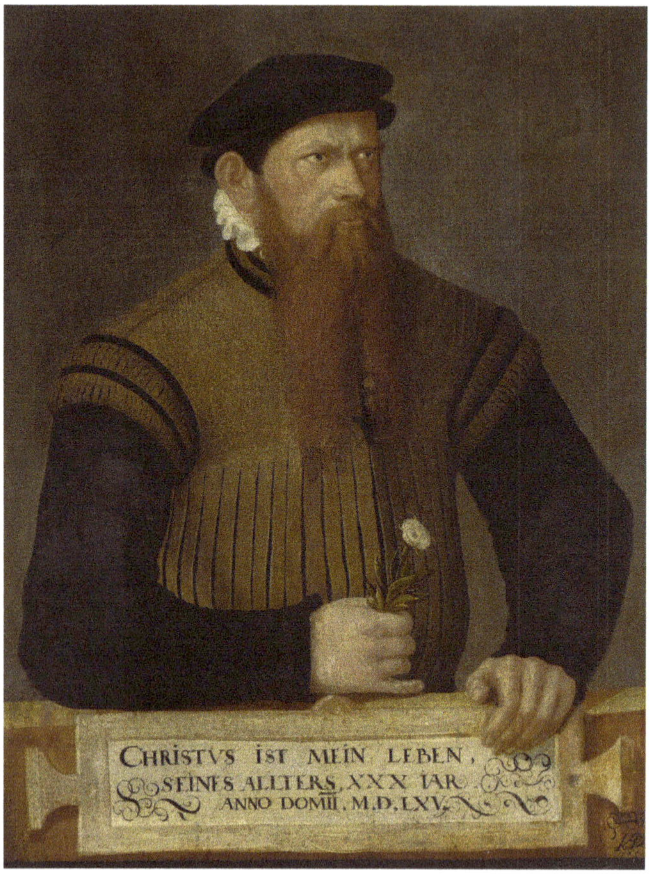

4. Portrait of a Bearded Man, 1565
Jost Amman, Kunstmuseum Basel
Ankauf mit Mitteln des Birmann-Fonds 1906 Inv. 9

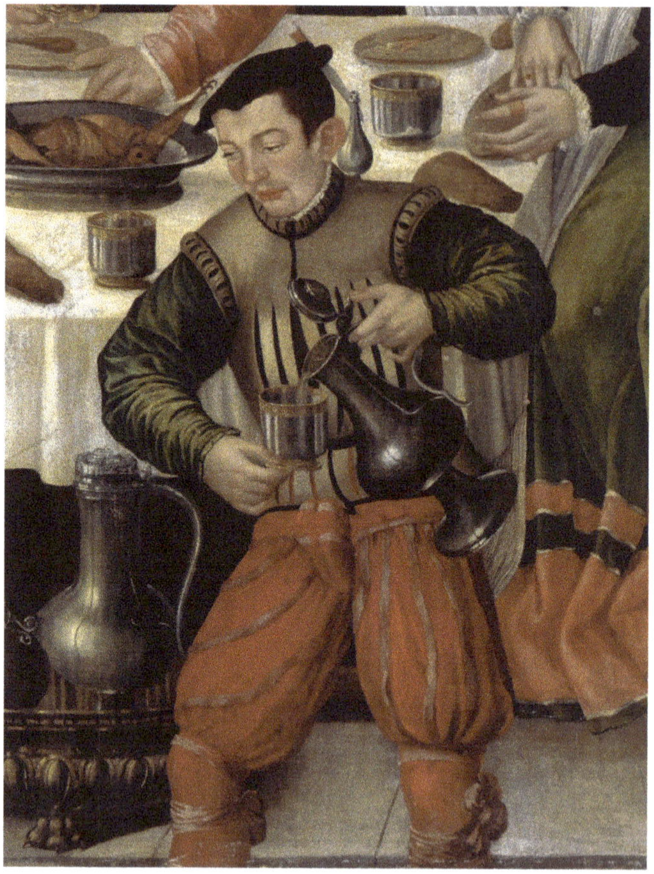

5. Detail of The Portrait of the Family of Hans Rudolf Faesch, 1559 Hans Hug Kluber, painted in Basel
Kunstmuseum Basel, Inv. 1936

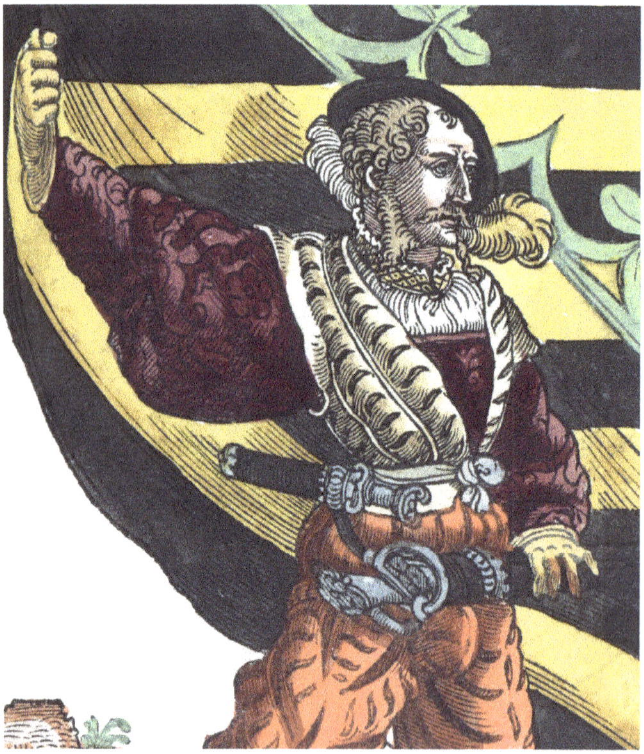

6. Detail of Plate 7. Sachsen, The Duchy of Saxony, depicting a ledergollar, leather vest.

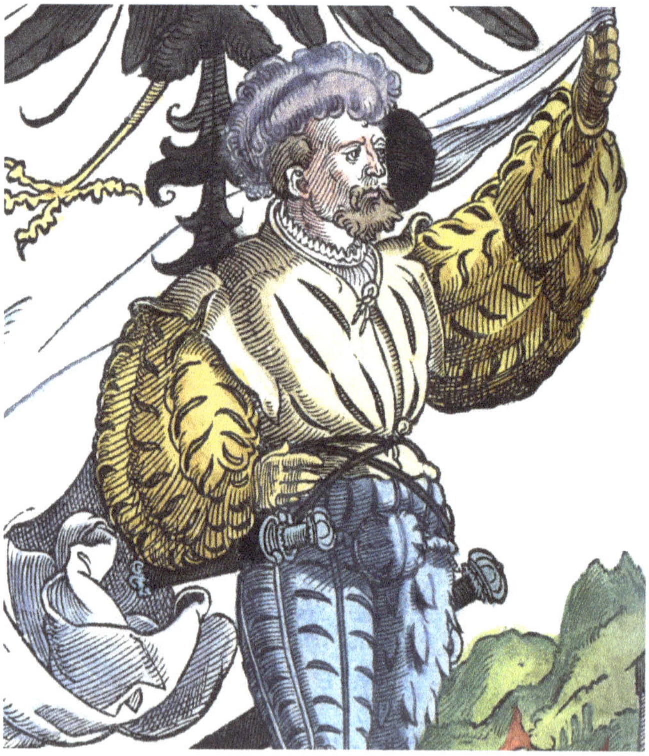

7. Detail of Plate 44. Schlerstat | Schlettstatt depicting a ledergollar, leather vest.

Hosen

The hosen (pants) in the woodcuts are interesting to observe and study as they depict a variety of styles in the transitional period of the 1540s. The 1540s were the beginning of the puffed pludderhosen period, which would culminate in the extreme styles of the 1560s and 1570s (fig.12, 13). The series of woodcuts in this manuscript proves that the older styles of hosen, which were more closely fitted (fig.8, 9), were still being created and worn at the same time as the looser, puffier styles (fig.10, 11). It is interesting to note that this transition and shift in fashionable hosen styles did not happen immediately but rather was a slower process than might be expected by only using portraiture as a reference.

The different rear leg seam positions and codpiece styles may be the result of different regional preferences or social indications, which would have been obvious to the viewer at the time but have since lost their meaning to the modern viewer.

Although the full message of their clothes has been lost, the Landsknecht continue to inspire tailors and makers in their quest for imaginative and playful clothing design and construction.

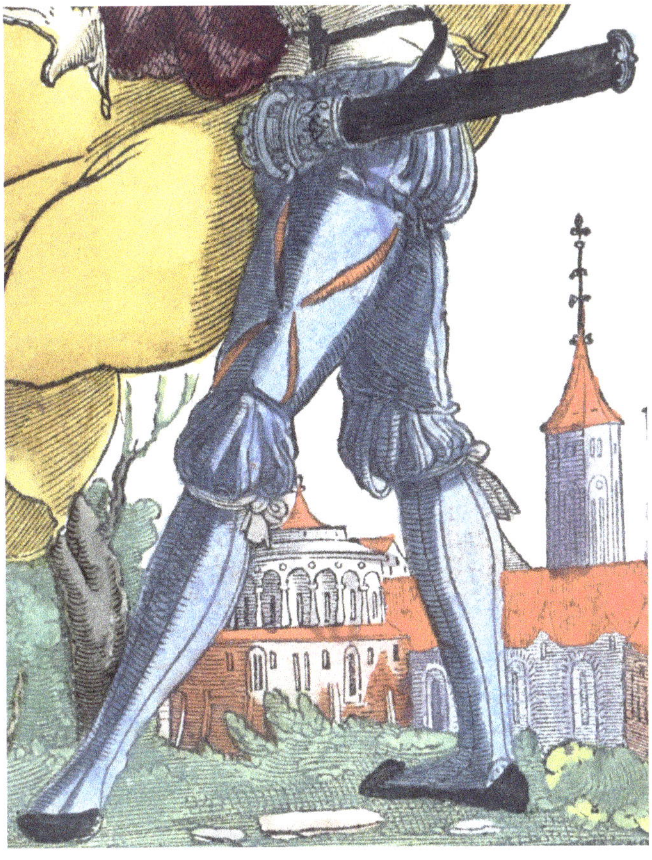

8. Detail of Plate 12. Lothringen | Lutring, Duchy of Lorraine

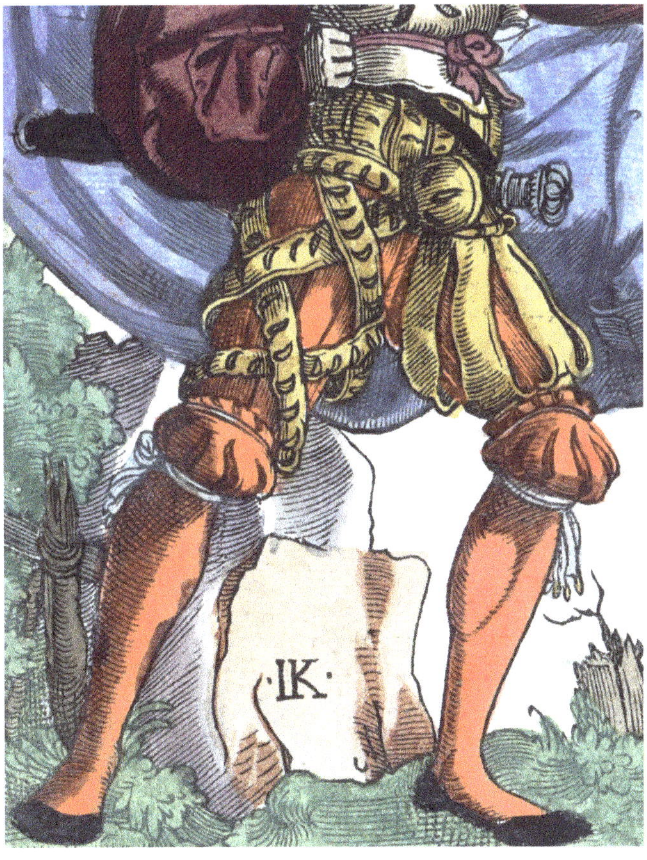

Fig 9. Detail of Plate 32. Aldenwalden | Alwalden

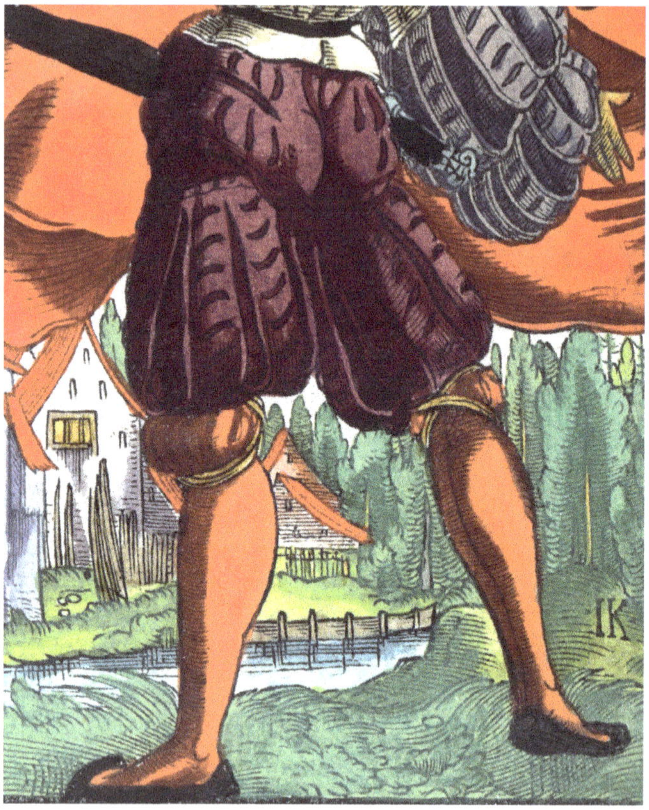

10. Detail of Plate 31. Thufis | Thussis, Thusis, Switzerland

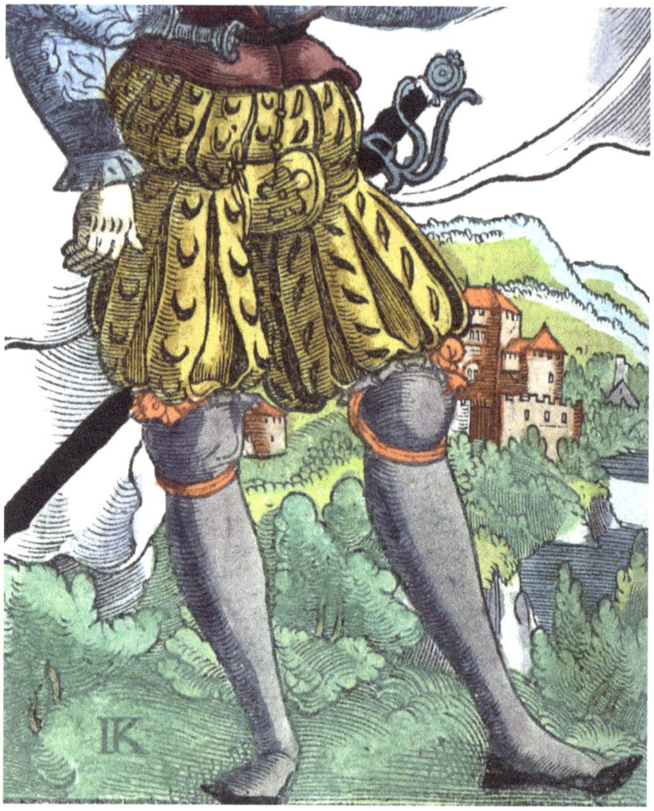

11. Detail of Plate 45. Cöln | Koln, Cologne

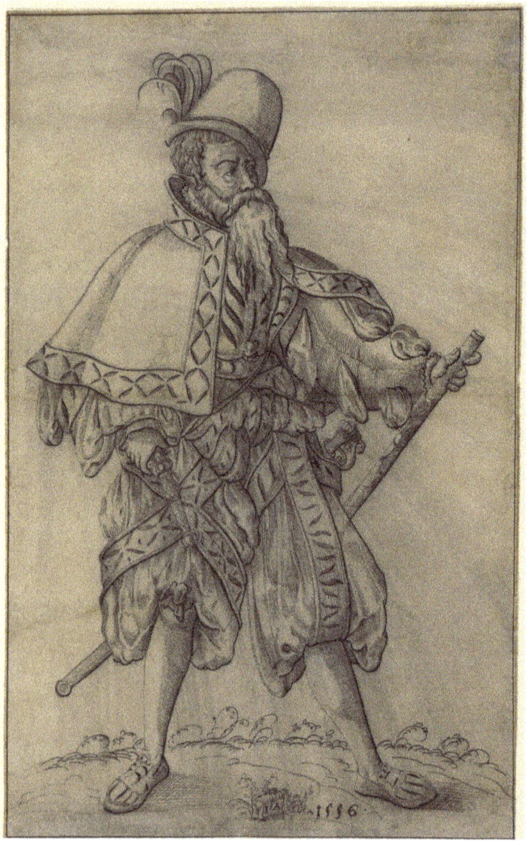

12. An Officer of the Rank of ‚Oberster Feldprofoss' in the Imperial Army, Jost Amman 1556, after a print by Virgil Solis The J. Paul Getty Museum, Los Angeles, 89.GA.15

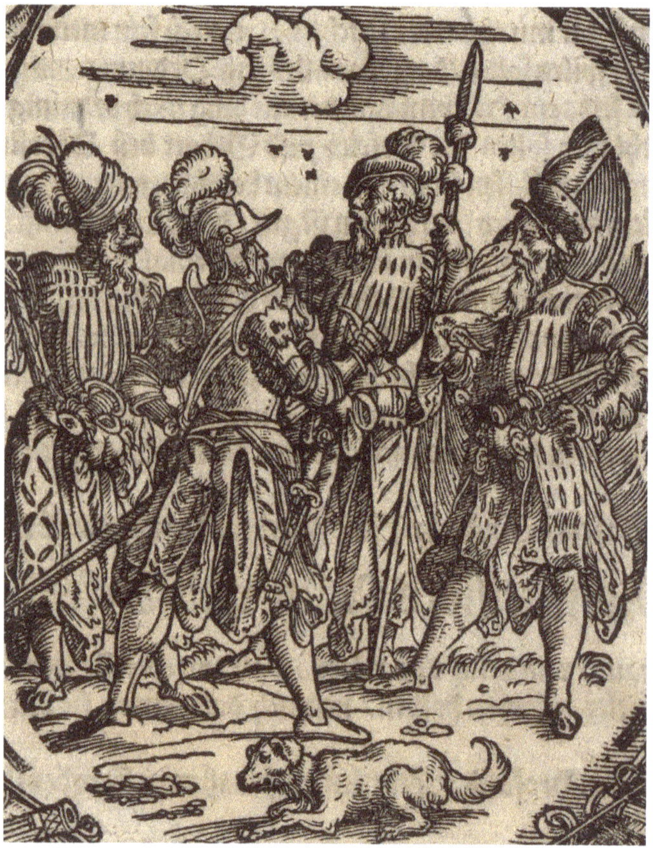

13. Four Standing Soldiers Jost Amman, 1573 Rijksmuseum, RP-P-OB-21.392

Flag Construction

The flags depicted in this book would have been constructed and decorated in several different ways. The flag making sections in 16th c tailors' manuals only mention the use of silk in flag making and focus mainly on pieced flags, but there are surviving 16th century flags in wool, silk or linen which use a variety of techniques. Cennini in his 15th c craftsman handbook "*Il Libro dell'Arte*" describes the process of preparing the fabric for flags, drawing the design and painting flags and other hangings. Among other items, these types of painted flags were most likely what Jakob Kallenberg specialized in painting. Many of the flags seen in the book would have been created using a combination of a variety of sewing and painting techniques.

Sewn or pieced designs are often seen in larger flags, and there are several excellent surviving examples of this type. The design was usually made solely through piecing with no painted elements. This same construction method was used in later 17th and 18th c silk and cotton flags, and most modern flags are also made this way. A white and red wool ship's flag dated 1570-1630 from Spain or the Netherlands is a rare survivor (fig. 14). Its construction matches other pieced flags made of silk found in other collections, such as the large silk flag held in the museum in Enns, Austria which is discussed in detail in Drei Schnittbücher[1].

Another method commonly used to decorate a flag was painting. This fragment of a painted ship flag, also from the Netherlands, is made from a fine linen (fig. 15). Silk flags were also painted, as this fine 1594 example from Sweden demonstrates (fig 16). The flag was made with a light blue silk taffeta, and then painted with gold, silver and other colored paints.

Another method of ornamentation was embroidery and appliqué, in which the design is painted with needle and threads of silk and metal instead of paint. This embroidered Spanish flag dates to 1596 and was made the Order of Our Lady of Mercy, depicting King David (fig.17). The flag is made of silk damask, decorated with velvet appliques and embroidered with silk and metal threads. The red and silver tassel trim is very similar to that seen on other extant flags.

However, these heraldic designs were not just used for flags but were also used for large displays for tournaments. An excellent example is this embroidered and appliquéd Imperial crest dated 1559 from Augsburg (fig.18, 19). Here the design is worked in a variety of methods of padded embroidery and raised appliqué, resulting in a design which is three dimensional.

---

1 Barich Katherine and Marion McNealy. Drei Schnittbücher: Three Austrian Master Tailor Books of the 16th Century. 2015 p.377-379

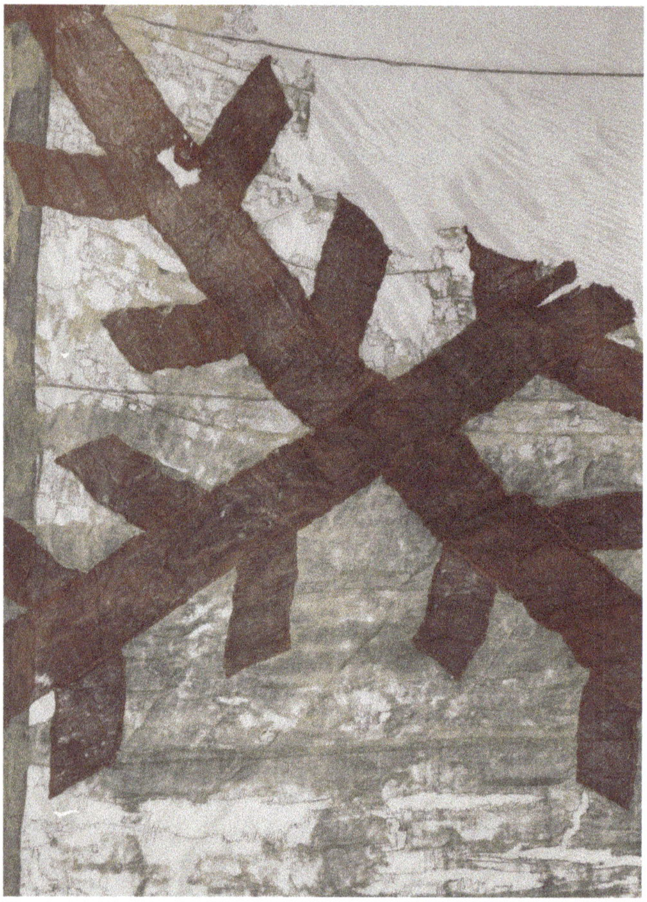

14. Fragment of a Ship Flag, 1570–1630
Pieced red and white wool
Rijksmuseum, NG-MC-1889-83-1

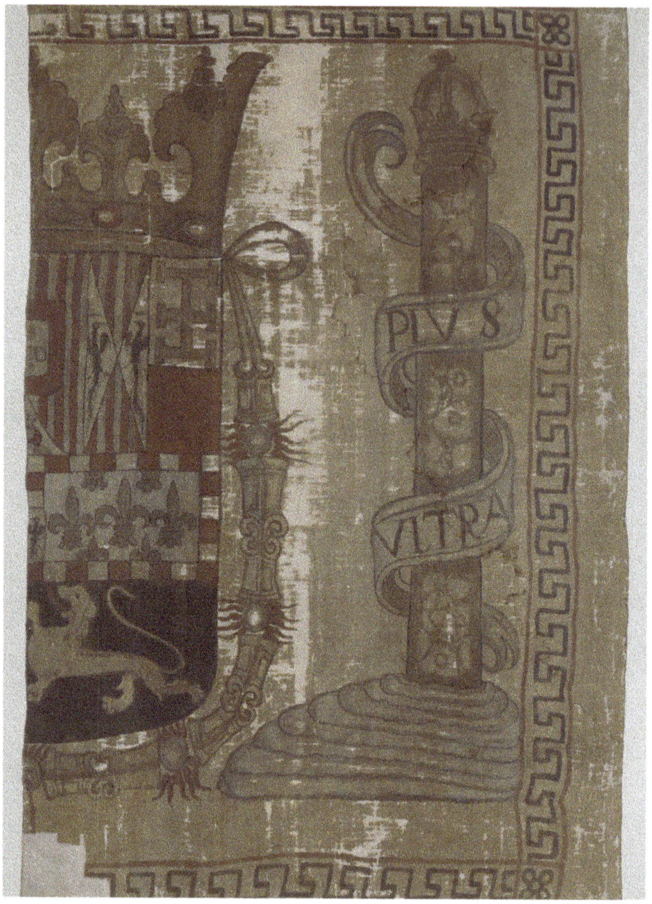

15. Fragment of a Ship Flag, 1570–1630
Painted Linen
Rijksmuseum, NG-MC-1889-83-3

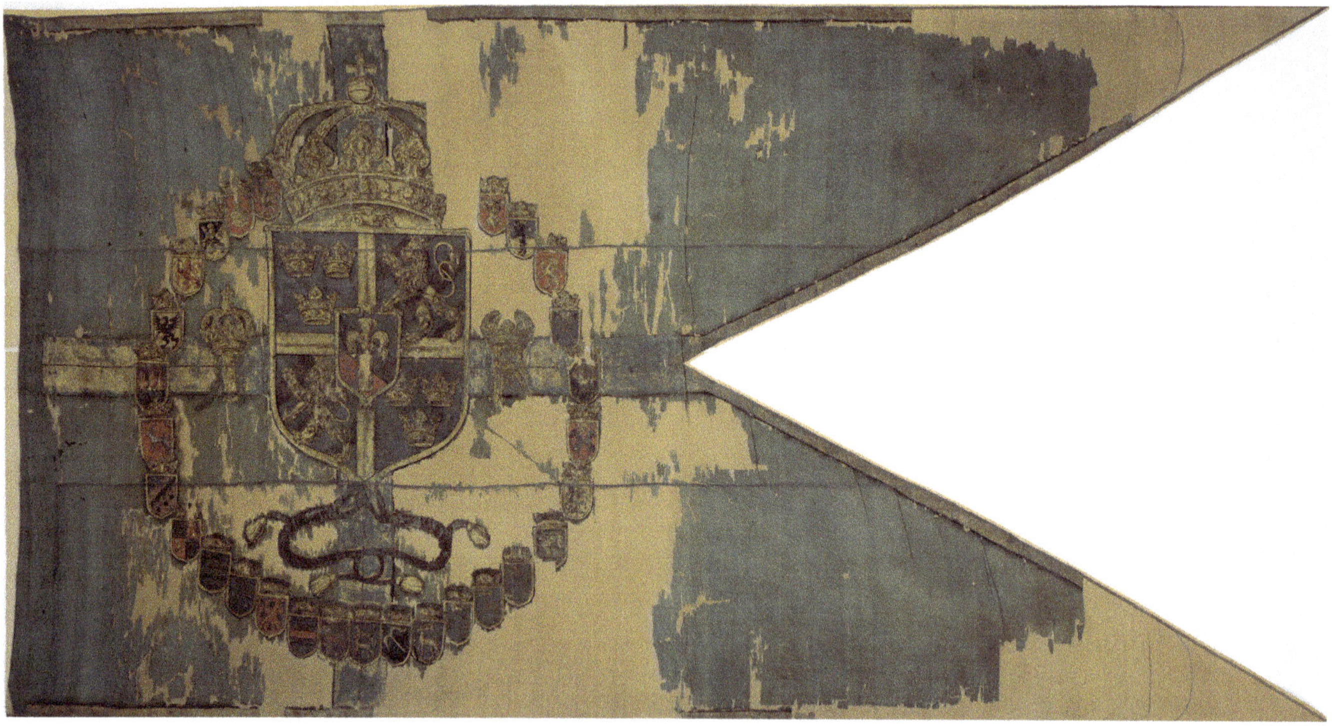

16. National banner from Johan III's funeral and son Sigismund's coronation in 1594. Sewn by court tailors Claes Bertilsson and Anders Svart, probably painted by Baptista van Uther Livrustkammaren, Livrustkammaren /SHM (CC BY 4.0) No. 1867: 55

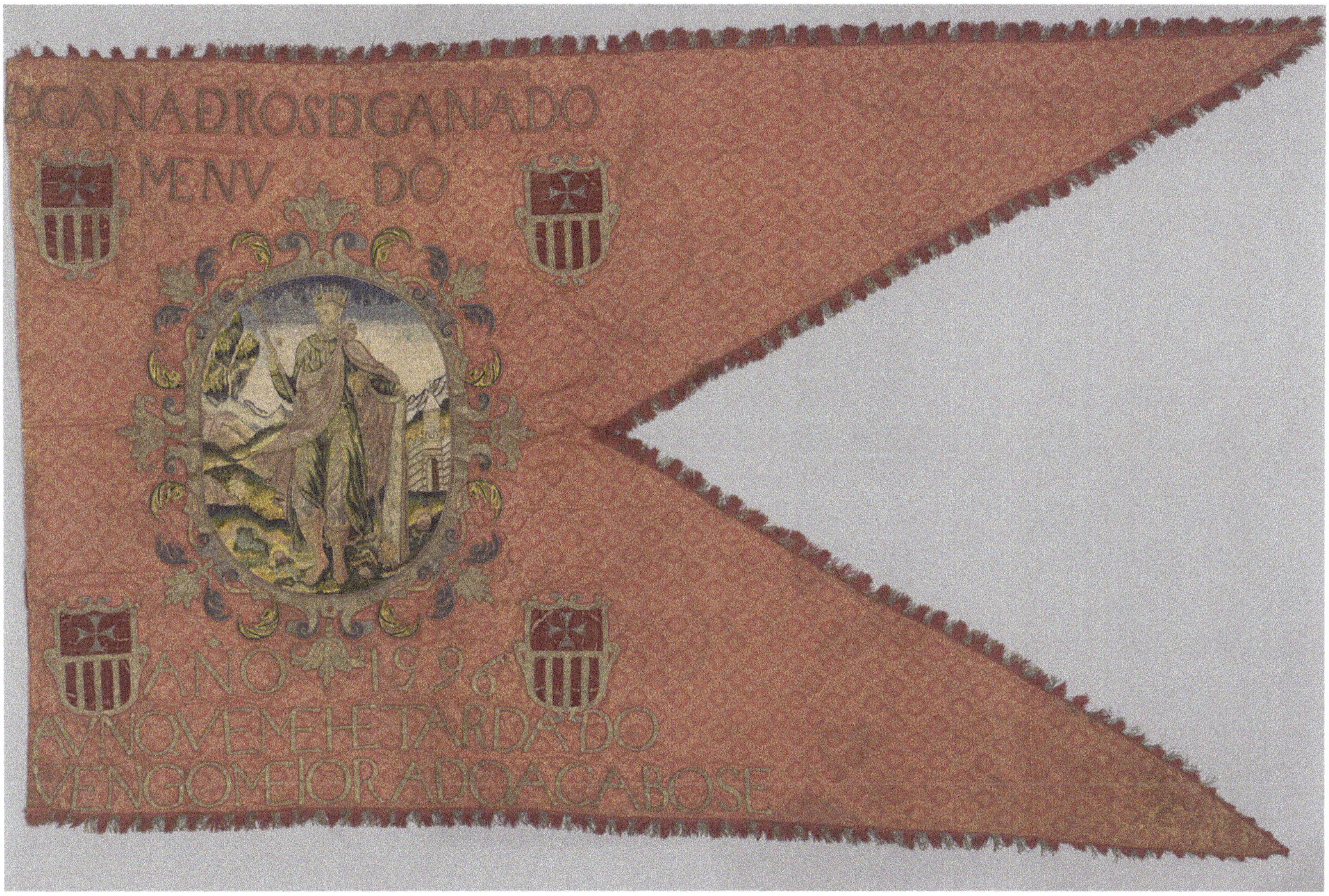

17. Banner, 1596, Spanish, Silk and metal thread
The Metropolitan Museum of Art Gift of Melvin Gutman, 1958, Accession Number: 58.172

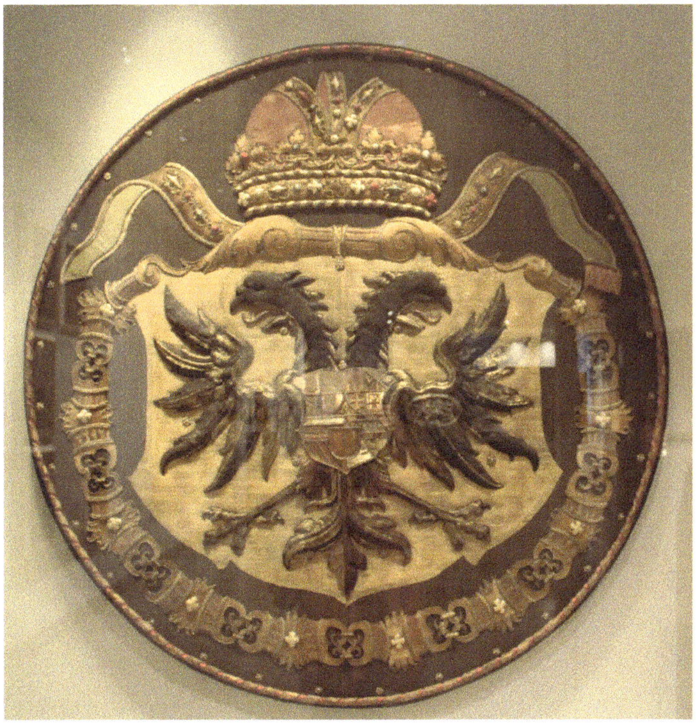

18. Imperial crest, 1559 Maximilian Museum Augsburg, Photo by Tina Vadász-Hain

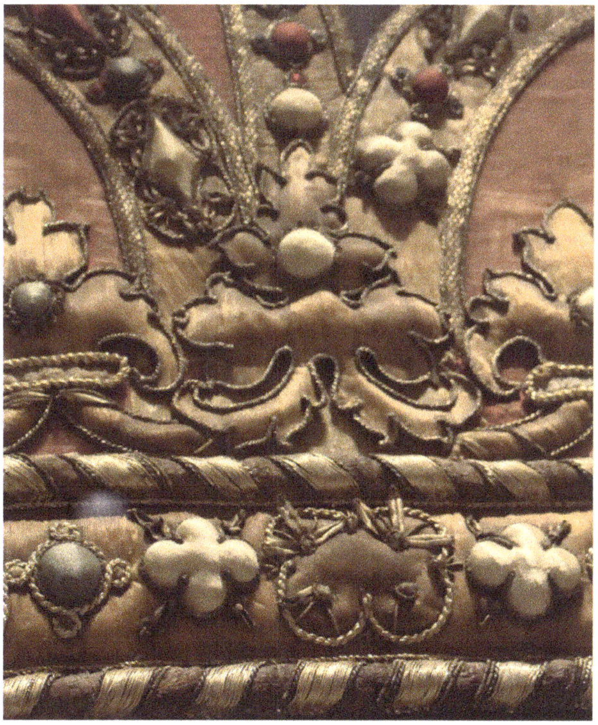

19. Detail of Imperial crest, 1559 Maximilian Museum Augsburg, Photo by Tina Vadász-Hain

Order of Precedence of the Imperial Diet of the Holy Roman Empire

1. Des Reichs Fahn |
   The Imperial Banner

Des Reichs Fahn.

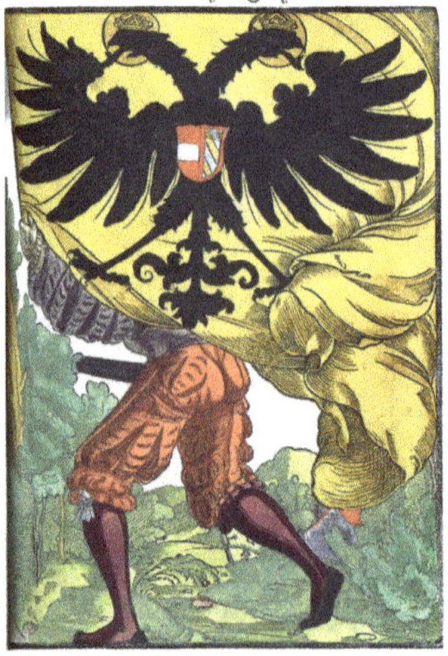

Flags of the Seven Electors are divided into two parts, Ecclesiastical (Catholic Church) and Secular.

Three Ecclesiastical Electors:

2. Menz |
   The Prince Archbishop
   of Mainz

Menz.

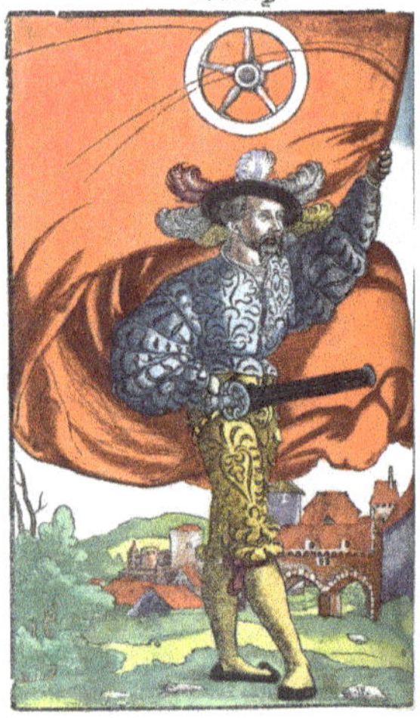

A white wheel
on a red field.

3. Cöln |
   The Prince Archbishop
   of Cologne

Cöln.

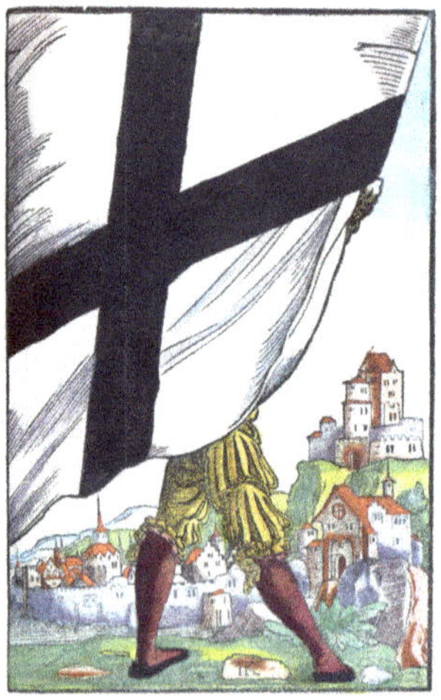

A black cross
on a white field.

4. Trier |
   The Prince Archbishop
   of Trier

Trier.

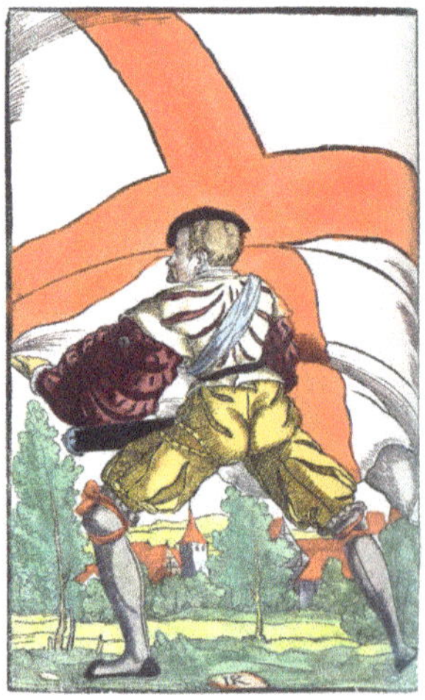

A red cross
on a white field.

Four Secular Electors:

5. Behenm | Böhmen, The King of Bohemia, Imperial Cup Bearer. Area of modern Czech Republic

Beheym.

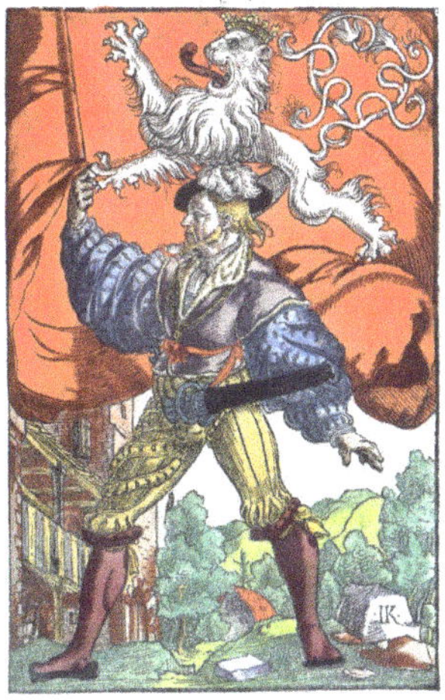

A white lion on a red field facing left.

6. Pfalz | Pfaltz The Count Palatine of the Rhine, Imperial Steward

Pfaltz.

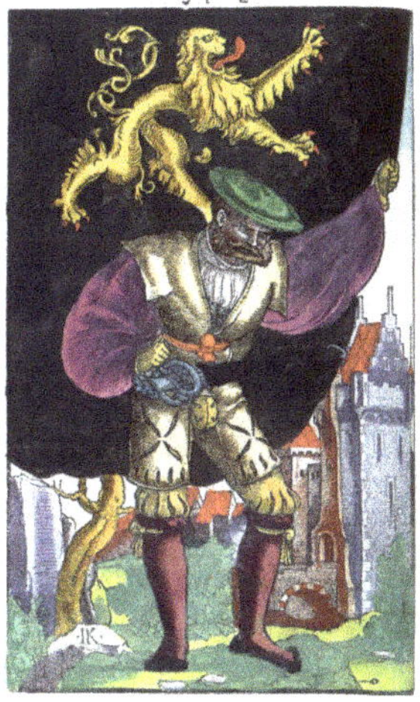

A gold lion on a black field with red tongue and claws.

7. Sachssen | Sachsen, The Duchy of Saxony, Imperial Marshal

Sachssen.

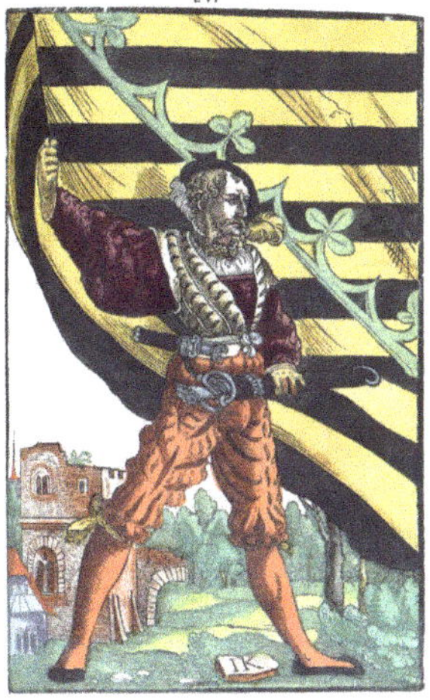

Gold and black stripes with a ducal coronet in green across it diagonally.

8. Brandenburg | The Margrave of Brandenburg, Imperial Chamberlain

Brandenburg.

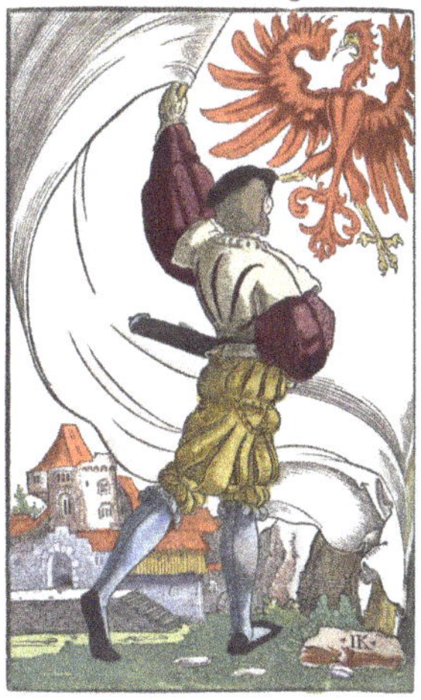

Red eagle on a white field with gold feet.

The Groups of Four of the Imperial Constitution

The Four Seuln – Duchies:

9. Braunschwenck | Braunschweig (Duchy of Brunswick-Lüneburg)

Braunschweyck.

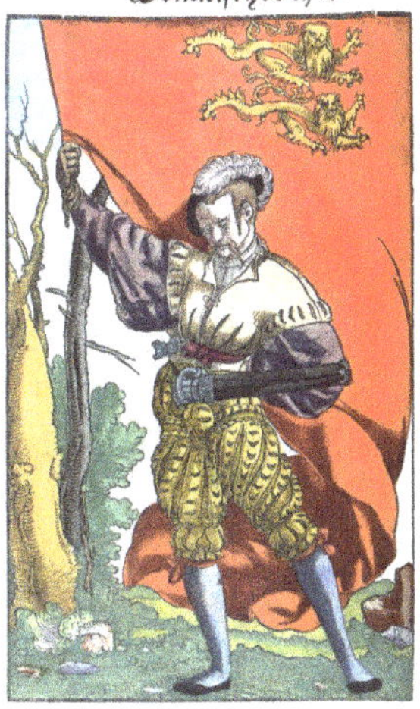

Two gold lions on a red field.

10. Bairn | Bayern (Duchy of Bavaria) [Mislabled as Deckendorff]

Deckendorff.

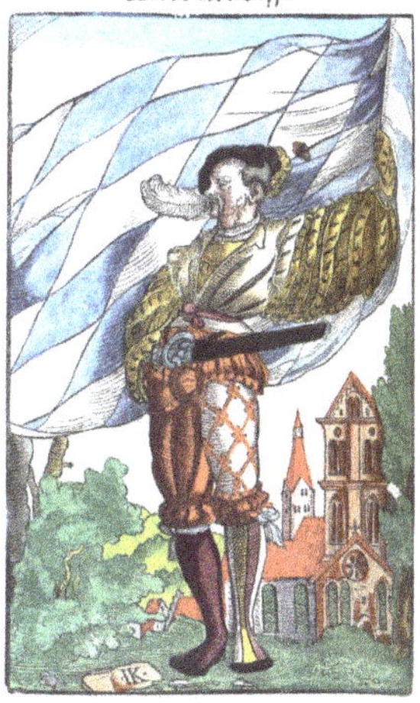

Blue and white diamonds.

11. Schwaben | Swaben (Duchy of Swabia)

Schwaben.

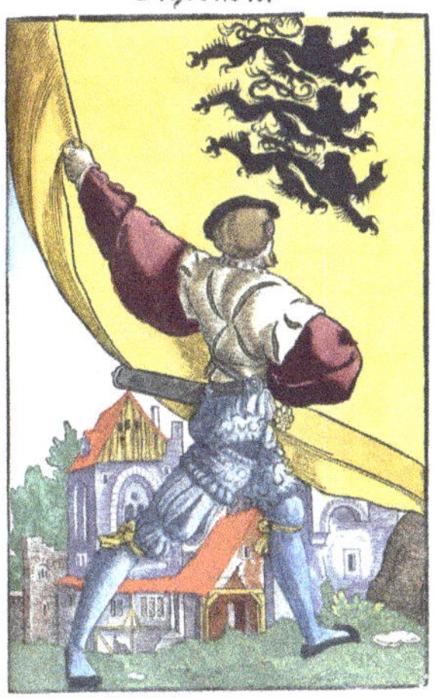

Three black lions on a gold field.

12. Lothringen | Lutring (Duchy of Lorraine)

Lothringen.

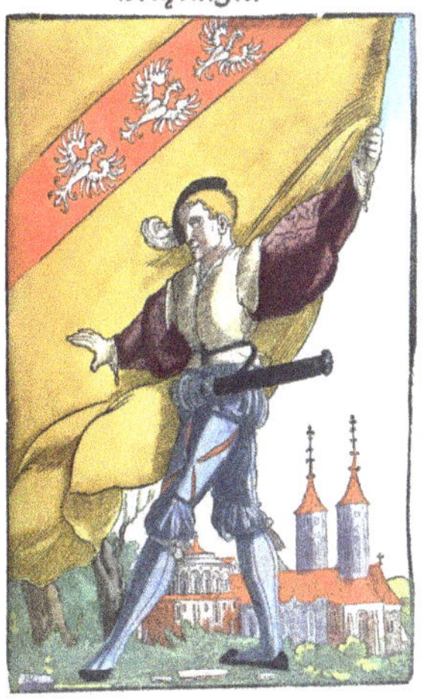

Three white eagles, on a red diagonal banner, on a gold field.

The Four Marggraven – Margkgraffen:

13. Mörhern | Mähren, Moravia

Mörhern.

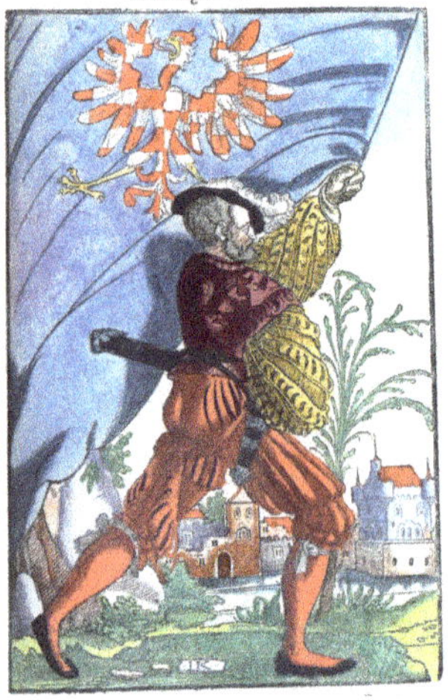

An eagle, in red and white check, with gold feet, on a blue field.

14. Brandenburg | Brandenburg

Brandenburg.

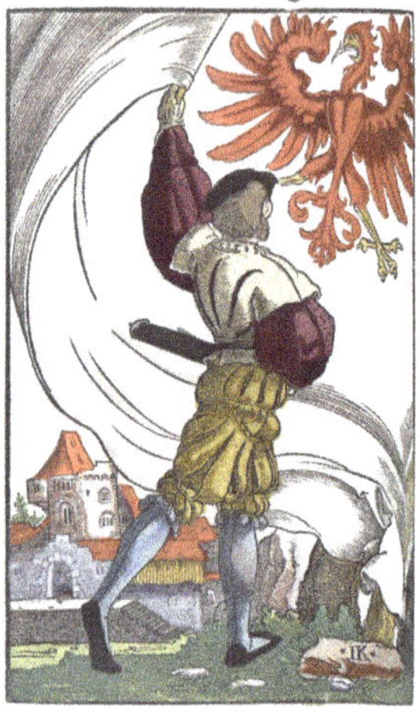

Red eagle on a white field with gold feet.

15. Meichssen | Meißen (Mislabeled as Landaw | Landau in der Pfalz)

Landaw.

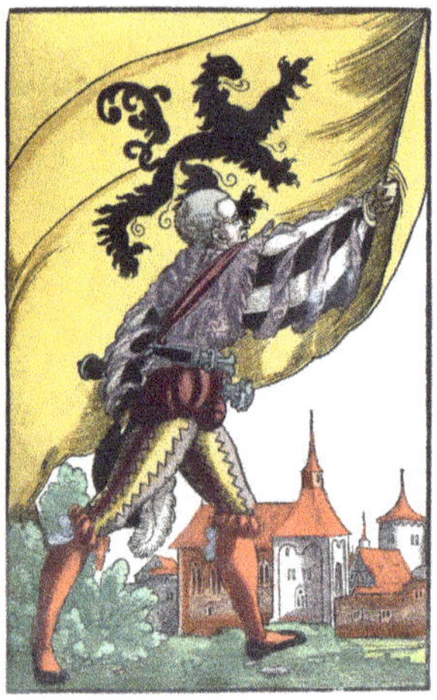

Black lion on a gold field.

16. Baden | Baden

Baden.

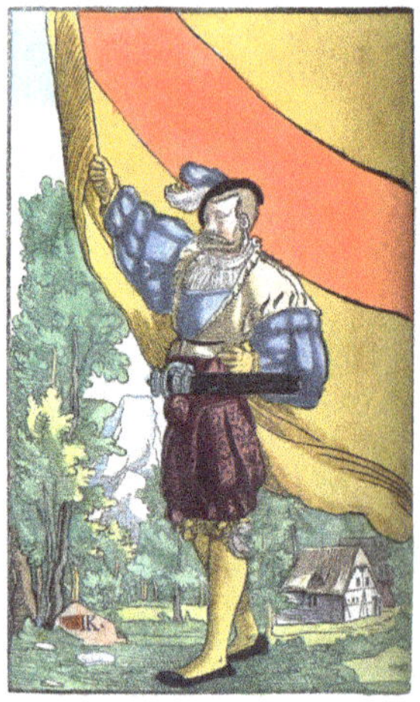

Upper part of the flag in yellow, in the middle red, and then lower part in yellow.

The Four Landsgraffen:

17. Düringen | Düring, Thüringen, Thuringia

Düringen.

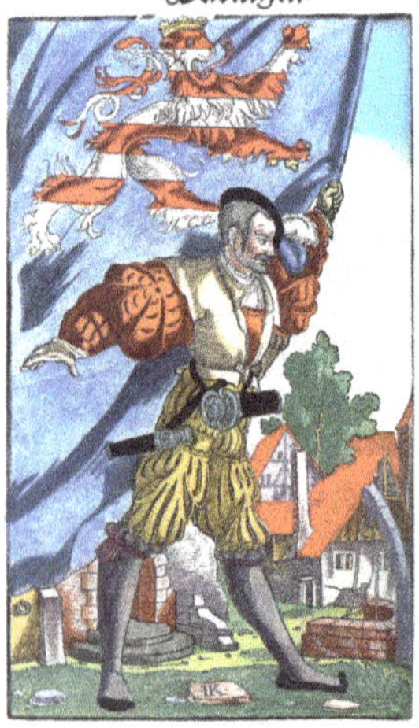

A lion on the upper part with a gold crown, striped in red and white, on a blue field.

18. Leuchtenburg | Leuchtenberg

Leuchtenburg.

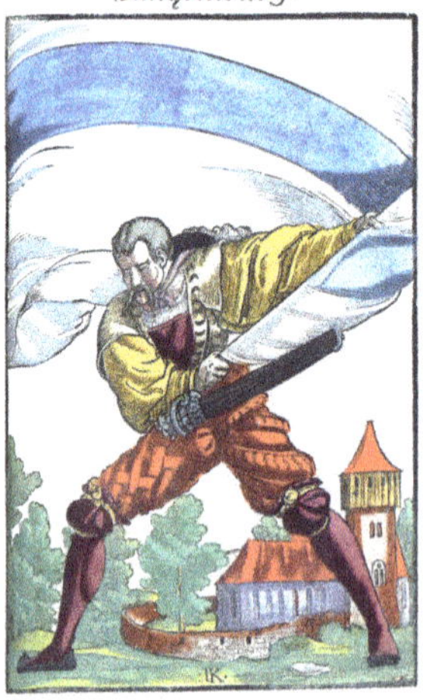

White on the upper part, blue in the middle and white on the lower part.

19. Hessen | Hessen, (Mislabeled as Düringen | Thuringia)

Düringen.

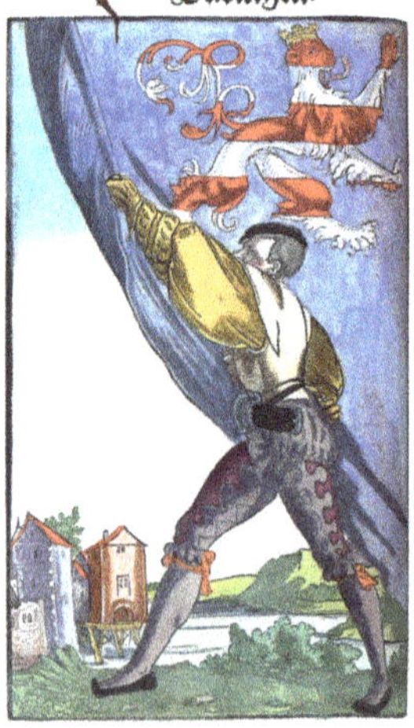

On the upper part, a lion with a gold crown on a blue field. The body is in red and white stripes starting with the head in red.

20. Edelsasz | Edelsaß, Landgrafschaft Niederelsass, Bailiffs of Lower Alsace

Edelsaß.

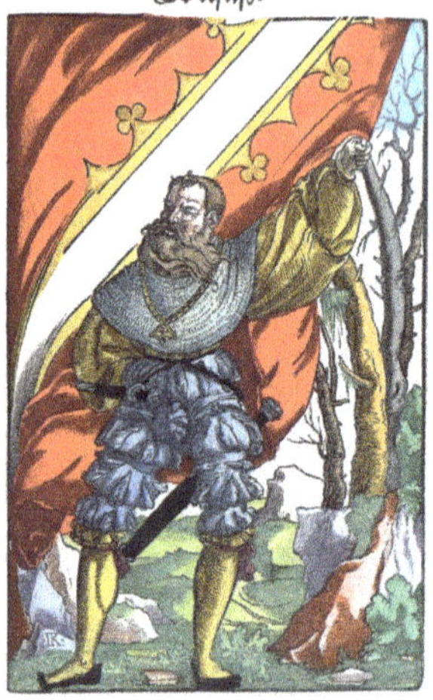

A diagonal white stripe bordered with two gold lines on a red field.

The Four Burggraffen – Burgraviate:

21. Nürenberg Burggraffen | Nürnberg (Nuremberg) Burgraviate Flag

Nürenberg.

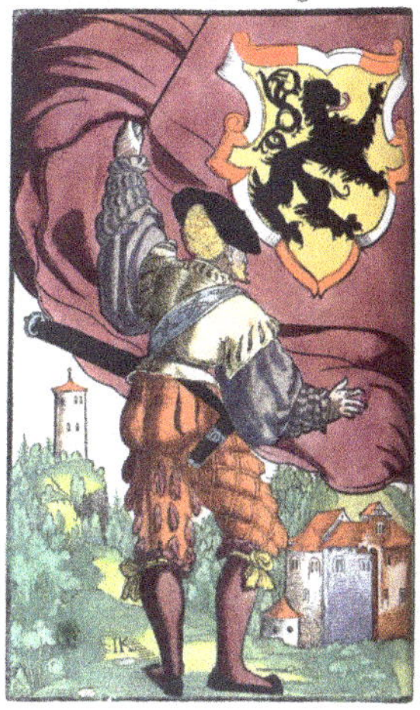

A shield with black lion on a gold field, with the outside of the shield bordered in a red and white stripe, on a dark red flag.

22. Maidburg | Maidburg, Hardegg-Maidburg

Maidburg.

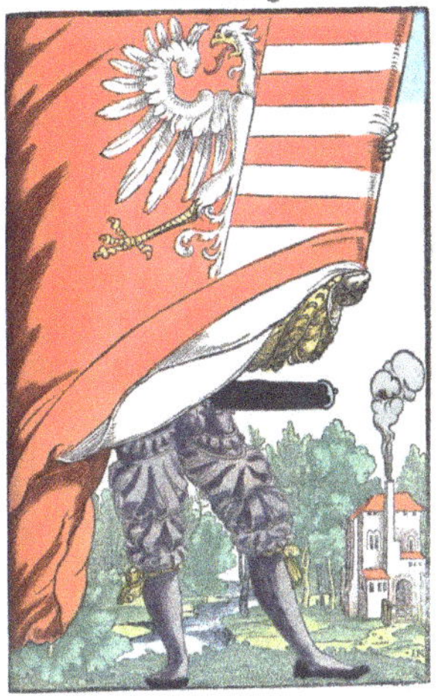

A white half eagle on red field and the other half of the flag striped in red and white.

23. Reneck | Reineck

Reneck.

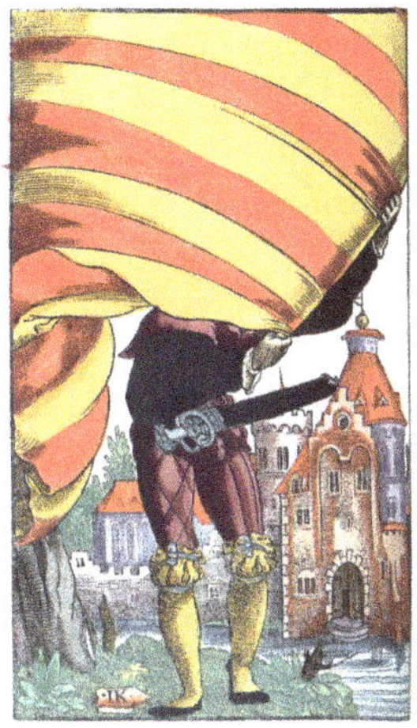

Flag striped in gold and red, starting with gold.

24. Stramberg | Stromberg

Stramberg.

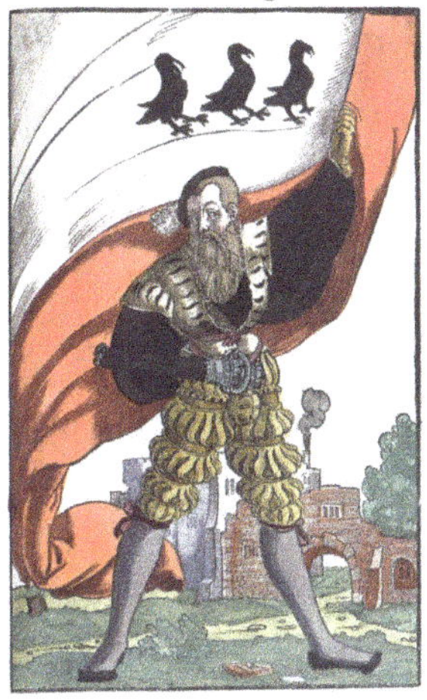

Three black ravens or bird in a white field with a red stripe on the bottom.

The Four Graffen:

25. Cleue | Grafen von Kleve, Count of Cleves

Cleue.

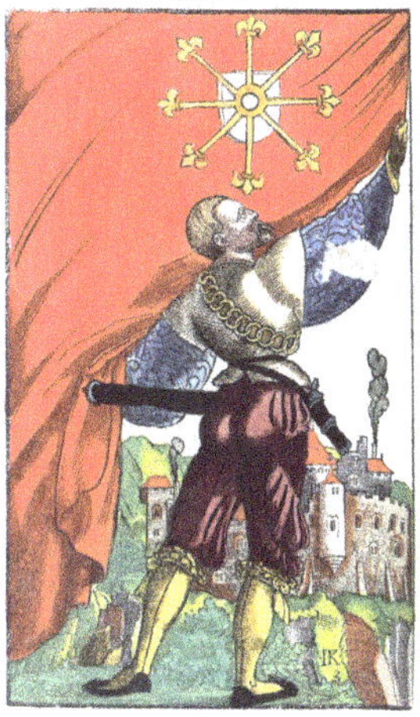

A red field in the middle a white shield with gold gilgen.

26. Saphoy | Grafen von Savoyen, Count of Savoy

Saphoy.

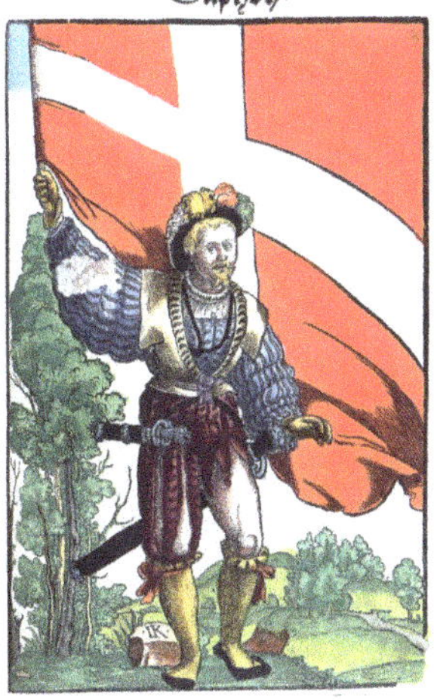

A white cross on a red field.

27. Schwarzburgk | Swartzburg, Schwarzburg

Schwarzburgk.

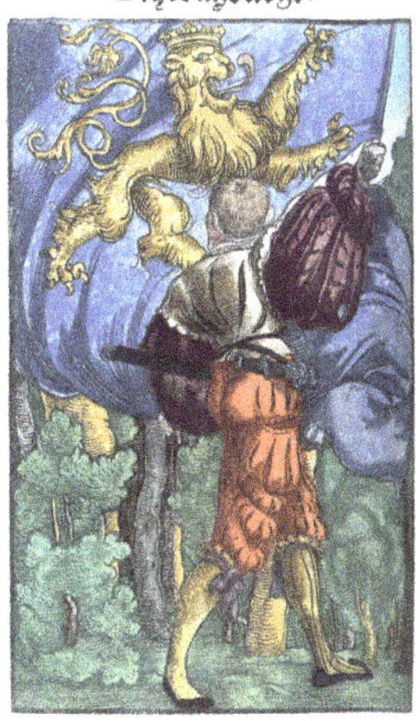

A gold lion on a blue field.

28. Zilli | Celje, Slovenia

Zilli.

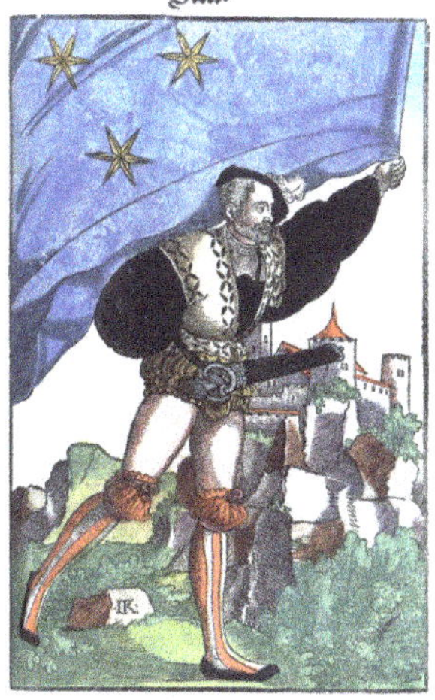

Three gold stars on a blue field.

The Four Semper Frenen:

29. Lintpurg | Limpurg

Lintpurg.

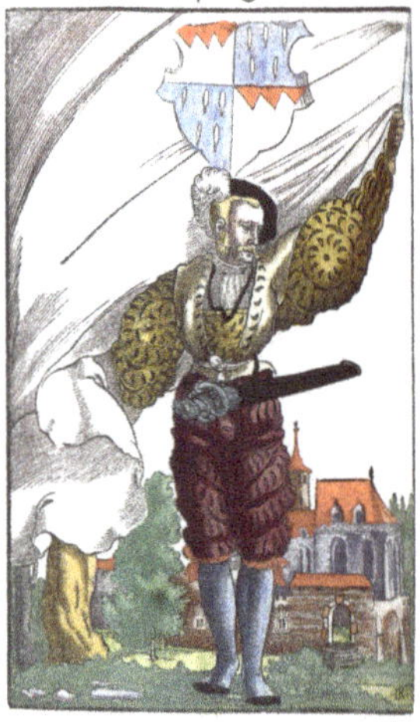

A shield on a white field. The shield is divided into four parts, the upper right and lower left part blue with five white churns, the other part notched red on top with white below.

30. Westerburg | Westerburg

Weſterburg.

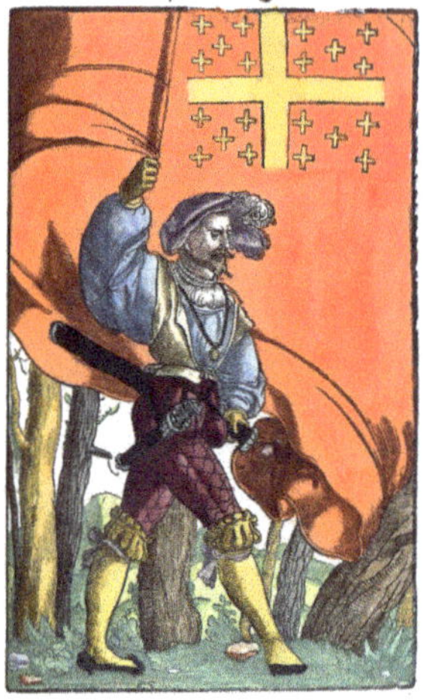

The large cross in gold, and the smaller crosses also in gold, on a red field.

31. Thufis | Thussis, Thusis, Switzerland

Thuſis.

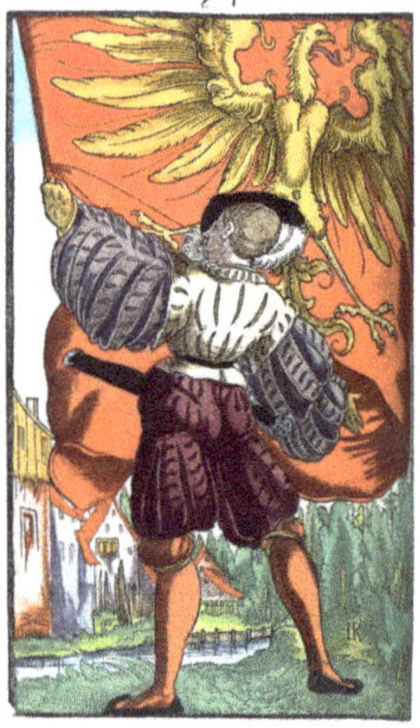

A golden eagle on a red field.

32. Aldenwalden | Alwalden

Aldenwalden.

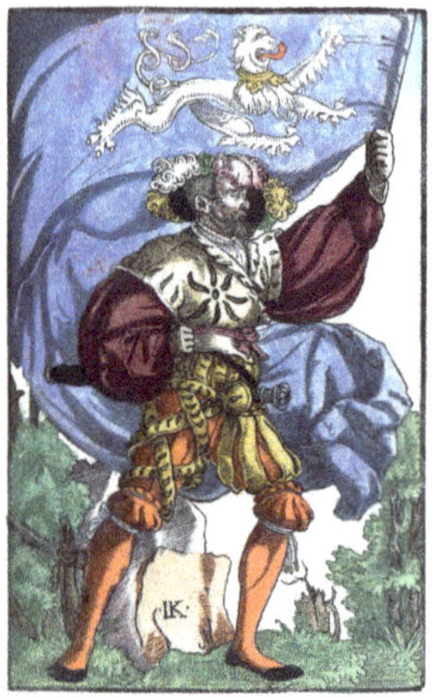

White lion in a blue field with a gold crown around his neck.

The Four Ritter – Knights:

33. Andelaw | Anndelaw, Andlau

Andelaw.

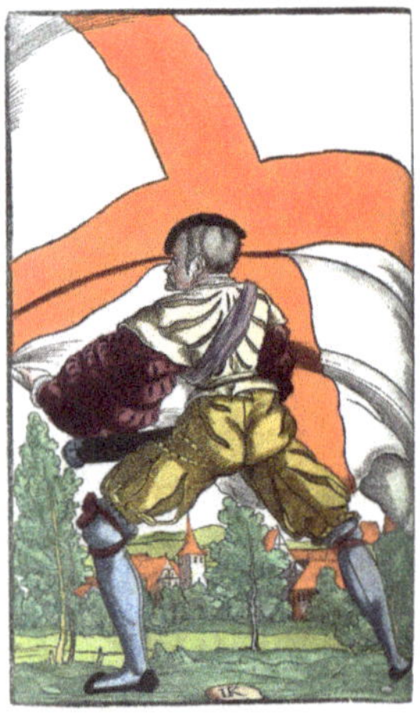

A red cross on a white field.

34. Weissenbach | Weissenbach

Weiſſenbach.

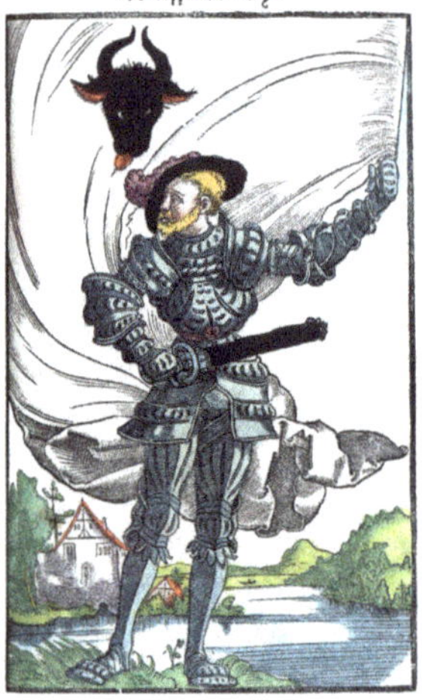

A black bull's head with red ears and tongue on a white field.

35. Frawenberg | Fraunberg

Frawenberg.

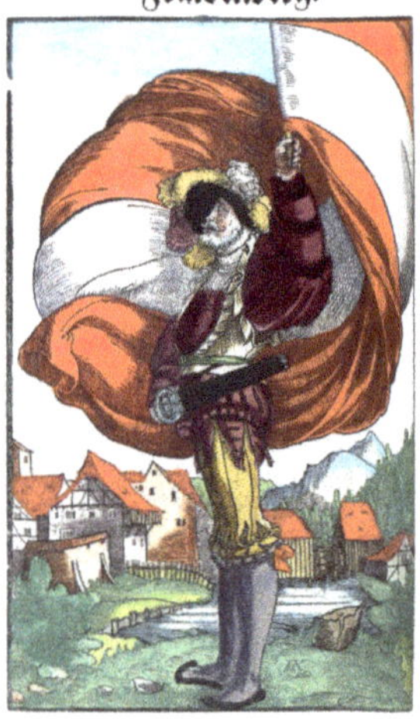

The middle stripe is white, with the top and bottom in red.

36. Strundeck | Strundeck

Strundeck.

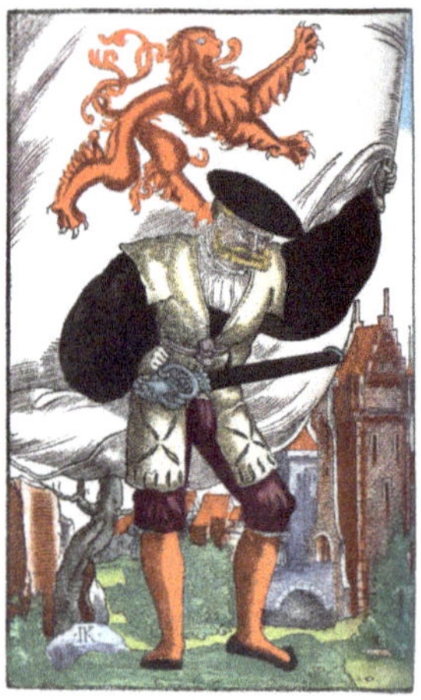

A red lion in a white field.

The Four Städte – Cities:

37. Augspurg | Augsburg

Augſpurg.

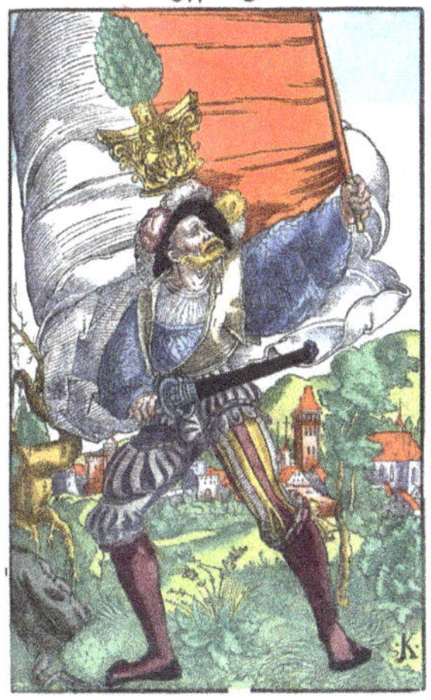

A bunch of green grapes on a gold capital, the flag half red and half white.

38. Merz | Metz, France

Metz.

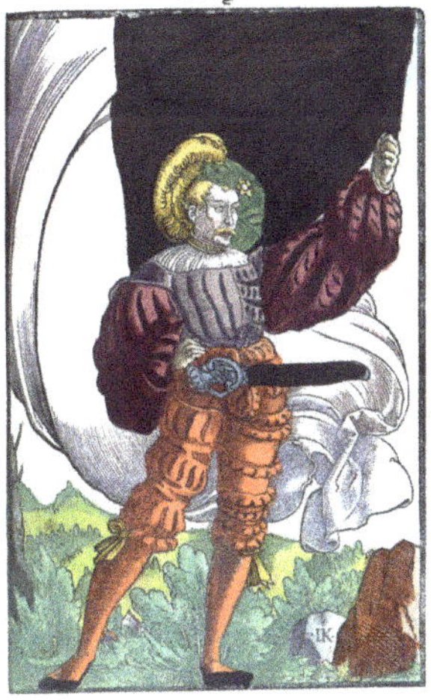

Half white and the other half black.

39. Ach | Aachen

Ach.

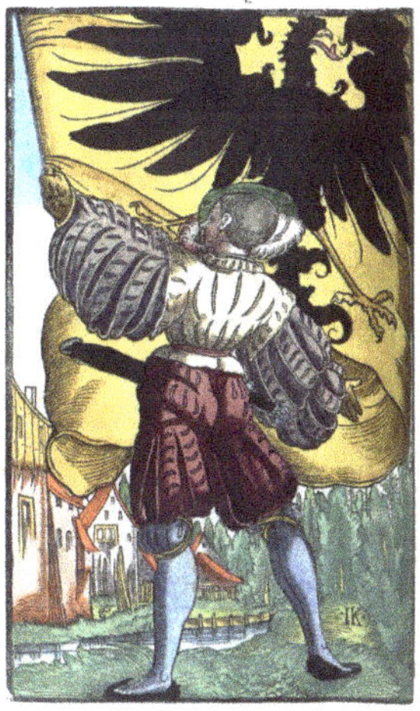

A black eagle on a golden field.

40. Lübeck | Lübeck

Lübeck.

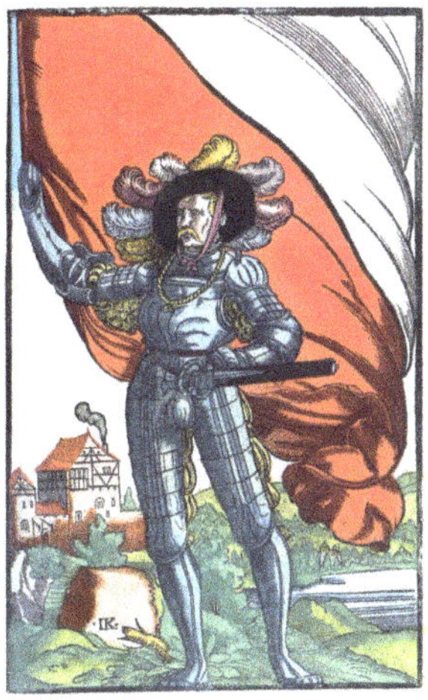

Upper part is white, and underneath is red.

The Four Dozffer:

41. Bamberg | Bamberg

Bamberg.

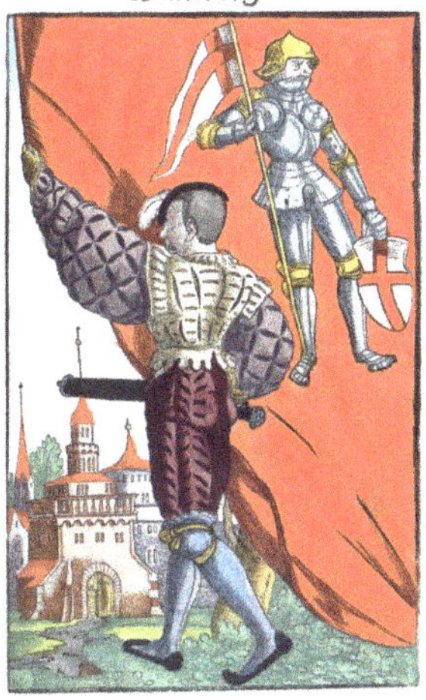

A red field with an armored man with golden helm, knees, and elbows holding a flag on a golden rod and shield, which both have a red cross on a white field.

42. Ulm | Ulm

Vlm.

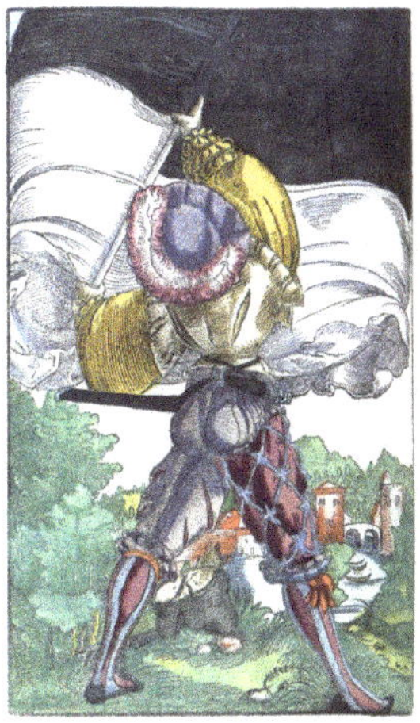

The upper part is black, lower part is white.

43. Hagnaw | Hagenau

Hagnaw.

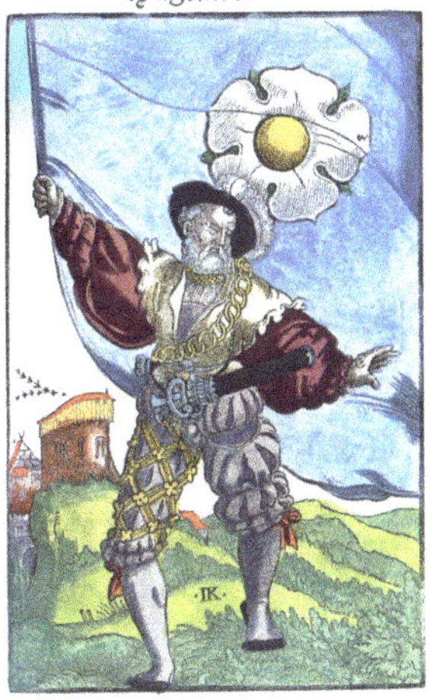

White rose with a gold center, and between each rose petal is a green spike, on a blue field.

44. Schlerstat | Schlettstatt

Schletſtat.

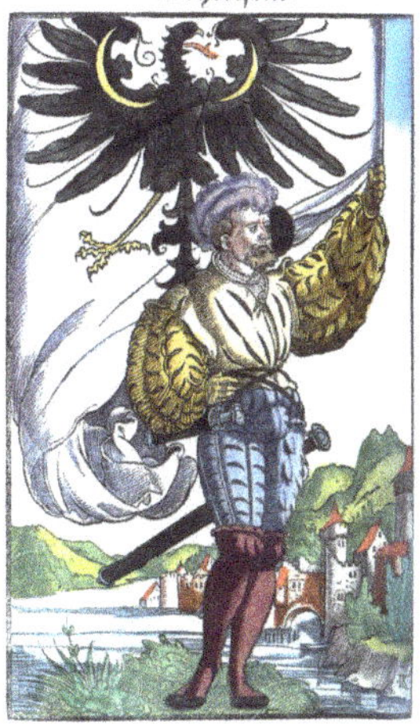

A black eagle on a white field.

The Four Bauern:

45. Cöln | Koln, Cologne

Cöln.

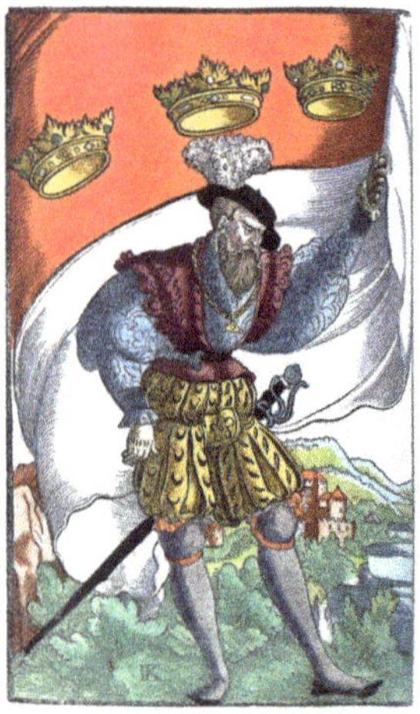

On the upper half three gold crowns on a red field, underneath white.

46. Regenspurg | Regensberg

Regenspurg.

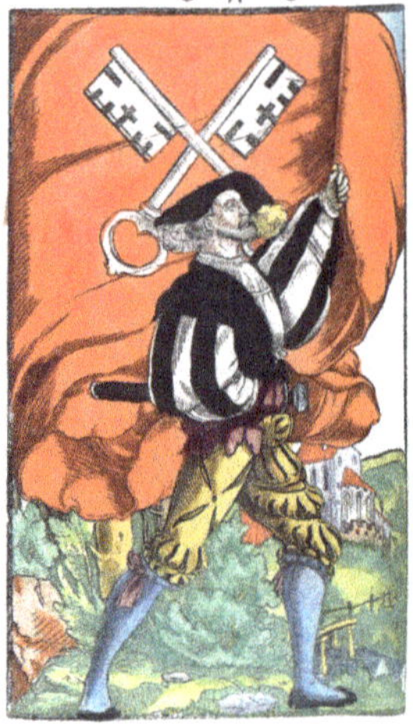

Two white keys crossed on a red field.

47. Costniz | Konstanz, Constance

Costnitz.

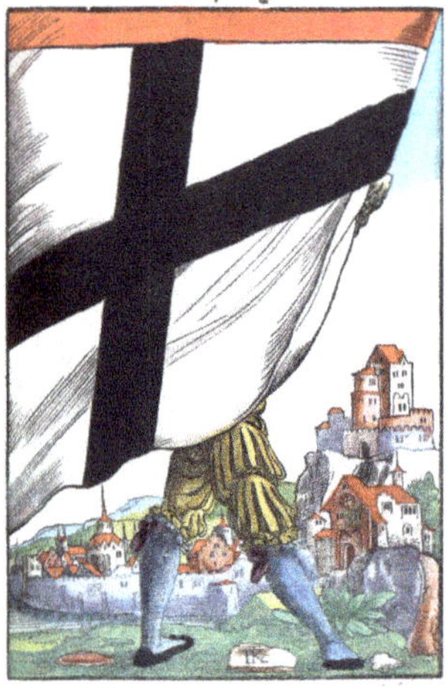

On the upper part, a small red strip and then a black cross on a white field.

48. Salzburg | Salzburg, Austria

Saltzburg.

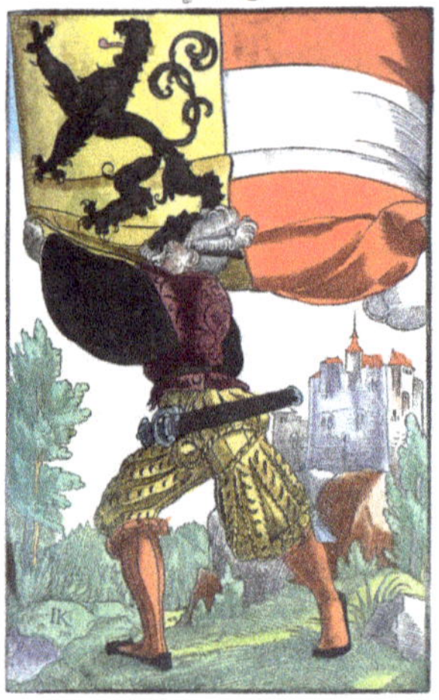

A black lion on a gold field, and the other half red, white, red

The Four Burgen:

49. Mendenburg inn Sachssen | Meyenburg

Meydenburg inn sachssen.

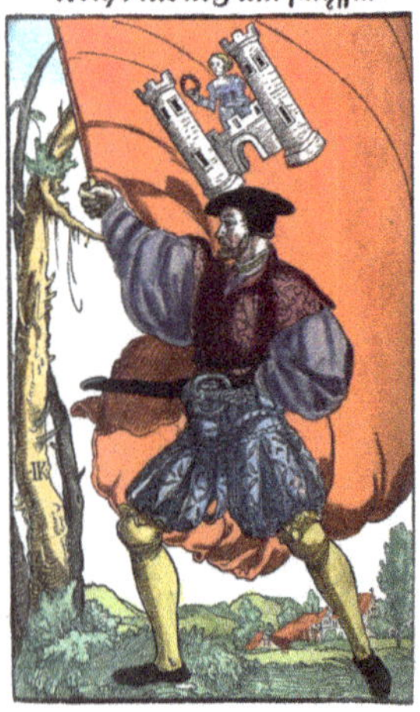

A white town on a red field with a young woman dressed in blue, gold hair holding a green wreath.

50. Lünzelburg | Luxemburg

Lützelburg.

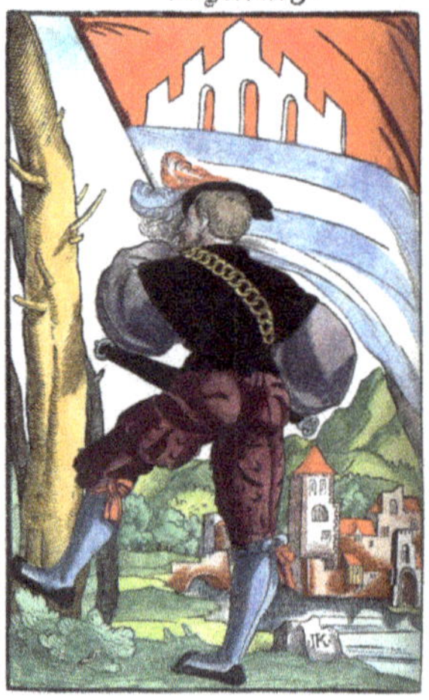

White battlements on a red field in the upper part, on the lower part, stripes blue, and white until the bottom of the flag.

51. Rottenburg am Necker | Rottenburg am Neckar

Rottenburg am Necker.

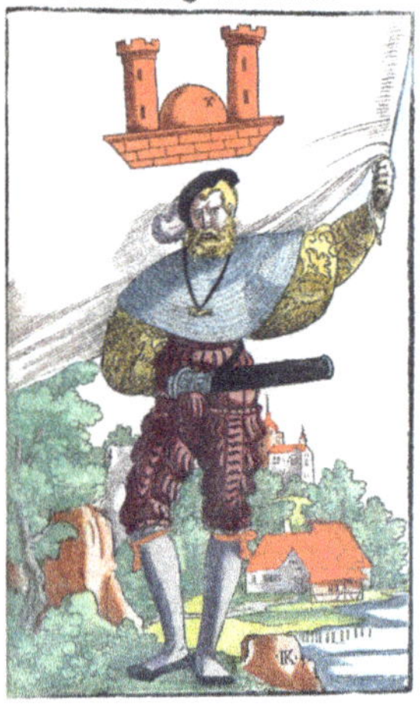

A red town on a white field.

52. Aldenburg | Altenburg

Aldenburg.

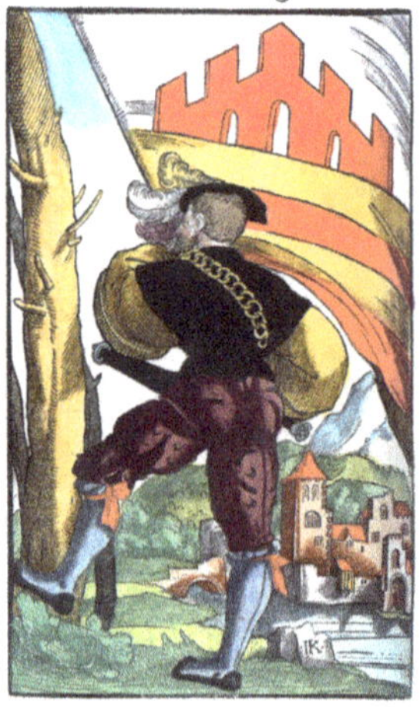

Red battlements on a white field in the upper part, on the lower part, stripes yellow and red until the bottom of the flag.

The Four Vicars

53. Brabandt |
Duchy of Brabant

Brabandt.

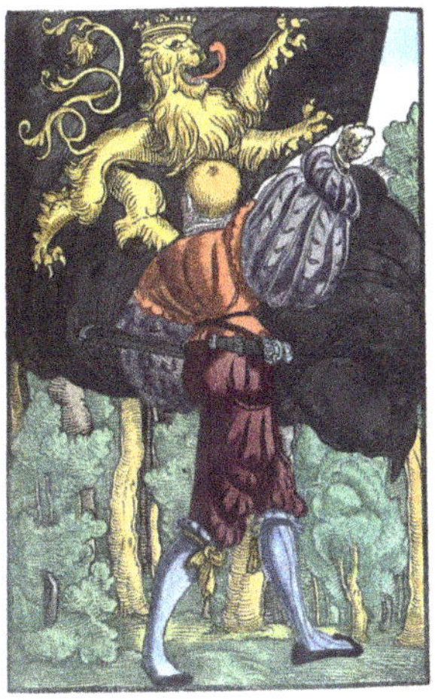

Yellow lion
on a black field.

54. Nider Sachssen |
Niedersachen, Lower Saxony

Nider Sachssen.

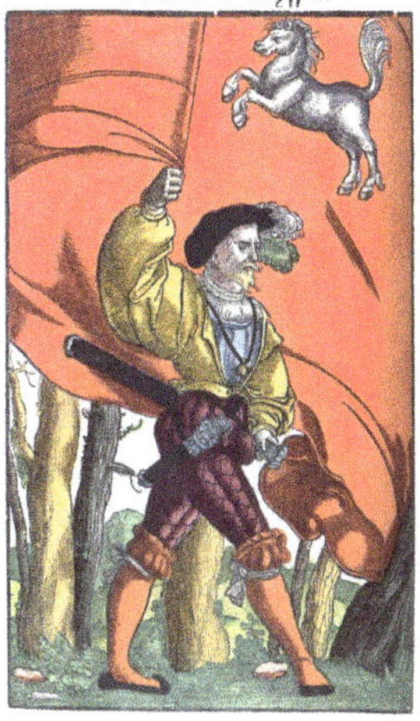

A white horse
on a red field.

55. Westerreich |
Westrich

Westerreich.

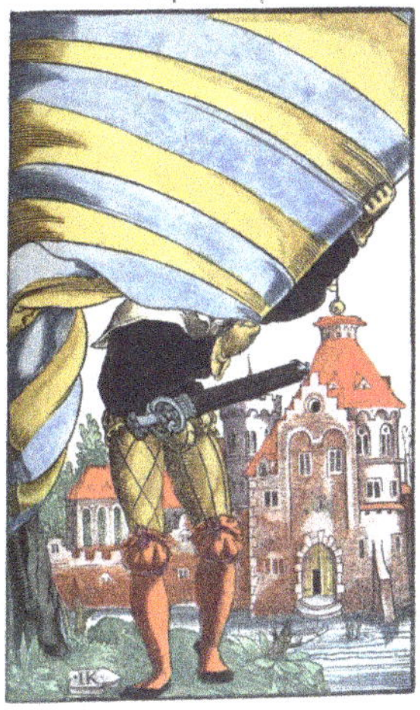

Yellow and blue stripes.

56. Schlessi |
Schlesien, Silesia

Schlessi.

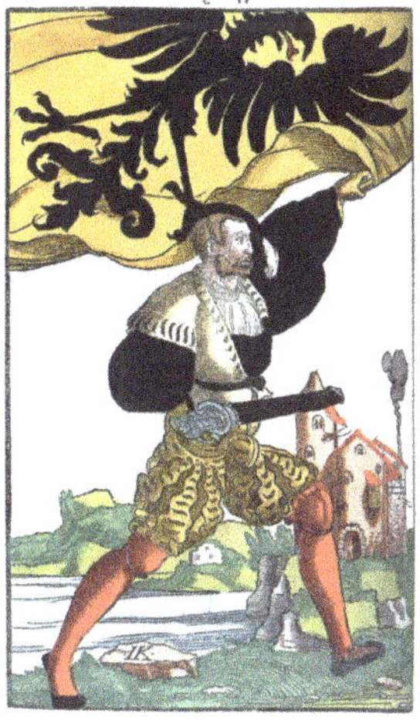

Black eagle
on a yellow field.

The Common States and Cities:

57. Offenburg |
Offenburg

Offenburg.

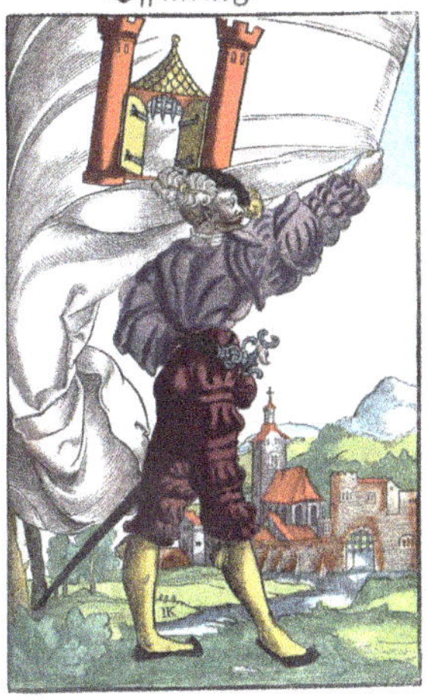

Red towers with
yellow roof and gates
with black bands,
all on a white field.

58. Oberzell |
Zell am Harmersbach

Oberzell.

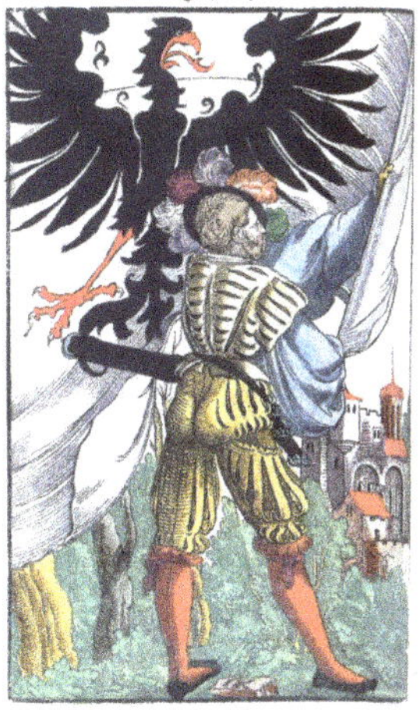

A black eagle on a white
field, with red beak
and foot.

59. Nerlingen |
Nördlingen

Nerlingen.

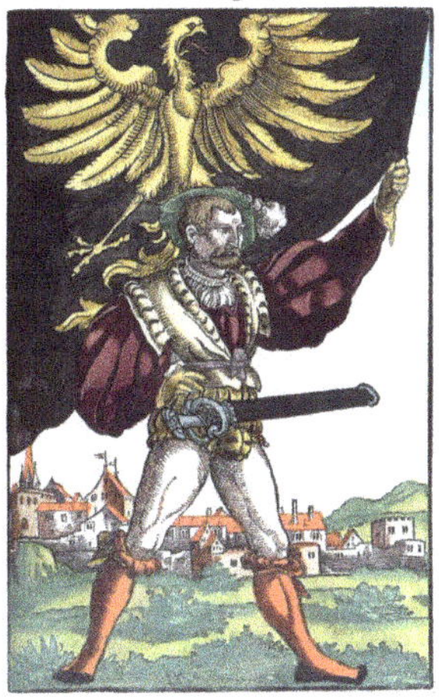

A yellow eagle
on a black field

60. Uberlingen |
Überlingen

Vberlingen.

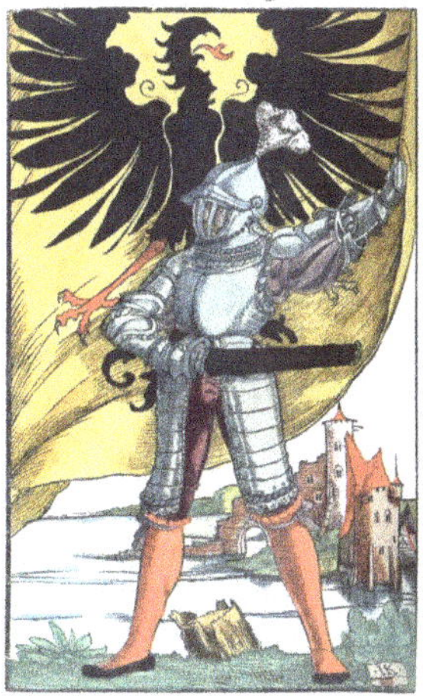

A black eagle on a yellow
field with a red tongue
and feet

61. Rosenhenm | Rosenheim

Rosenheym.

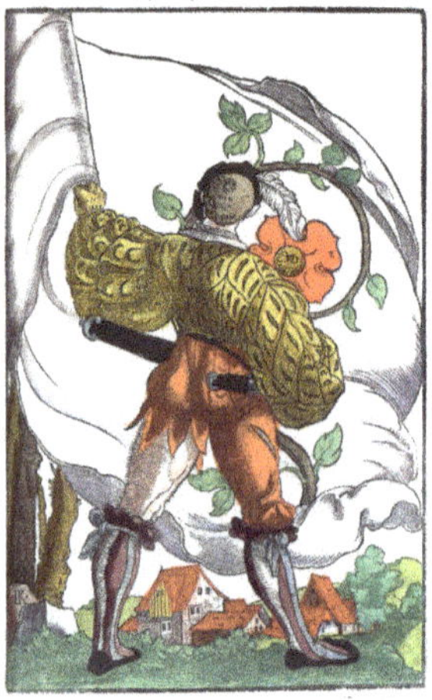

A red rose with a yellow center, on a white field.

62. Wimpfem | Bad Wimpfen

Wimpfem.

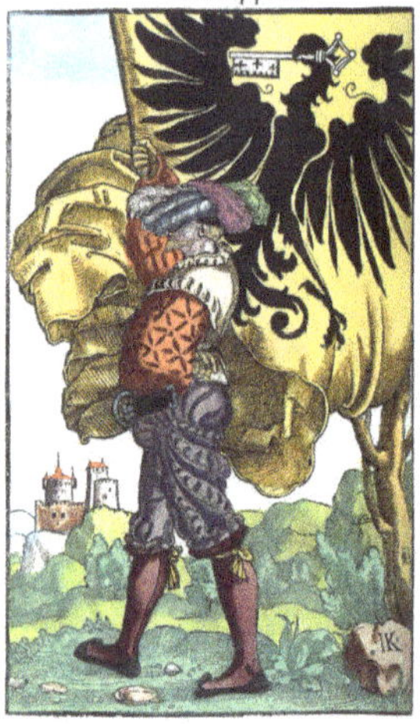

A black eagle with a red beak, holding a white key in its beak, on a yellow field.

63. Jenff | Genf, Geneva Switzerland

Jenff.

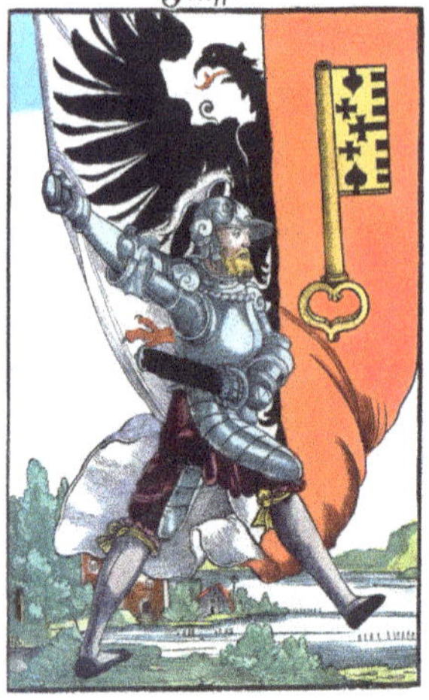

Yellow key on a red field, and the other half is a black eagle with a red beak and feet on a white field.

64. Toll | Toul, France

Toll.

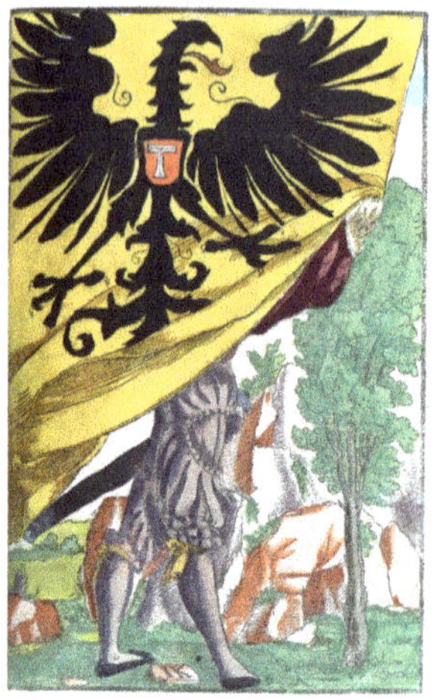

Black eagle with a red shield with the letter T on its body, on a yellow field.

65. Bisfantz | Besançon, France

Bysantz.

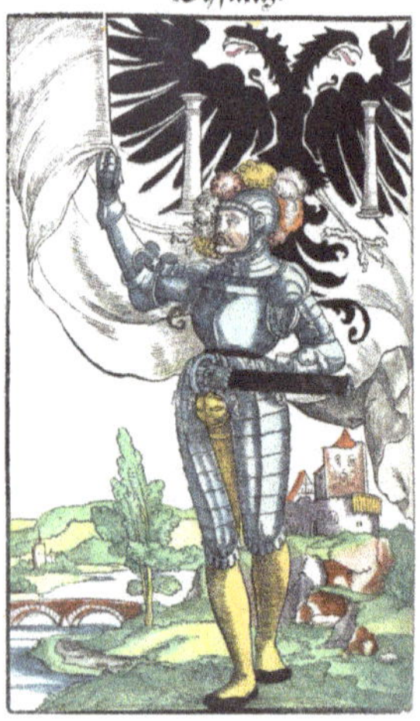

Black eagle on a white field, with two columns.

66. Villingen | Villingen

Villingen.

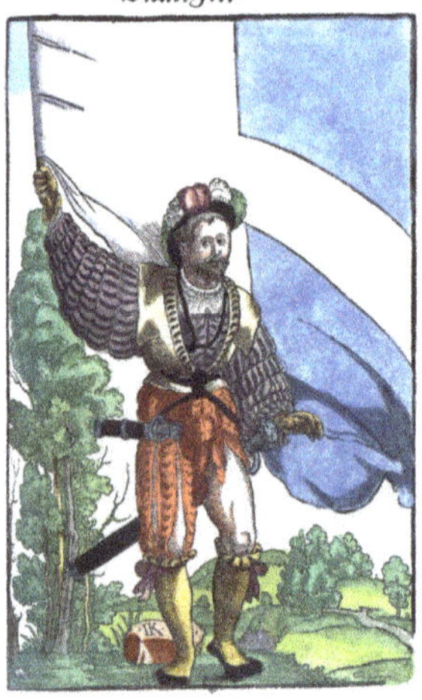

Half the flag white, the other half the top and bottom band blue, and the middle band white.

67. Herzenthal | Herentals, Belgium

Herzenthal.

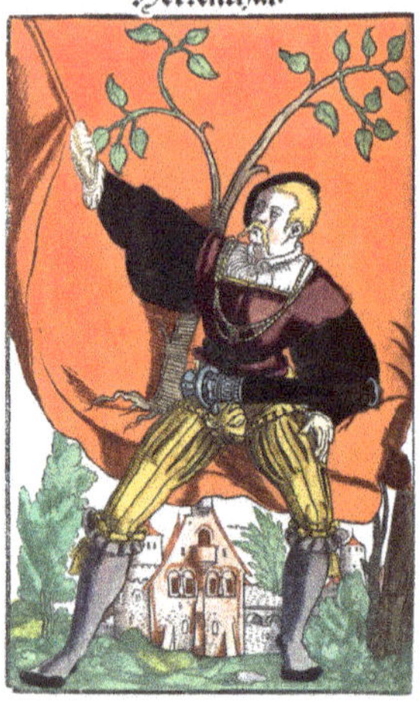

A tree branch with green leaves on a red field.

68. Benfelt | Benfeld, France

Benfelt.

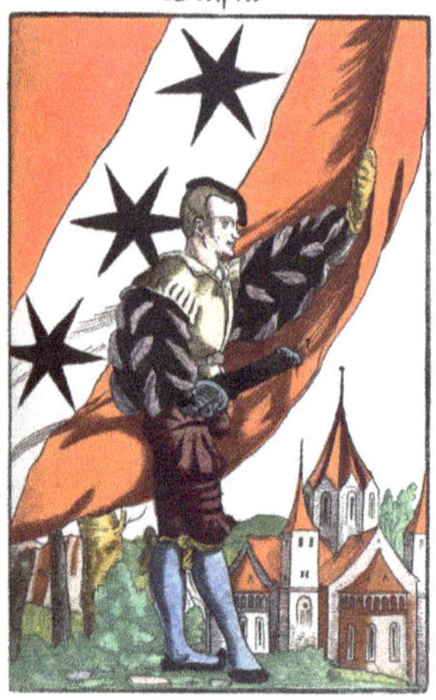

A white strip with three black stars with a red field on the sides.

69. Mülhaussen | Muhlhausen

Mülhaussen.

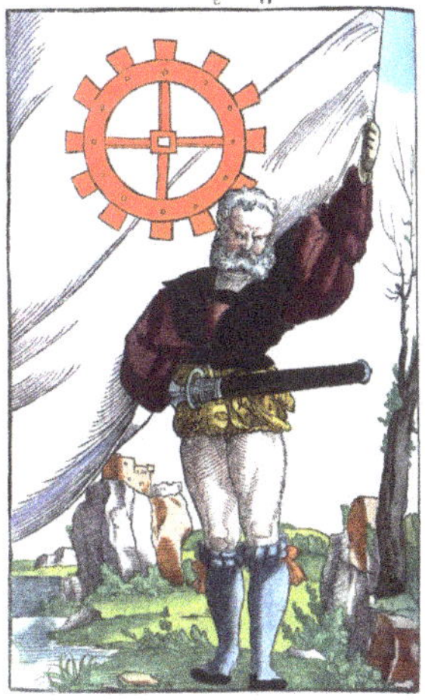

A red mill wheel on a white field.

70. Ole | Ala, Halla, Ahl am Etsch, in Trento Italy

Ole.

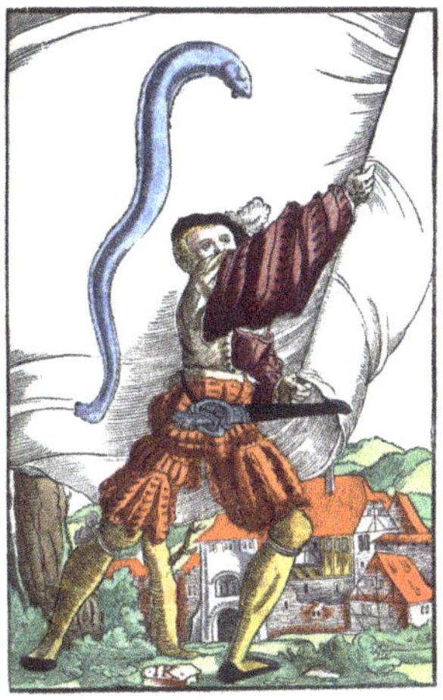

A blue eel on a white field.

71. Kolmar | Colmar

Kolmar.

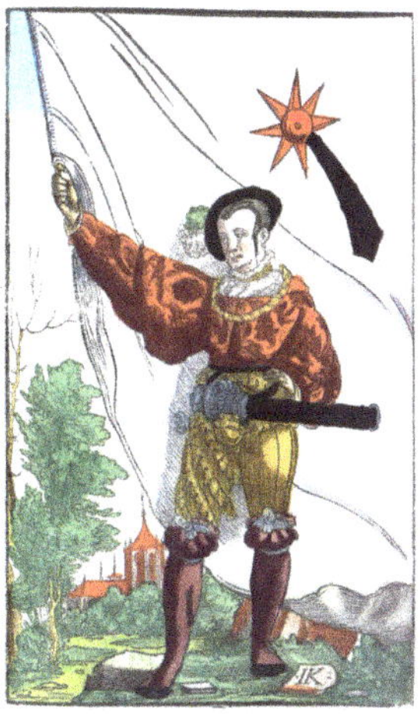

A mace with a red top and a black handle, on a white field.

72. Biberach | Biberach an der Riss

Biberach.

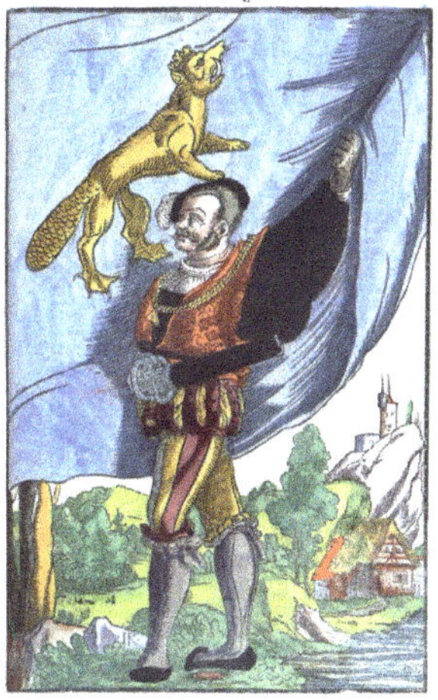

A yellow beaver wearing a crown, on a blue field.

73. Oppenheim | Oppenheim

Oppenheim.

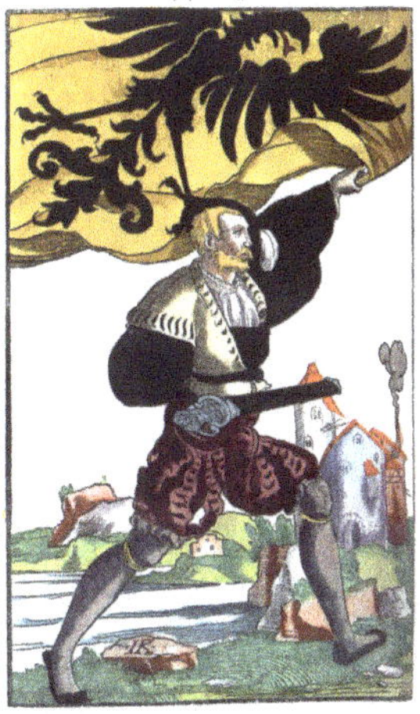

A black eagle on a yellow field

74. Breisach | Breisach

Breisach.

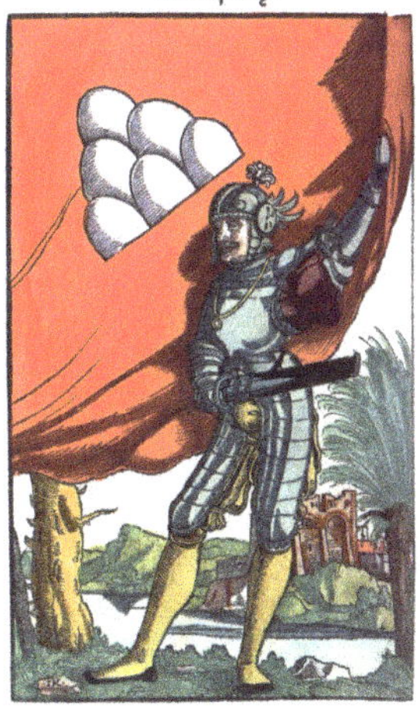

Six white mountains on a red field.

75. Schwebisch Gemind | Schwäbisch Gmünd

Schwebisch Gemind.

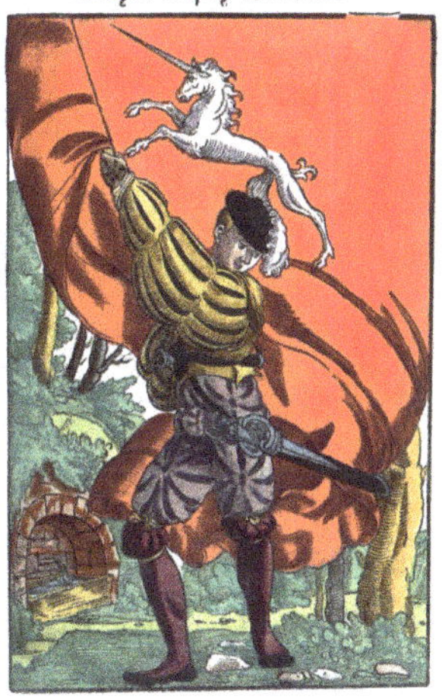

A white unicorn on a red field.

76. Keisers Lautern | Kaiserslautern

Keisers Lautern.

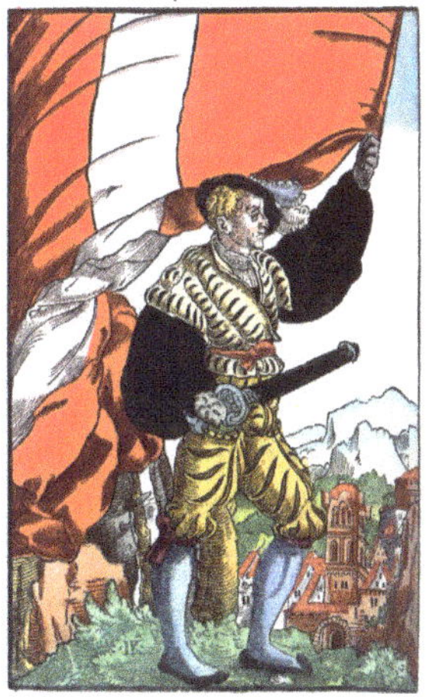

The middle part white, and the sides are red.

77. Memingen | Memingen

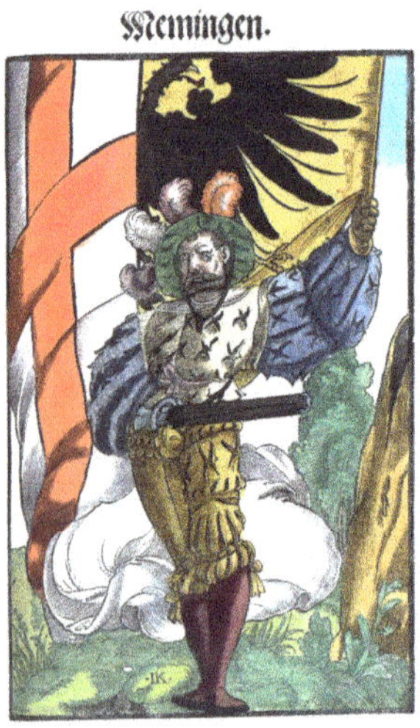

Half a black eagle on a yellow field on one side, the other side has a red cross on a white field.

78. Menz | Mainz

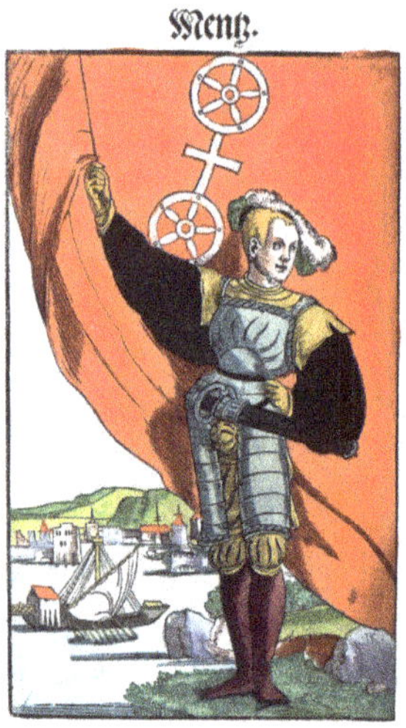

White wheels and cross on a red field.

79. Cobelenz | Koblenz

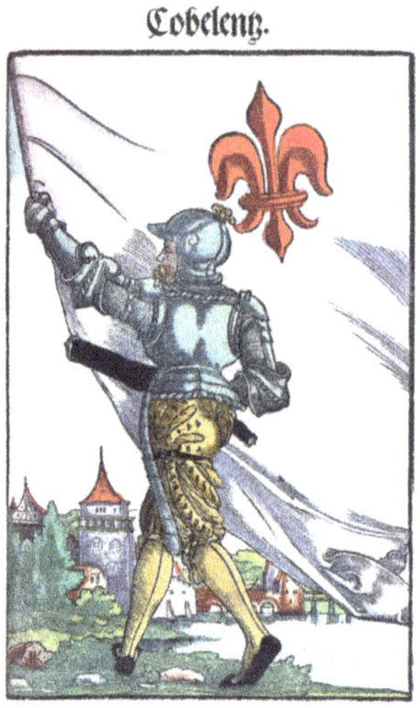

Red Fleur-de-lys on a white field.

80. Altdorff | Altdorf

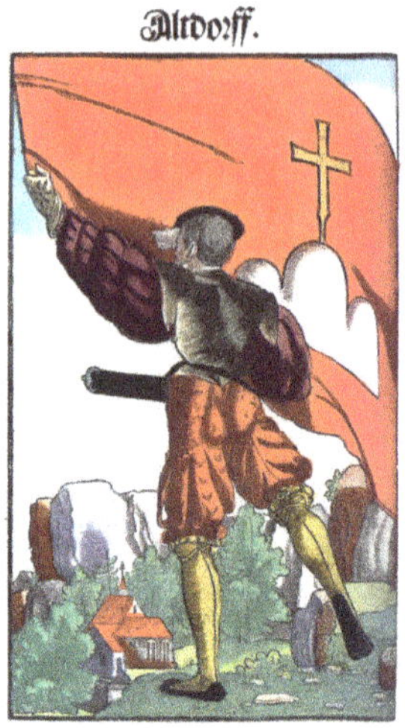

Three white mountains with a yellow cross on a red field.

81. Löuen | Leuven, Belgium

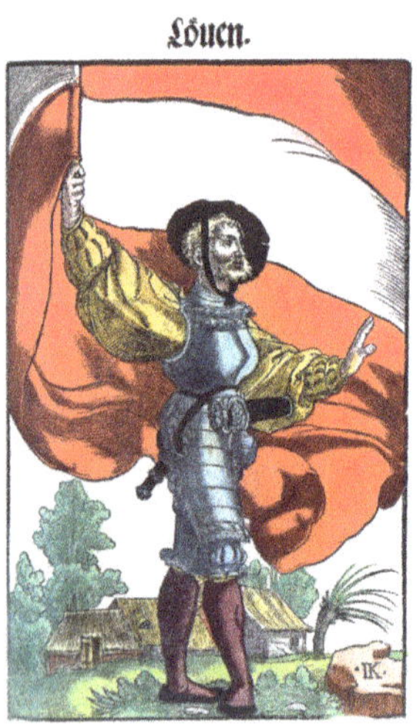

Middle field is white, the top and bottom bands in red.

82. Erdfurt | Erfurt

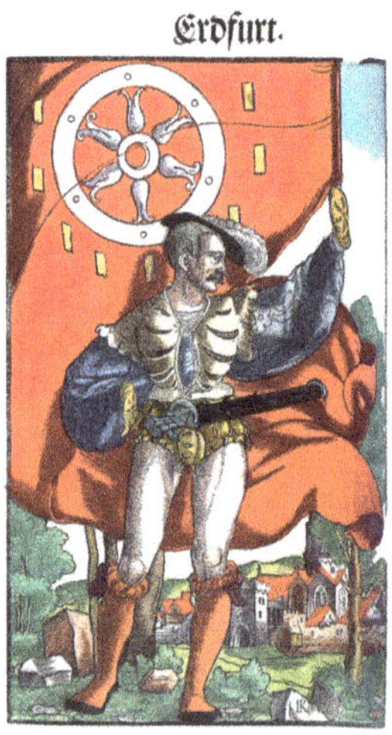

A white wheel surrounded by yellow stems, on a red field.

83. Lindaw | Lindau am Bodensee, Lindau by Lake Constance

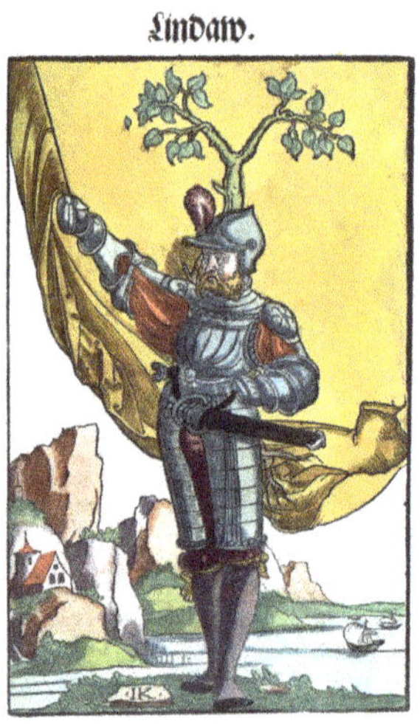

A Linden tree branch with green leaves, on a yellow field.

84. Halle in Sachssen | Halle/Saale in Saxony

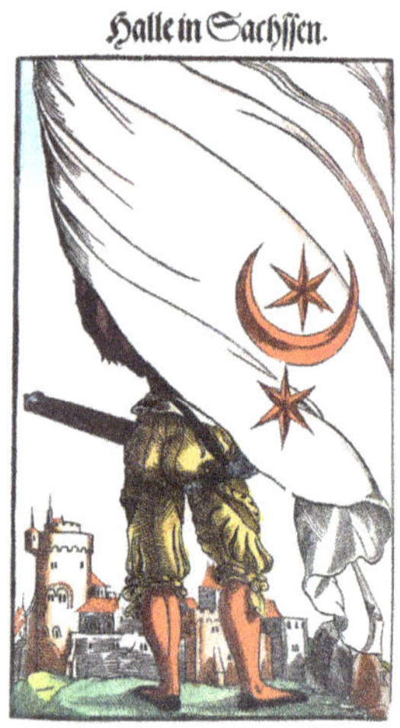

A red moon and stars on a white field.

85. Dircken | Turckheim

Dircken.

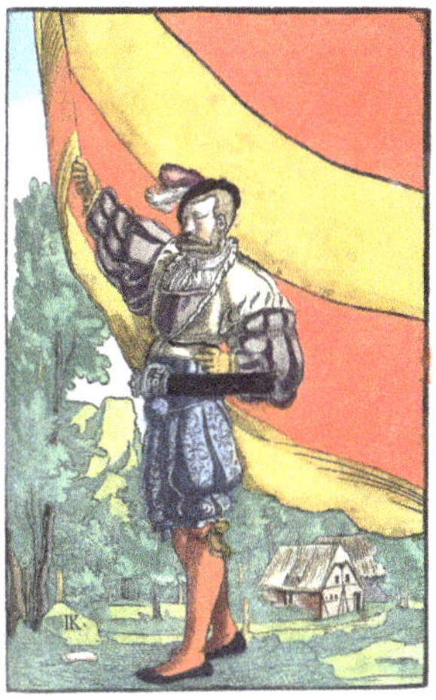

Four stripes, the first and third red, the second and fourth yellow.

86. Newenburg | Neuenburg am Rhein

Newenburg.

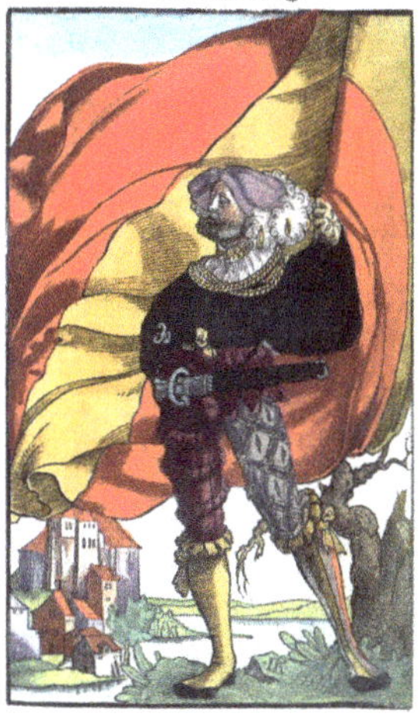

The middle part yellow, the top and bottom red.

87. Lantzhut | Landshut

Lantzhut.

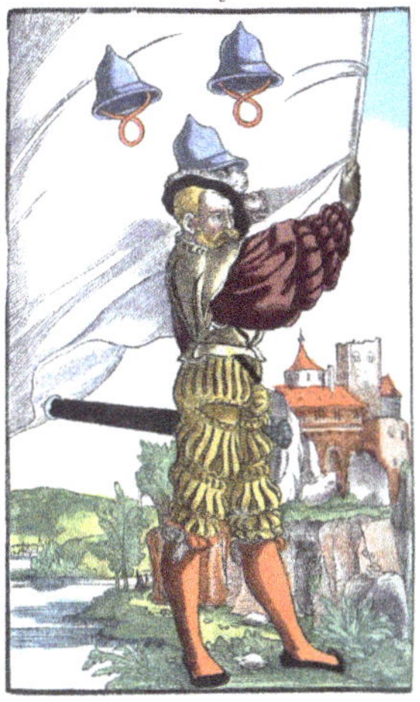

Three blue helmets with red straps on white field.

88. Schweinfurt | Schweinfurt

Schweinfurt.

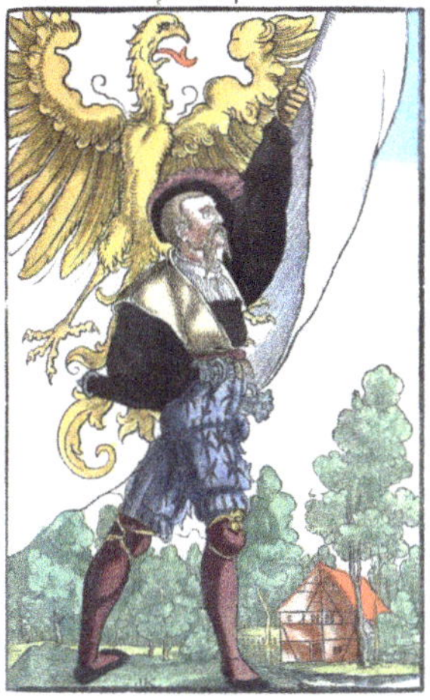

Yellow eagle on a white field.

89. Leichtennaw | Lichtenau

Liechtennaw.

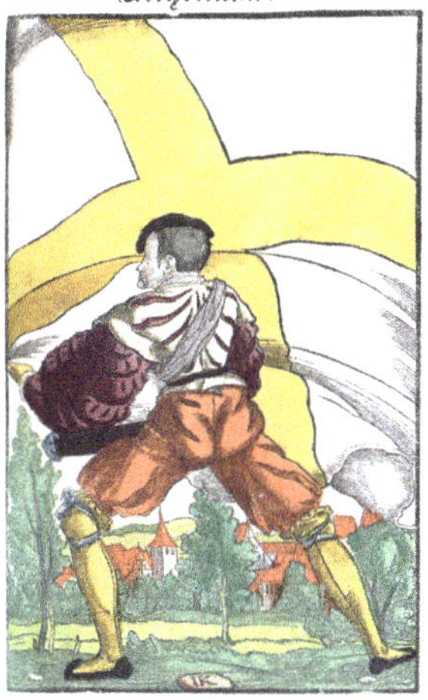

Yellow cross on a white field.

90. Mastricht | Maastricht, Netherlands

Mastricht.

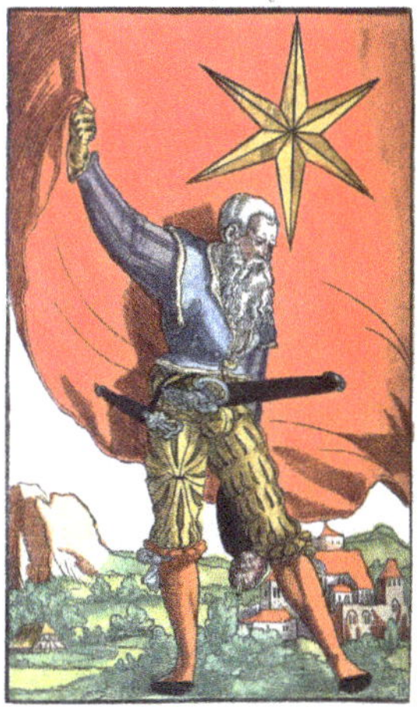

Yellow star on a red field.

91. Halberstat | Halberstadt

Halberstat.

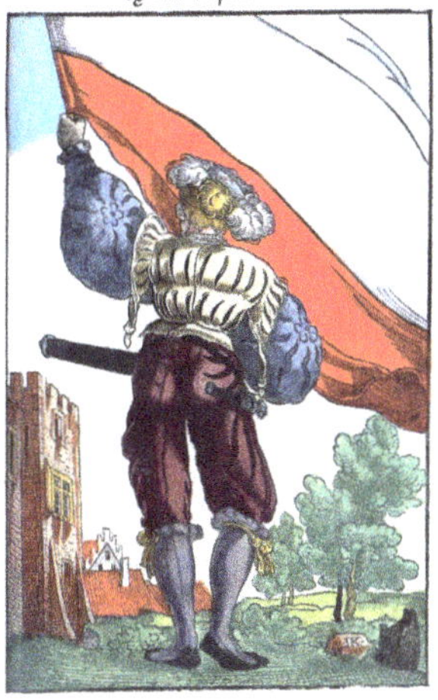

Top half white and the bottom half red.

92. Lagingen | Lauingen

Lagingen.

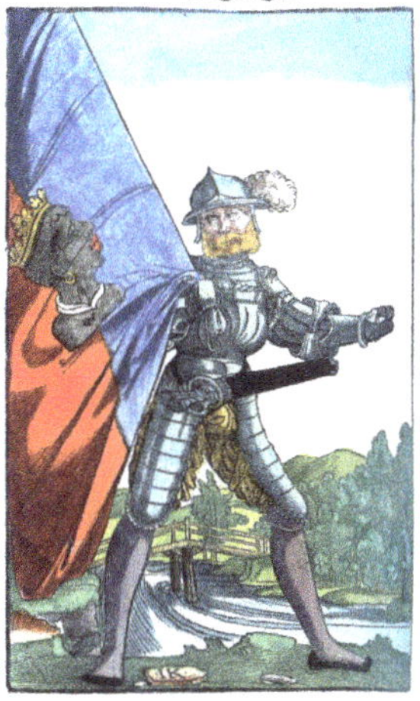

The head and shoulders of an African man, with red lips, gold earrings and crown, white necklace, on a field of blue and red, blue is closest to the pole.

93. Straszburg | Strasbourg, France

Straßburg.

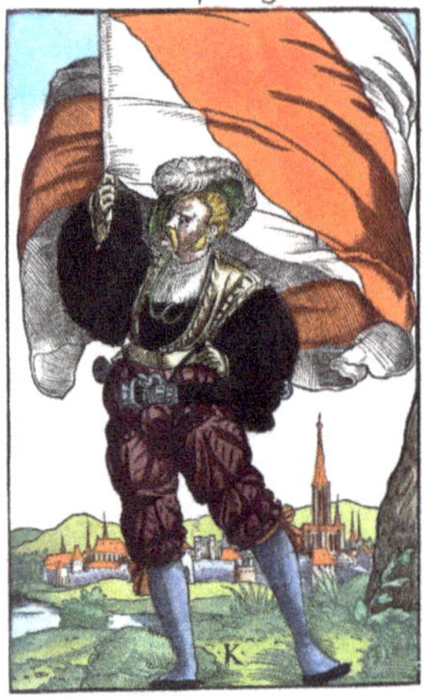

White above and below, red in the middle.

94. Franckfurt | Frankfurt

Franckfurt.

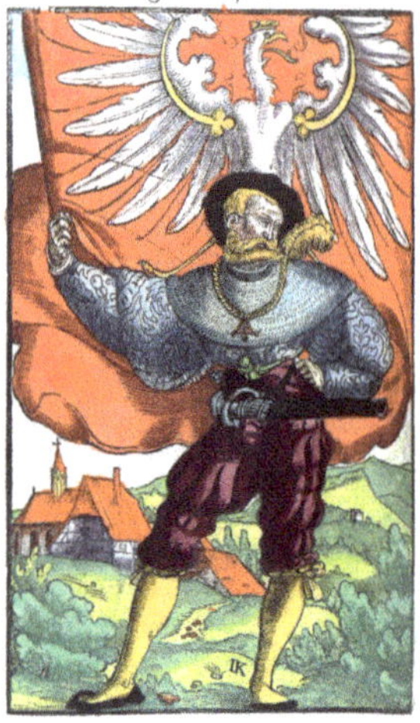

On a red field, a white eagle with a green tongue and claws, and a gold crown, beak, wing stripes and feet.

95. Buchhorn | Friedrichshafen

Buchhorn.

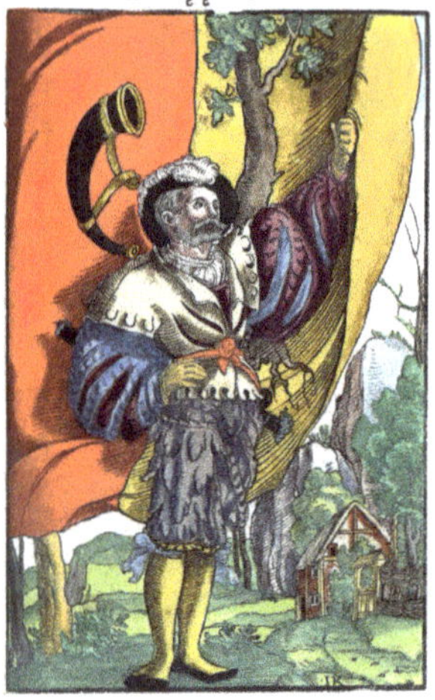

On a yellow field, a tree with green leaves. The other half is a red field with a black horn with yellow belt and fittings.

96. Gengen | Giengen an der Brenz

Gengen.

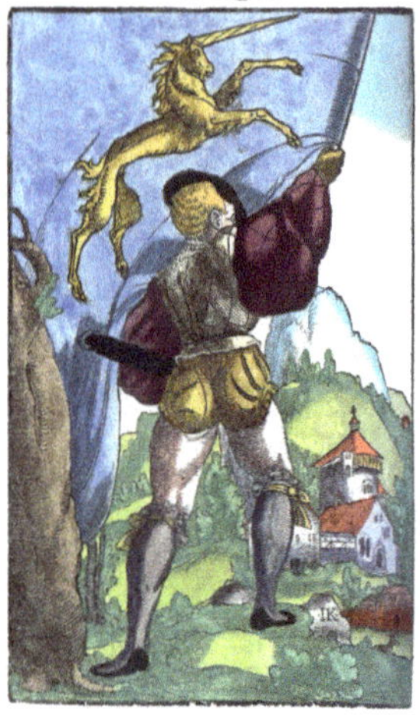

Yellow unicorn on a blue field.

97. Halle in Schwaben | Schwäbisch Hall

Halle in Schwaben.

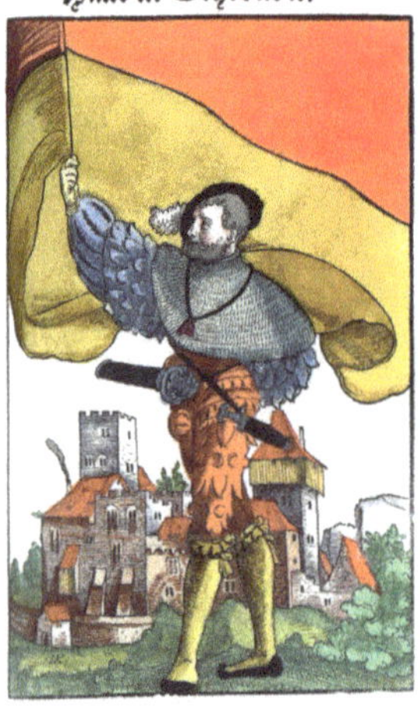

Top half in red, bottom half in yellow.

98. Gengenbach | Gengenbach

Gengenbach.

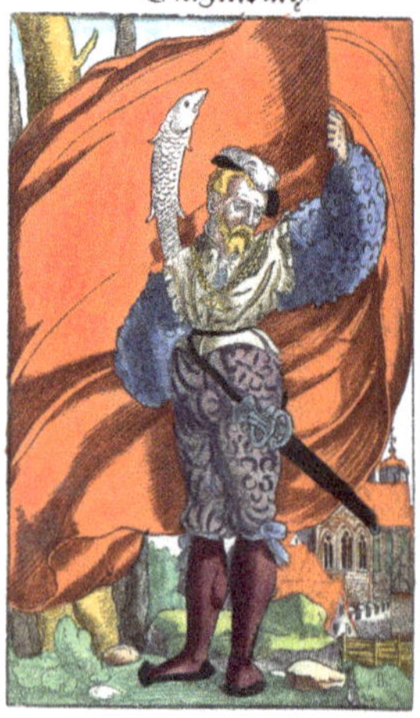

White fish on a red field.

99. Keiserberg | Kaysersberg

Keiserberg.

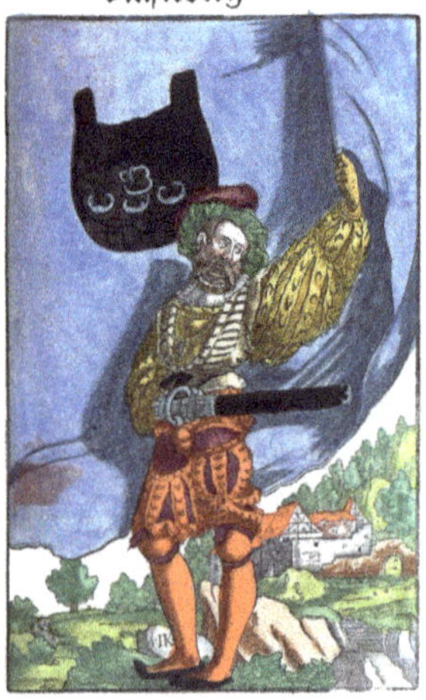

Black purse on a blue field.

100. Raffenspurg | Ravensburg

Raffenspurg.

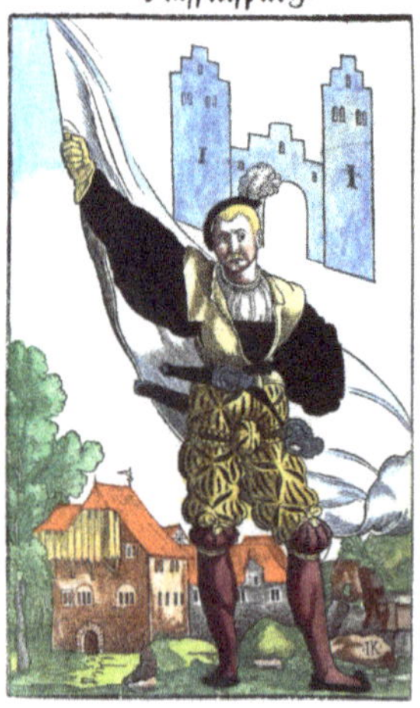

Blue castle on a white field.

101. Pfullendorff | Pfullendorf

Pfullendorff.

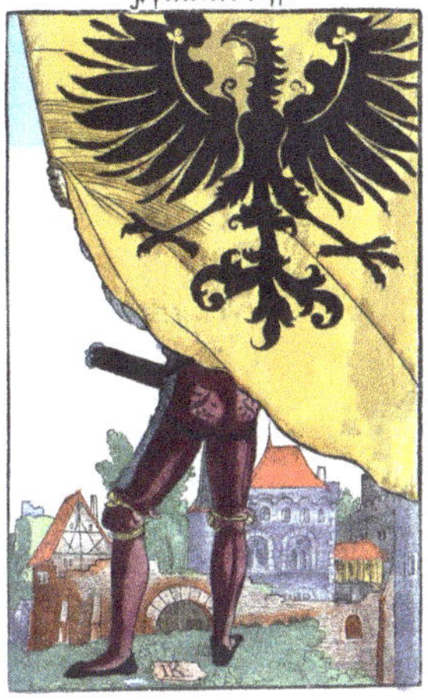

Black eagle with gold beak and stripes on the wings on a yellow field.

102. Issnach | Isny im Allgäu

Jßnach.

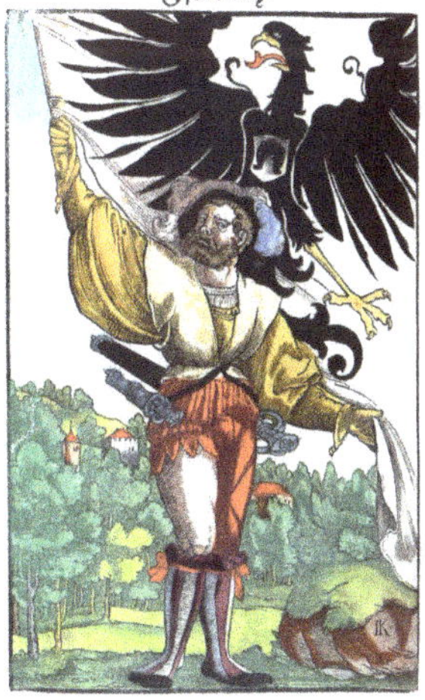

Black eagle on a white field, with a black horseshoe with white nails on a white shield on the eagle's chest.

103. Nürenberg | Nürnberg, Nuremberg Flag of the Imperial City

Nürenberg.

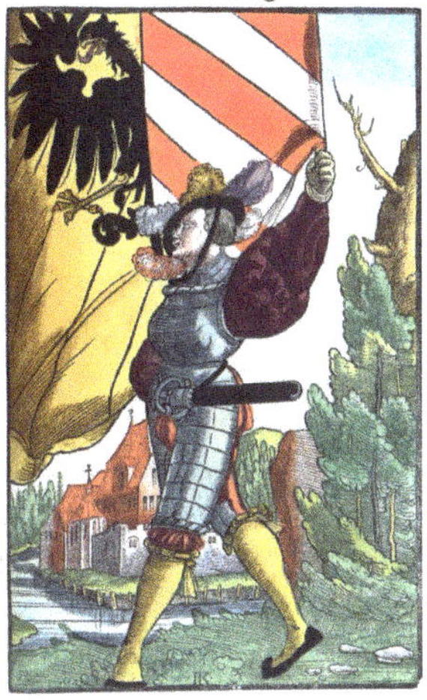

Black lion on a yellow field on one half of the flag, with white and red stripes on the other side.

104. Esslingen | Esslingen

Eßlingen.

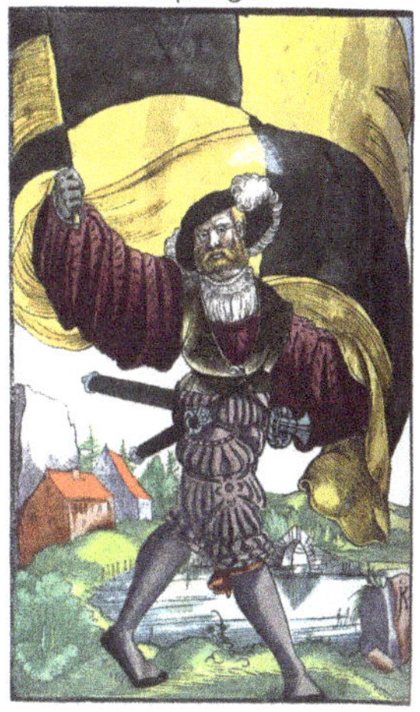

The flag is described in the list as a black eagle on a yellow field. However the depicted flag is yellow and black check, and might either be a battle flag, or a mislabeled flag.

105. Wirzburg | Wurzburg

Wirtzburg.

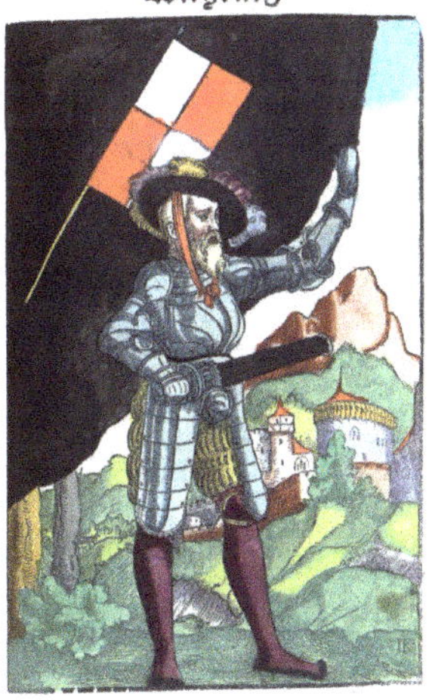

A white and red check flag with a yellow pole on a black field.

106. Kauffbeüren | Kaufbeuren

Kauffbeüren.

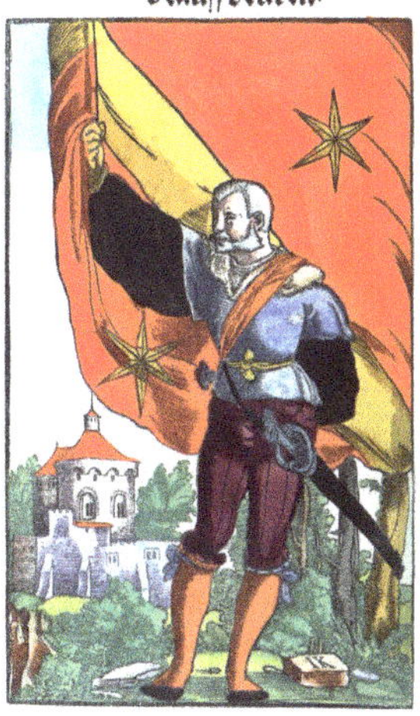

Two yellow stars on a red field, with a yellow stripe down the middle.

107. Rottenburg an der Thauber | Rothenburg ob der Tauber

Rottenburg an der Thauber.

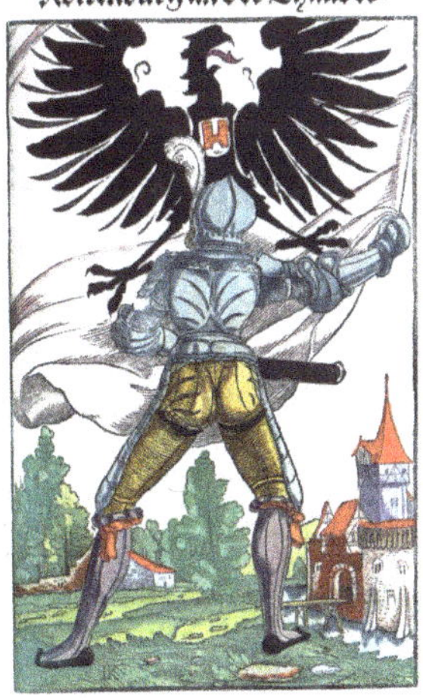

A black eagle on a white field, with a red castle on a white shield on the eagle's chest.

108. Speir | Speyer

Speir.

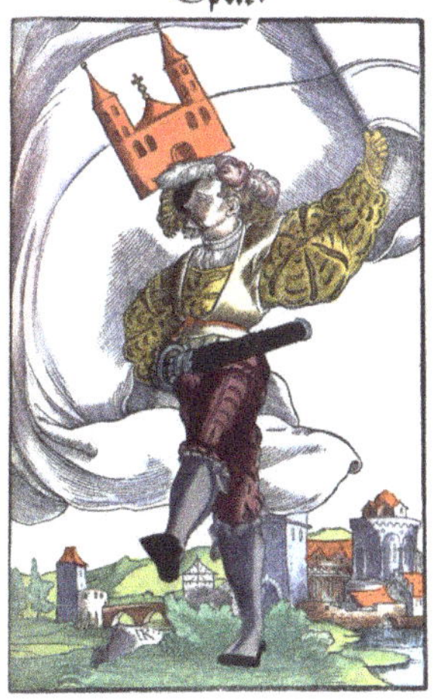

A red church on a white field.

109. Wurmbs | Worms

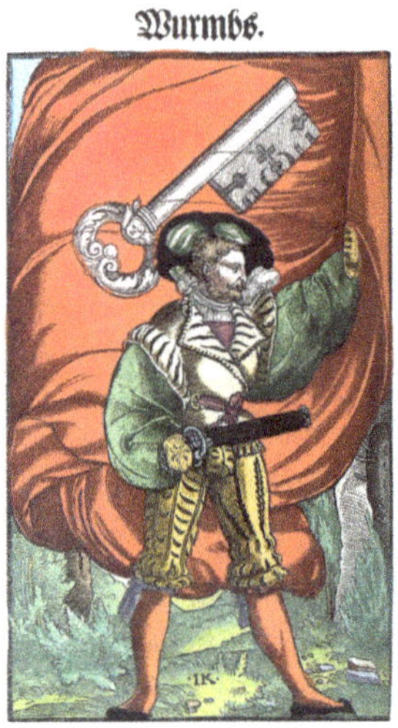

A large white key on a red field.

110. Sarburg | Sarrebourg, Lorraine France

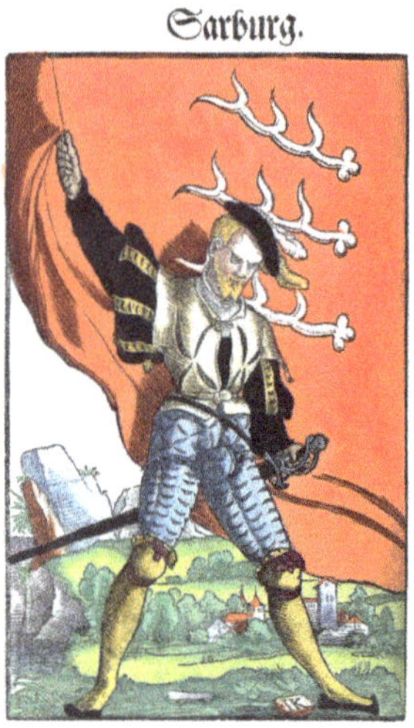

Three white stags horns on a red field.

111. Ritlingen | Reutlingen

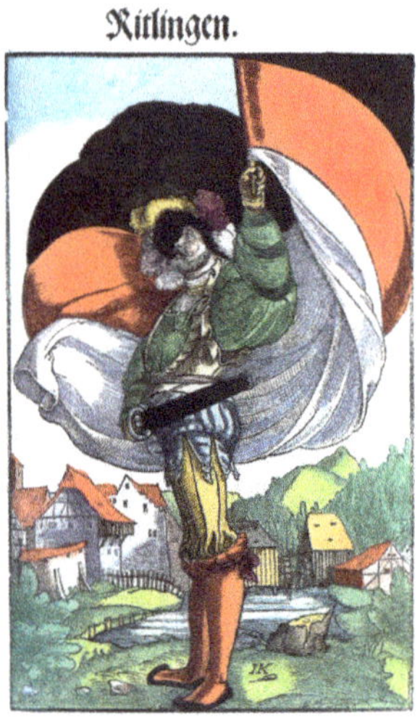

A flag with three horizontal stripes, the top black, the middle red, the bottom white.

112. Weissenburg das erst | Weißenburg in Bavaria

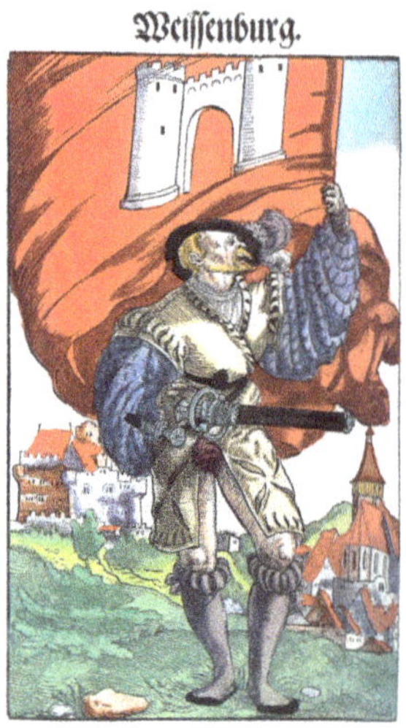

A white castle or gate on a red field.

113. Weisenburg das ander | Weißenburg in Alsace, Wissembourg, France

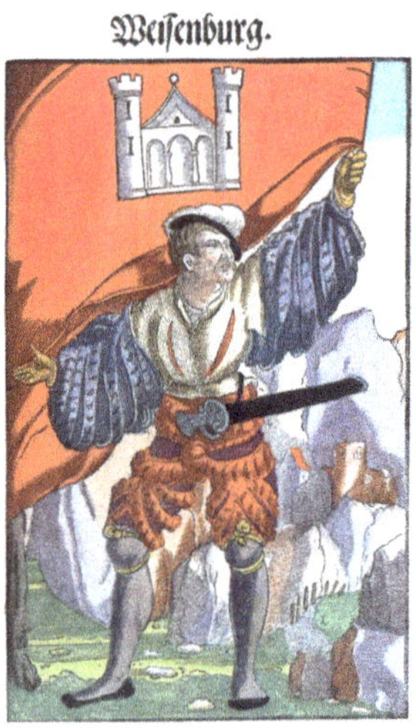

Like the first flag, but a different building.

114. Ingelnheim | Ingelheim am Rhein

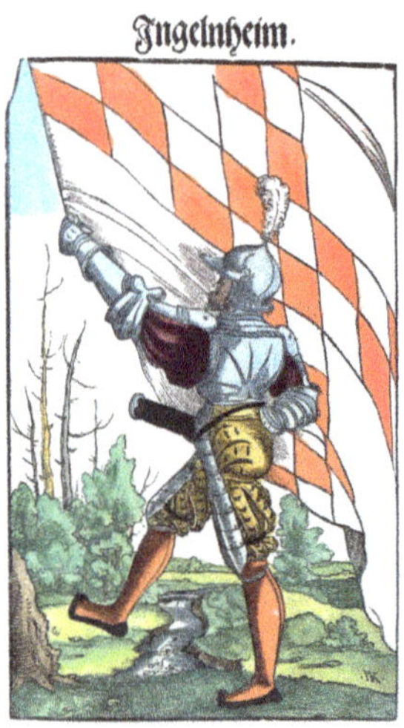

A cross made of two rows of red and white check, with the four corners in white.

115. Mechel | Mechelen, Belgium

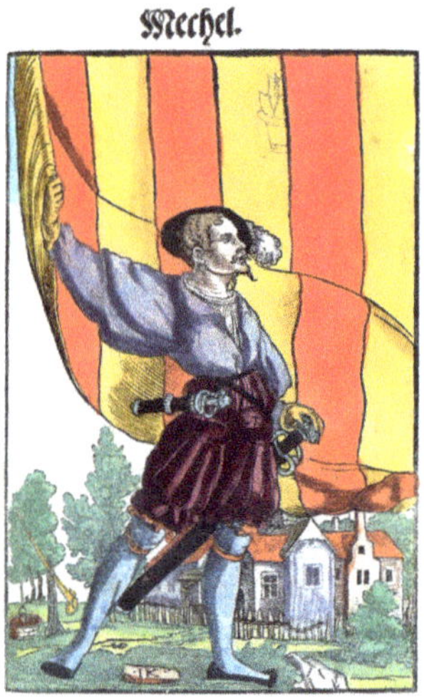

A flag with 7 vertical stripes, four yellow and three red, with yellow closest to the pole.

116. Lüttich | Liège, Belgium

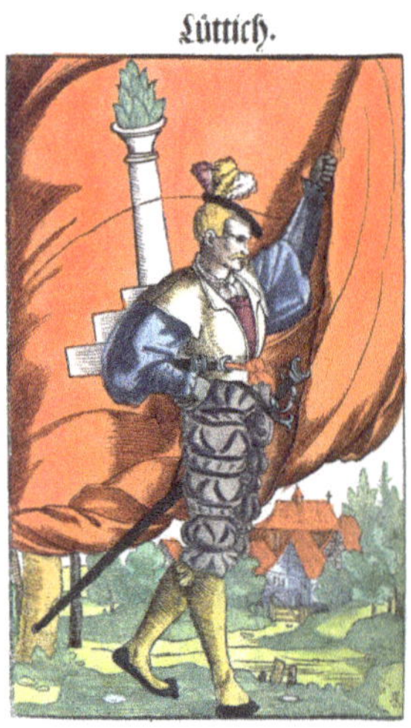

A white column on a red field.

117. Lübeck | Lübeck

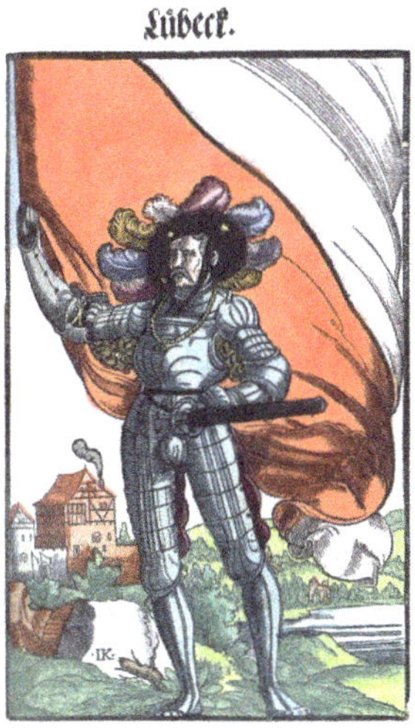

Two horizontal stripes, top is white, bottom is red.

118. Dinckelspühel | Dinkelsbühl

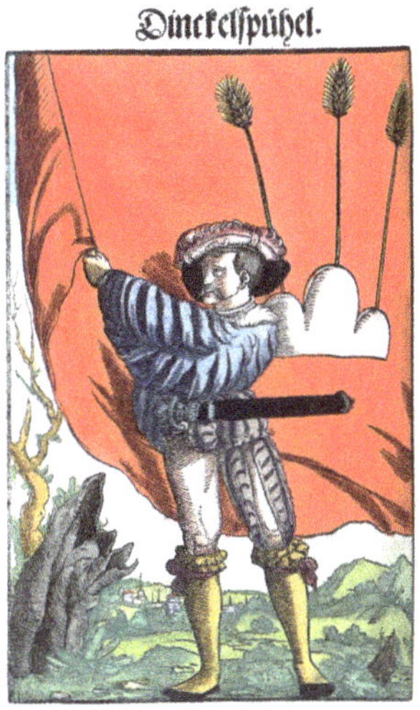

Three yellow stalks of wheat on three white mountains, all on a red field.

119. Trier | City of Trier

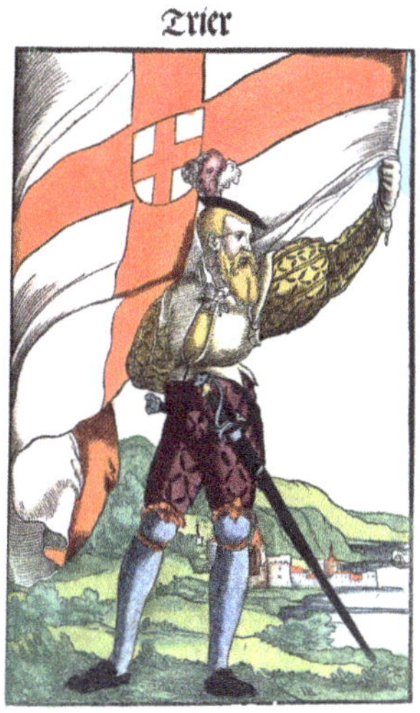

A red cross on a white field, with a shield with the same design in the center of the cross.

120. Kitzingen | Kenzingen

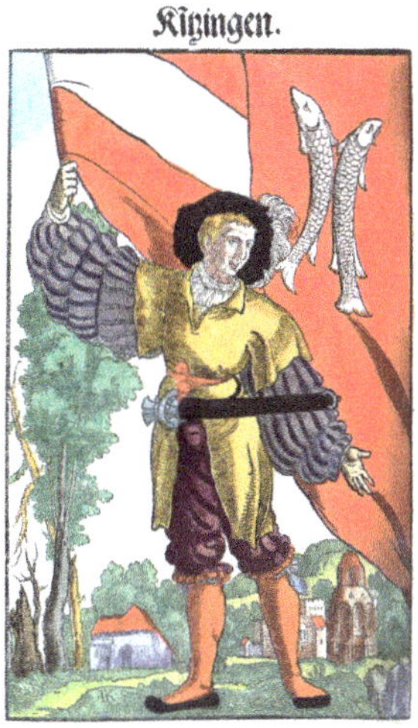

Closest to the pole, three horizontal stripes with white in the middle and red on top and bottom. On the other half of the flag, two white fish on a red field.

121. Wienssheym | Bad Windsheim

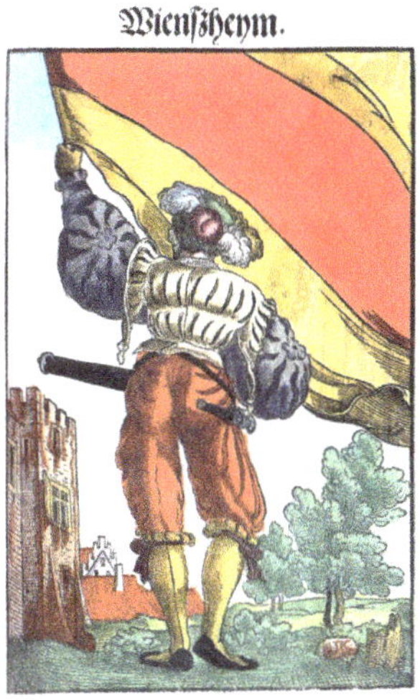

Windsheim is not included in the list of precedence, but its flag was included.

The original manuscript had 143 plates, and this edition has 144. Brandeburg is listed twice, as an Elector and as a Margrave however its flag only appeared once in the original manuscript. For clarity and ease of use, it has been included twice in this edition.

122.

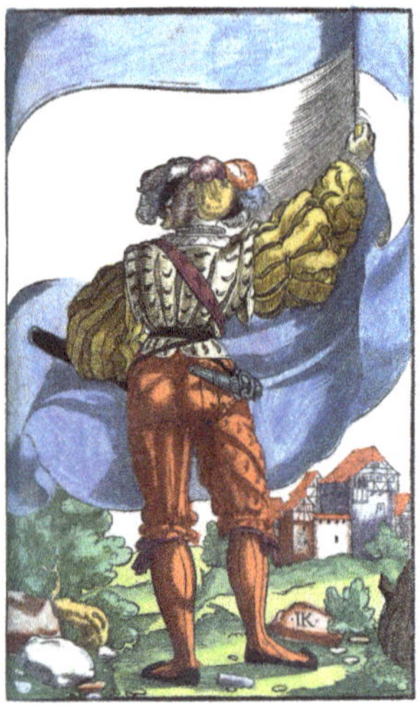

123.

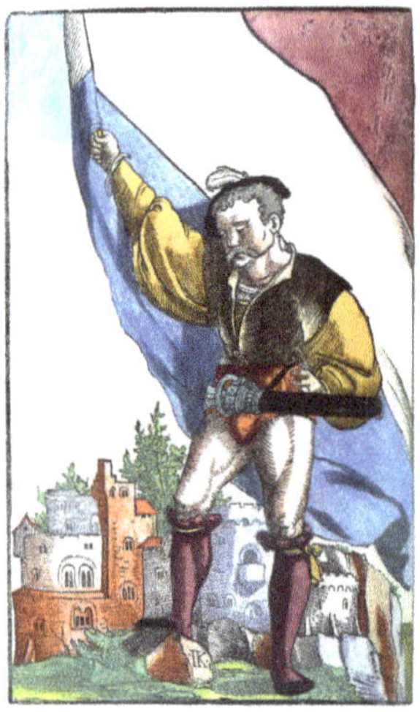

124.

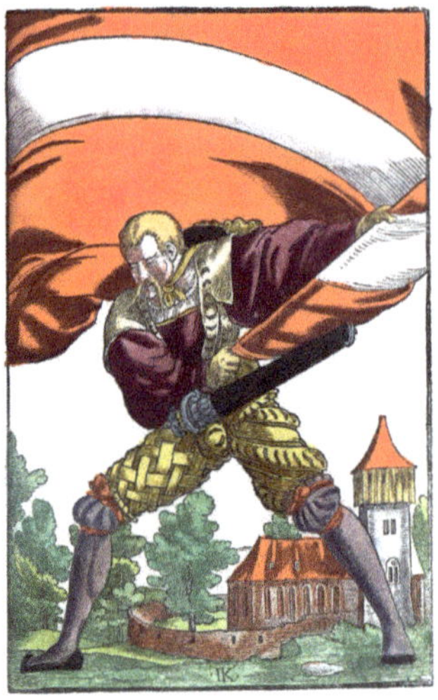

125.

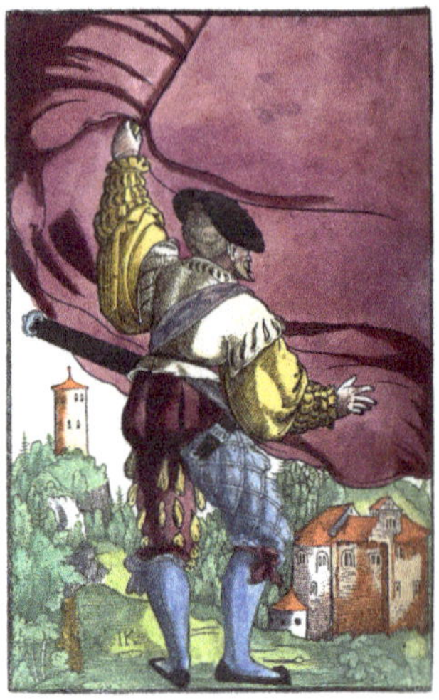

126.

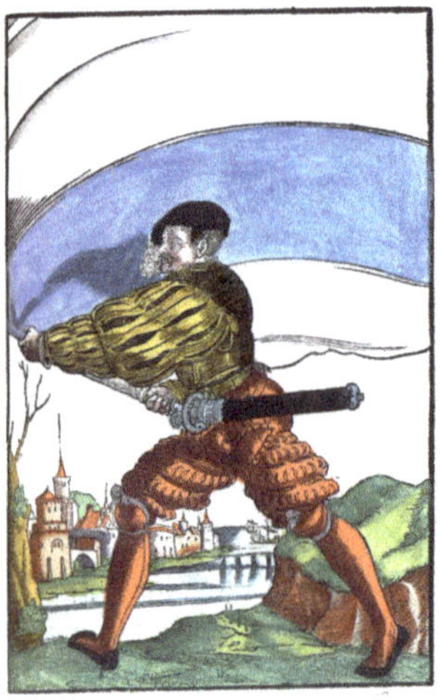

127.

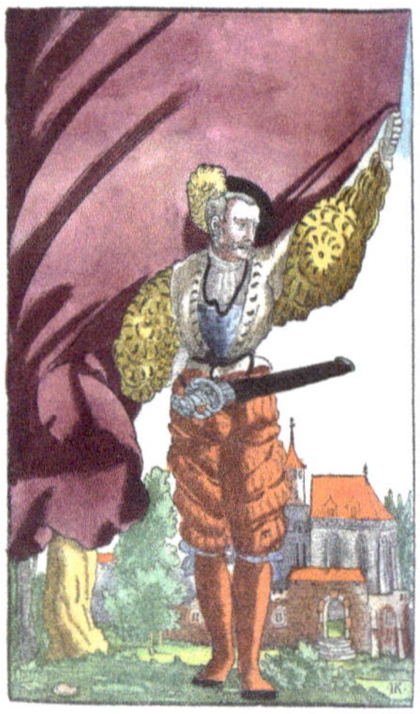

128.

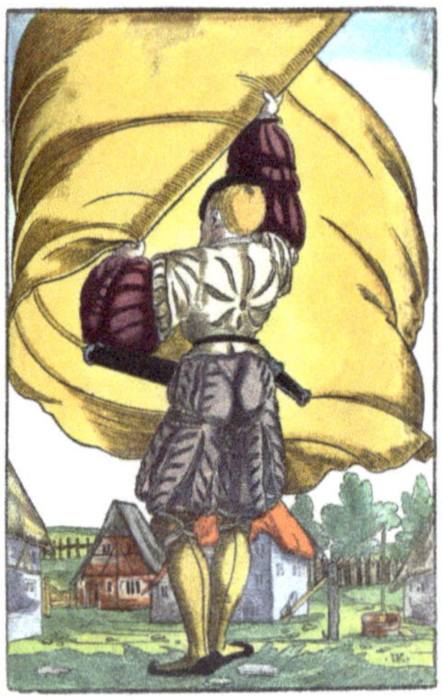

129.

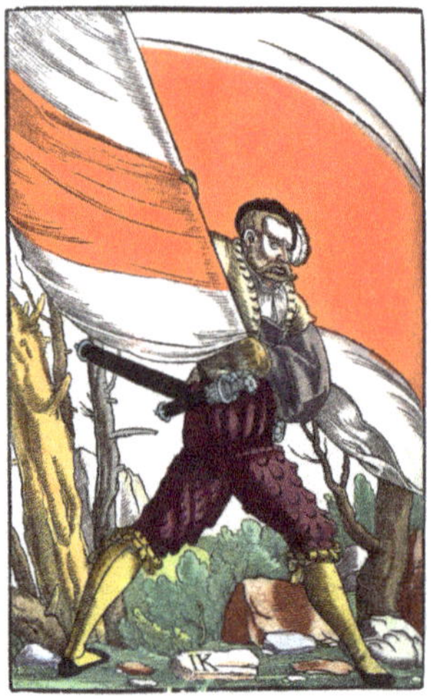

130.

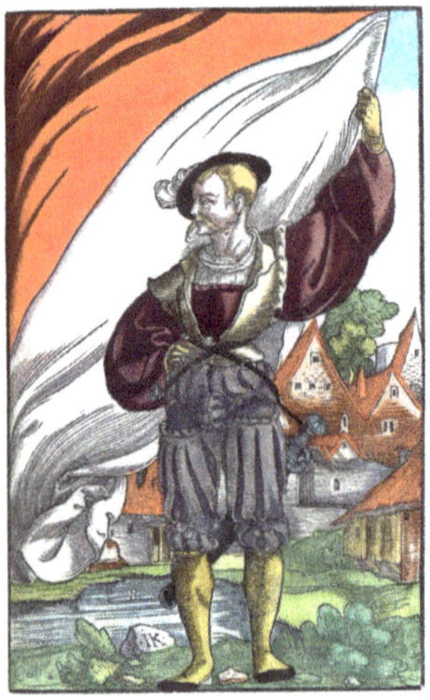

131.

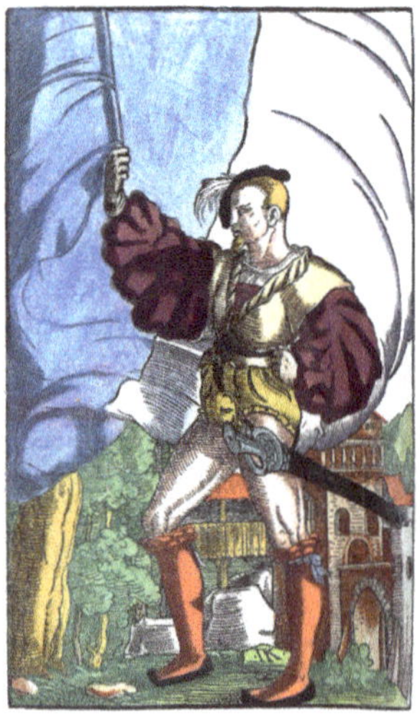

132.

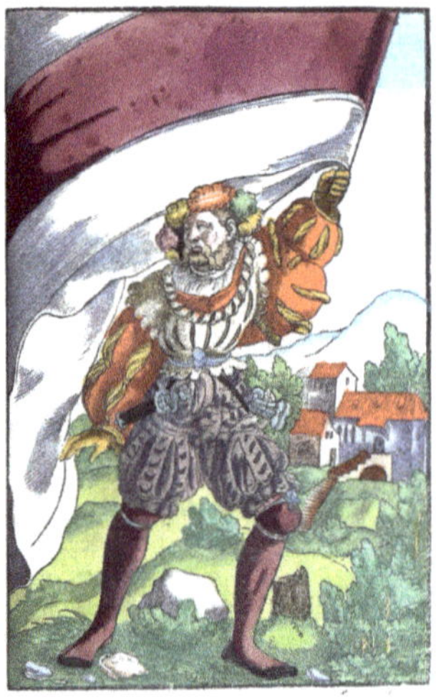

133.

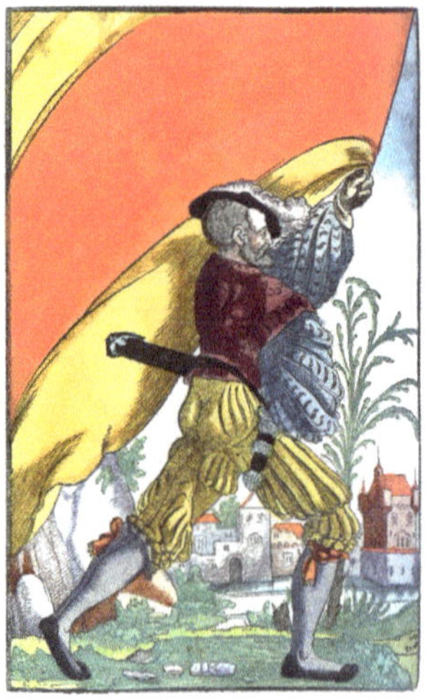

134.

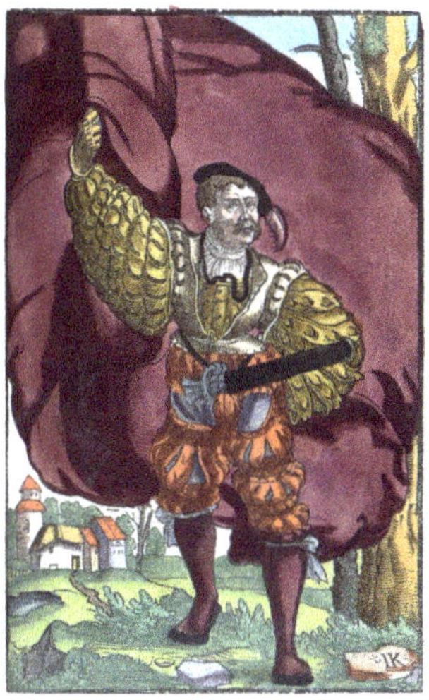

135.

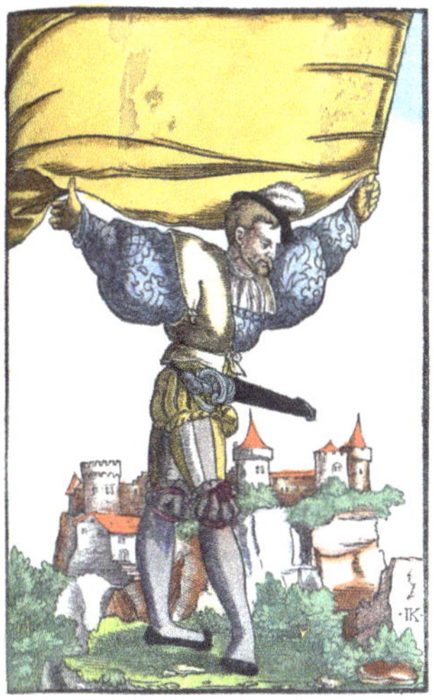

136.

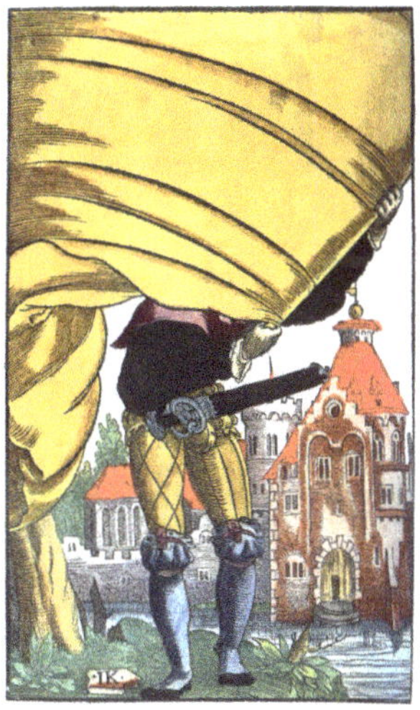

137.

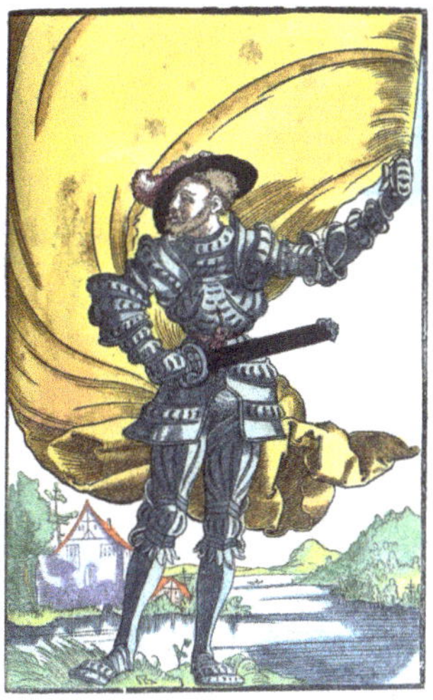

138.

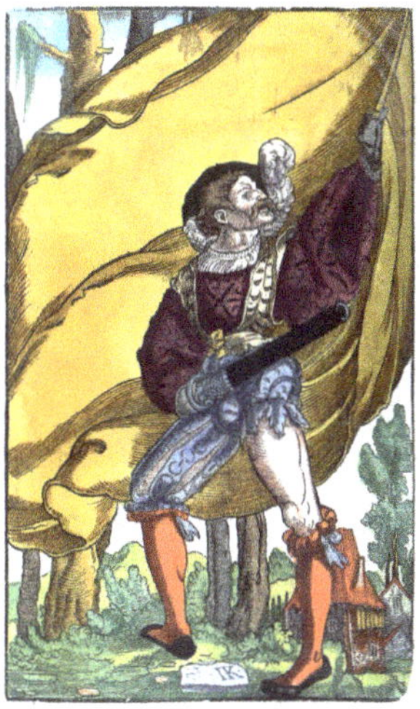

139.

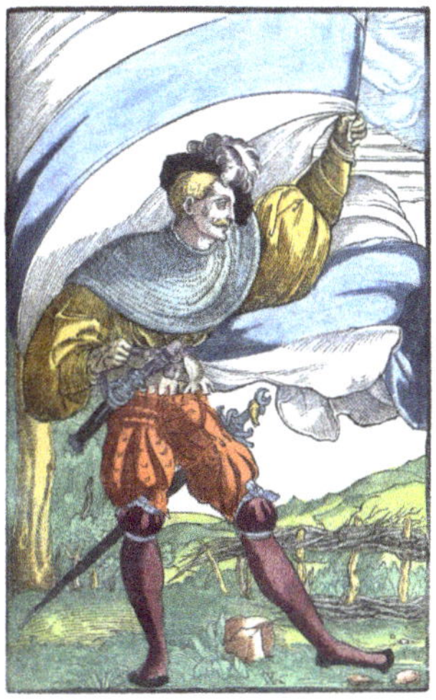

140.

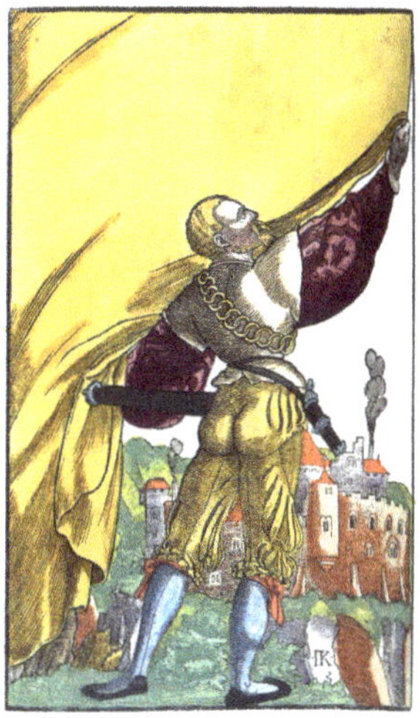

141..

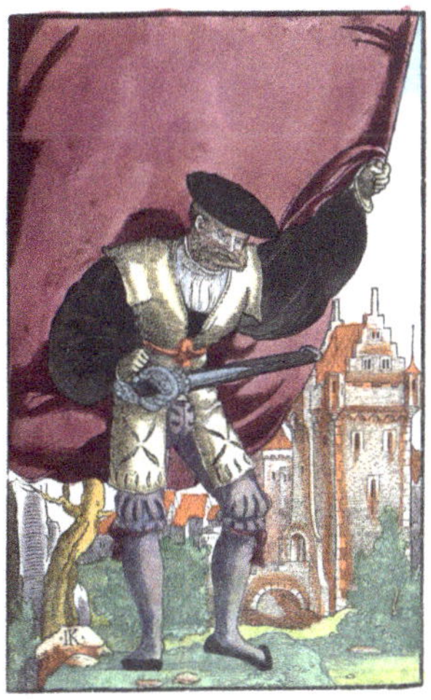

142.

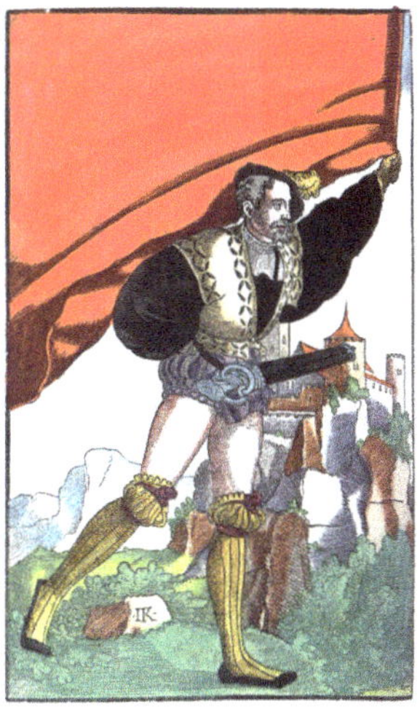

143.

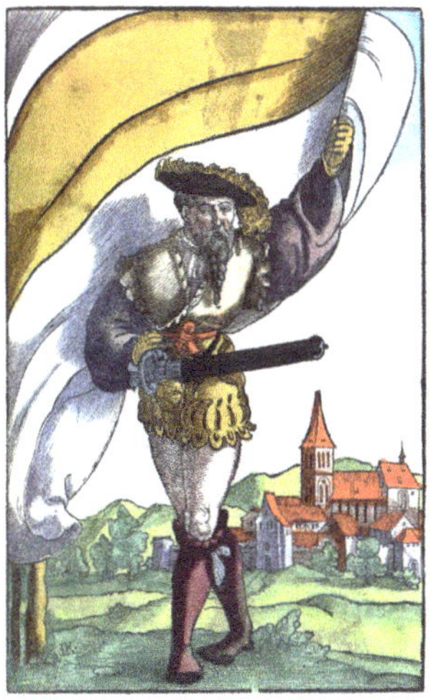

144.

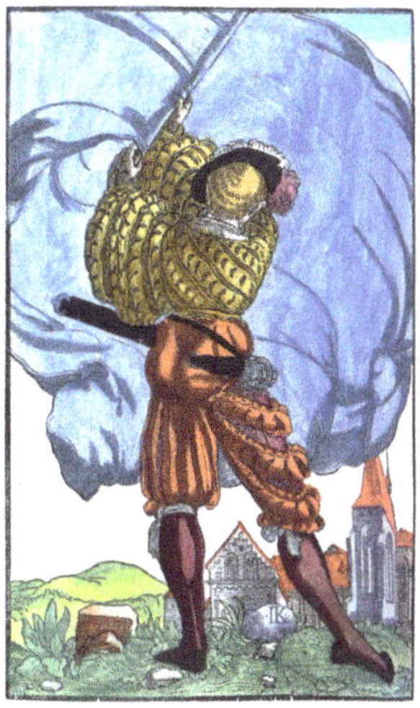

At the end of the book, there are twenty-three flags without heraldry, designed as blanks to allow for expansion or addition of the owners personal heraldry. In this manuscript they have been painted in a variety of designs, some of which look similar to battle flags seen elsewhere.

Conclusion

Heraldry was an important part of everyday life in early modern 16th century Germany. The book *Wappen des Heyligen Römischen Reichs Teutscher Nation* was published in Frankfurt by Cyriacus Jacob in 1545 as a handy guide to the flags of the Holy Roman Empire. Besides serving as a convenient heraldic reference guide, it may have been intended for a wide audience as a type of souvenir or patriotic memento for tourists or travelers to an Imperial Diet in 1545 or 1546, which was not held due to political issues and war.

However, for the modern reader, it provides a small glimpse into a brief moment in the past, before the religious wars, starting with the Schmalkaldic War in 1546-1547, would rip the lands and people of the Holy Roman Empire apart.

The Flags

of the

Holy Roman Empire

Plates 1 - 144

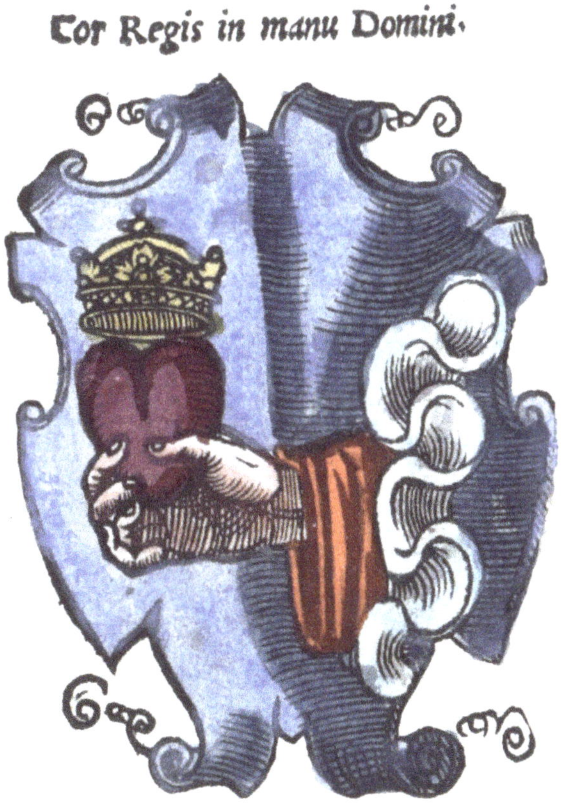

Des Reichs Fahn.

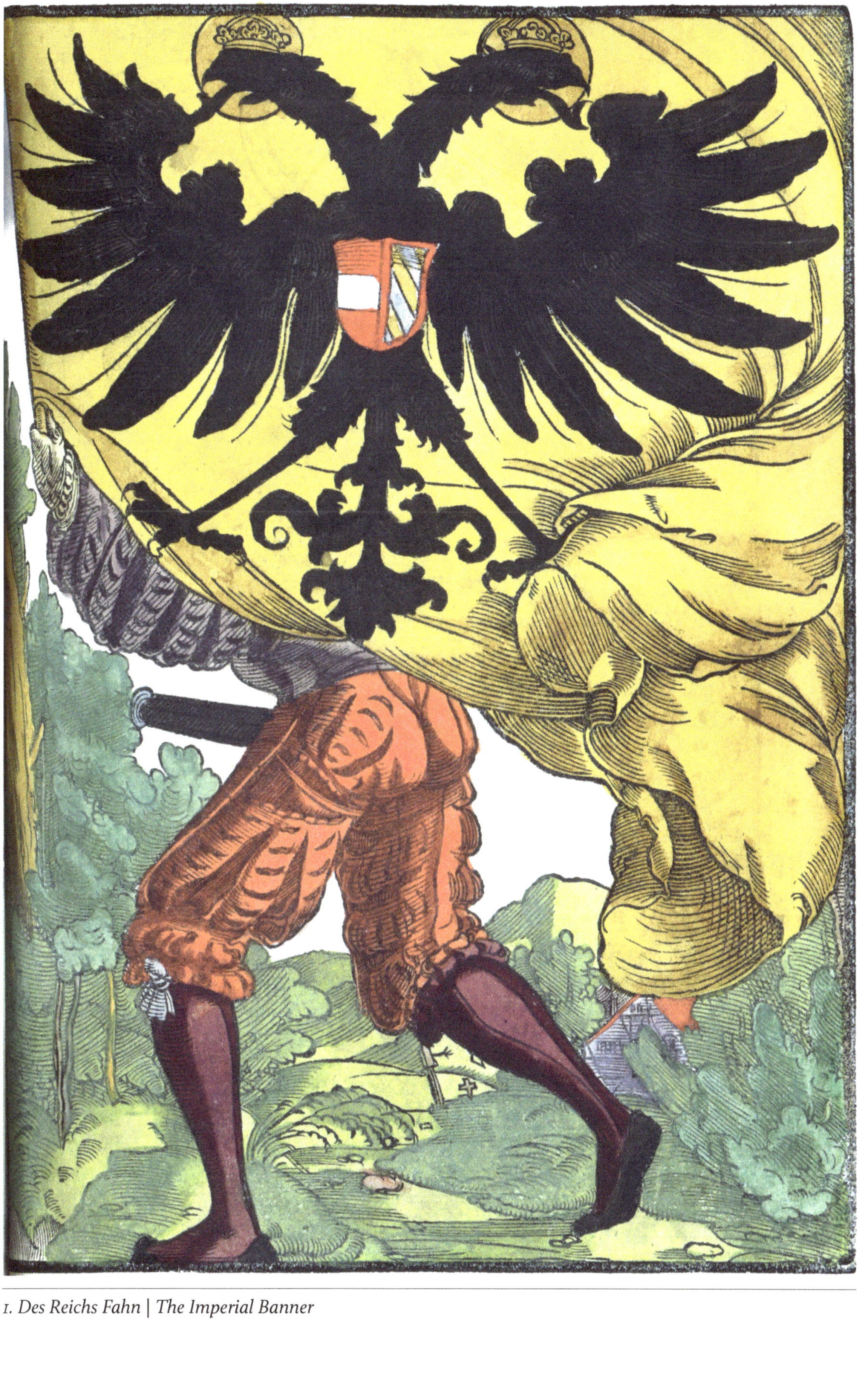

1. Des Reichs Fahn | The Imperial Banner

Menz.

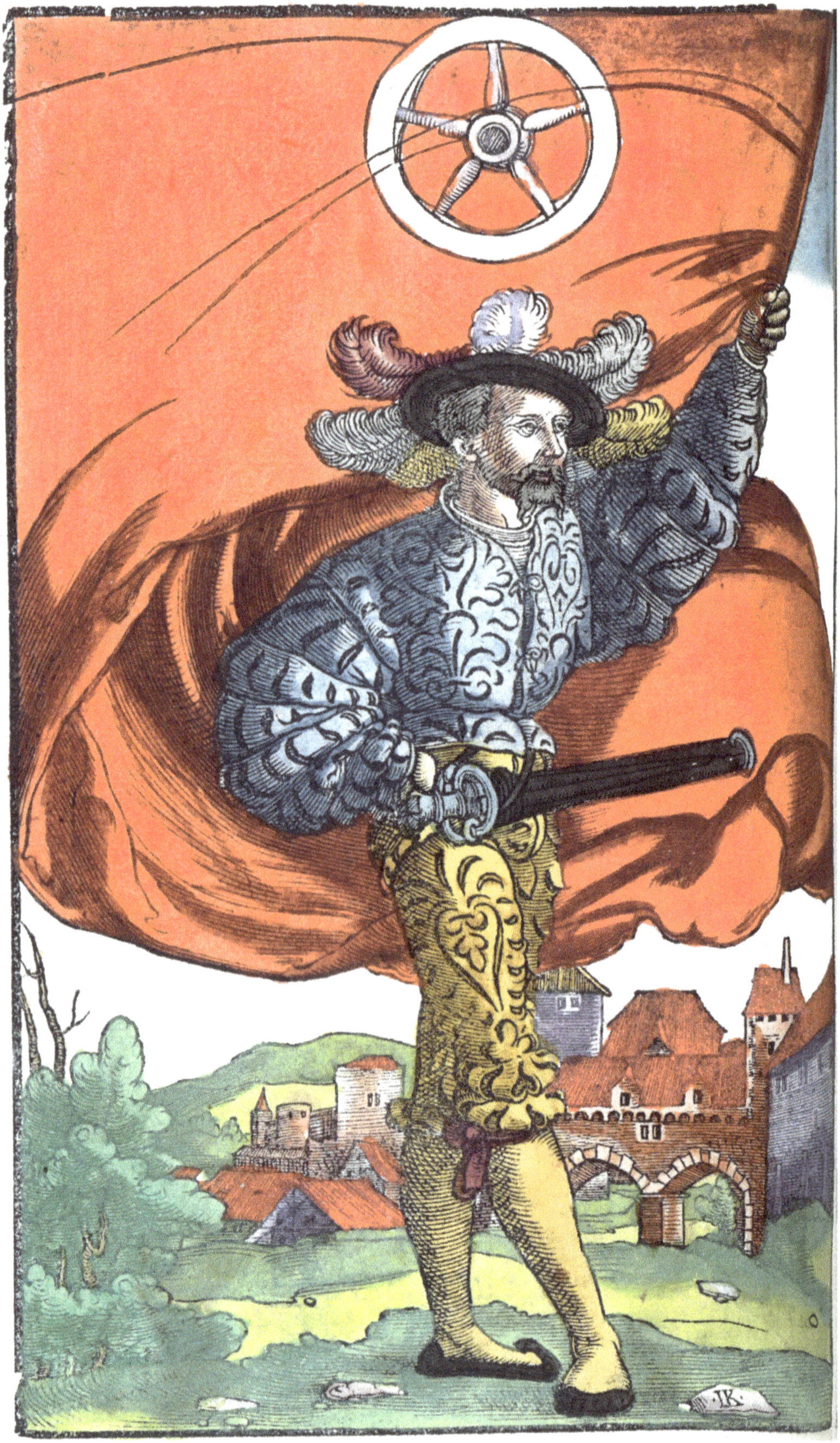

2. Menz | The Price Archbishop of Mainz

Cöln.

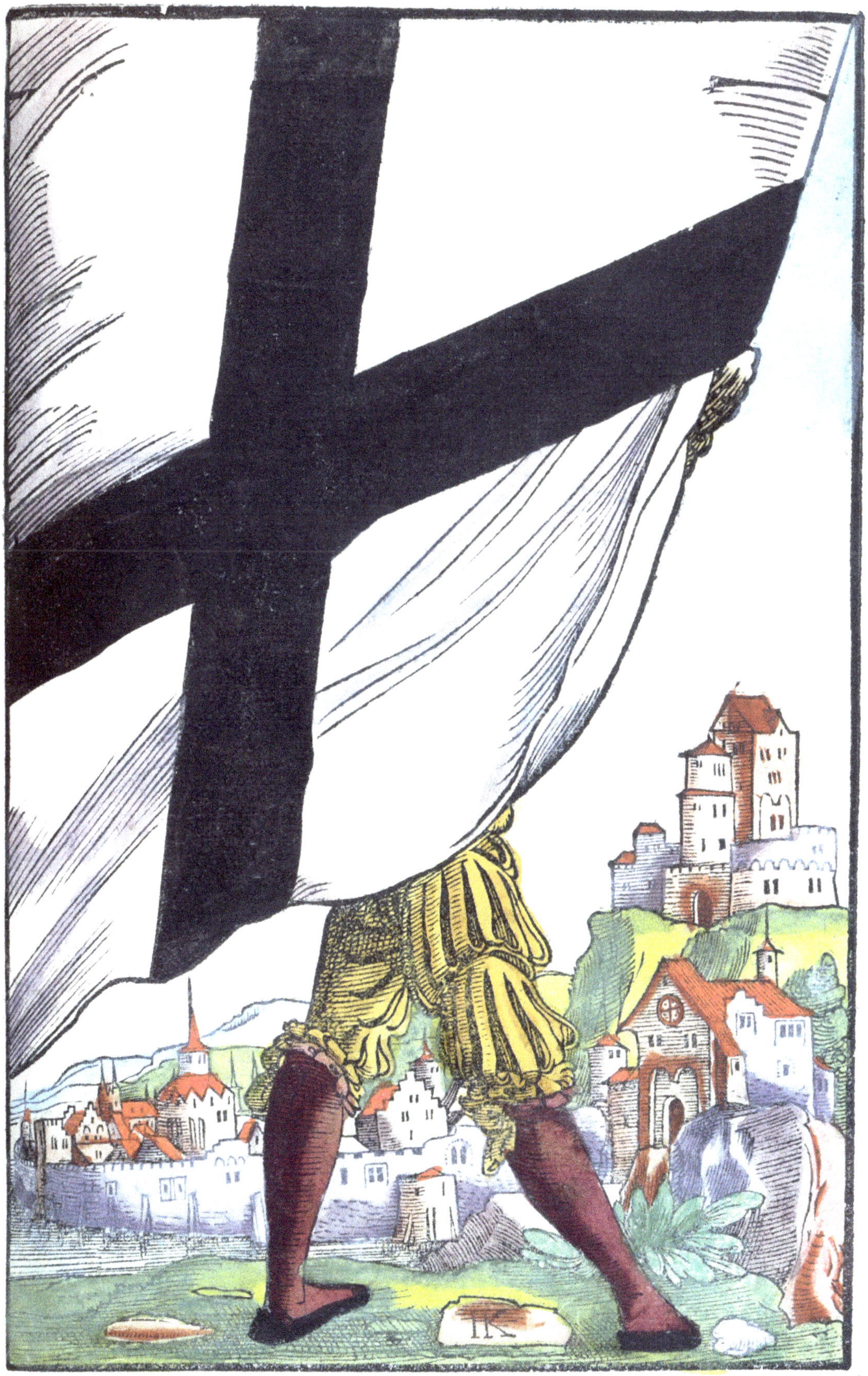

3. Cöln | The Prince Archbishop of Cologne

Trier.

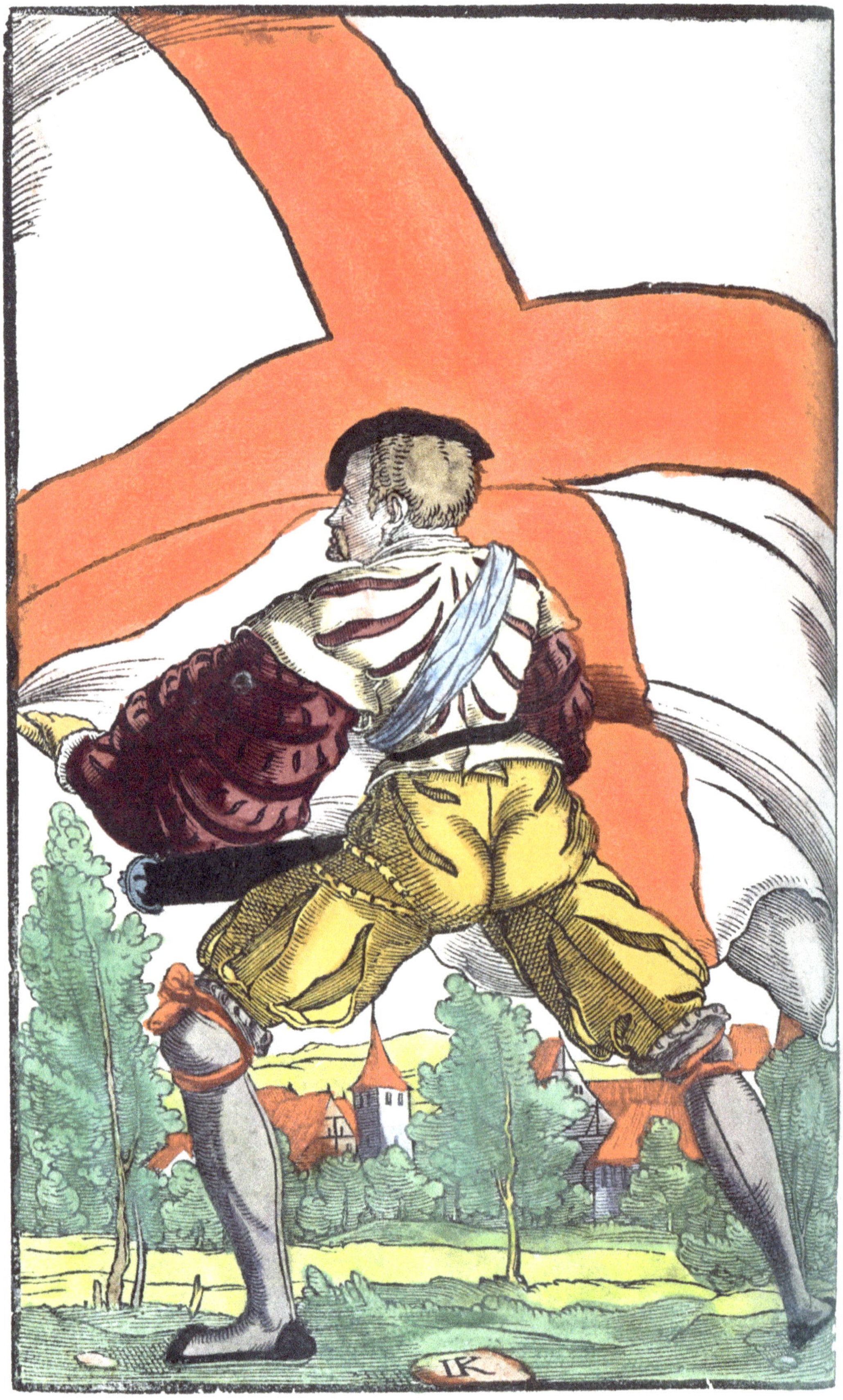

4. Trier | The Prince Archbishop of Trier

Behenm.

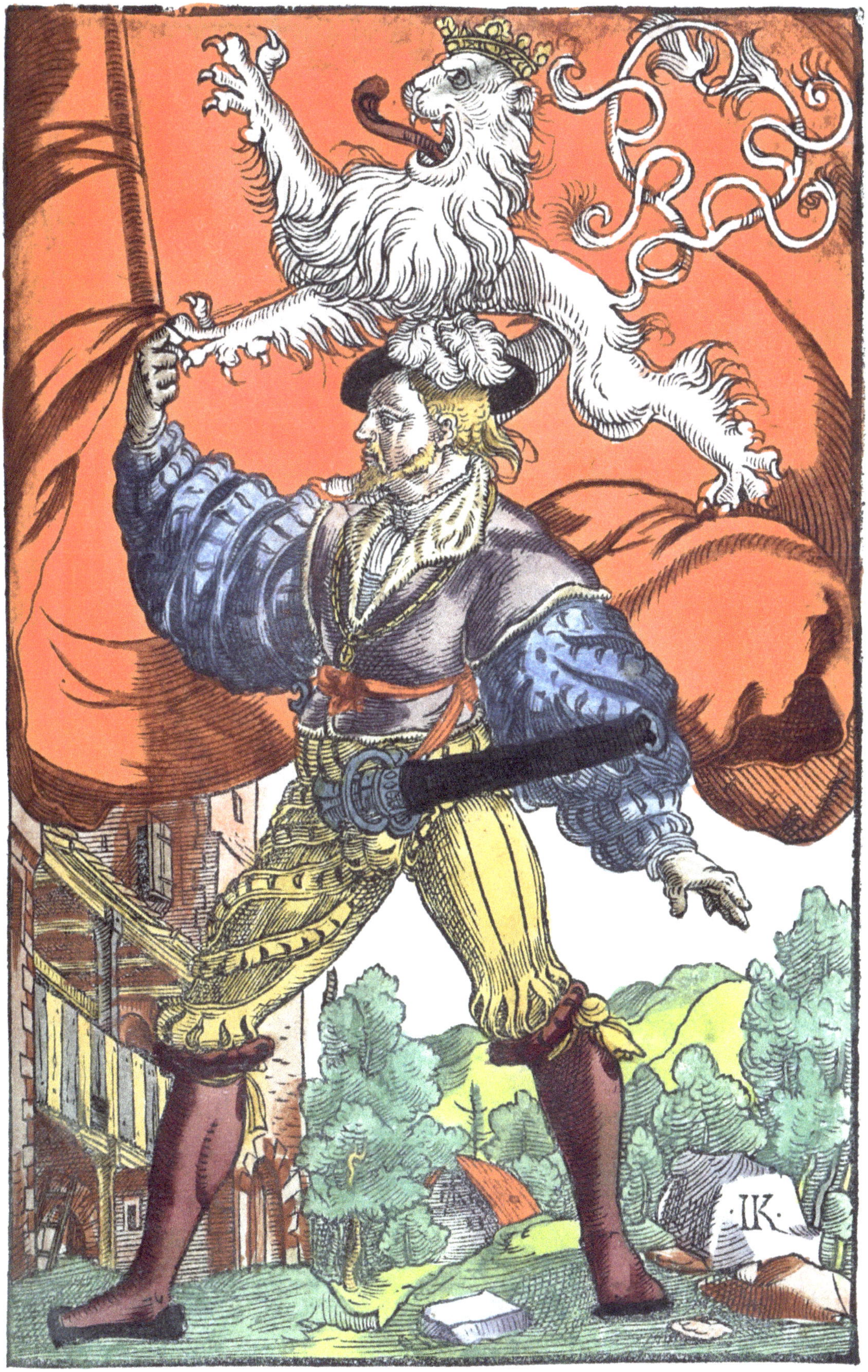

5. Behenm | Böhmen, The King of Bohemia, Imperial Cup Bearer. Area of modern Czech Republic

Pfalz.

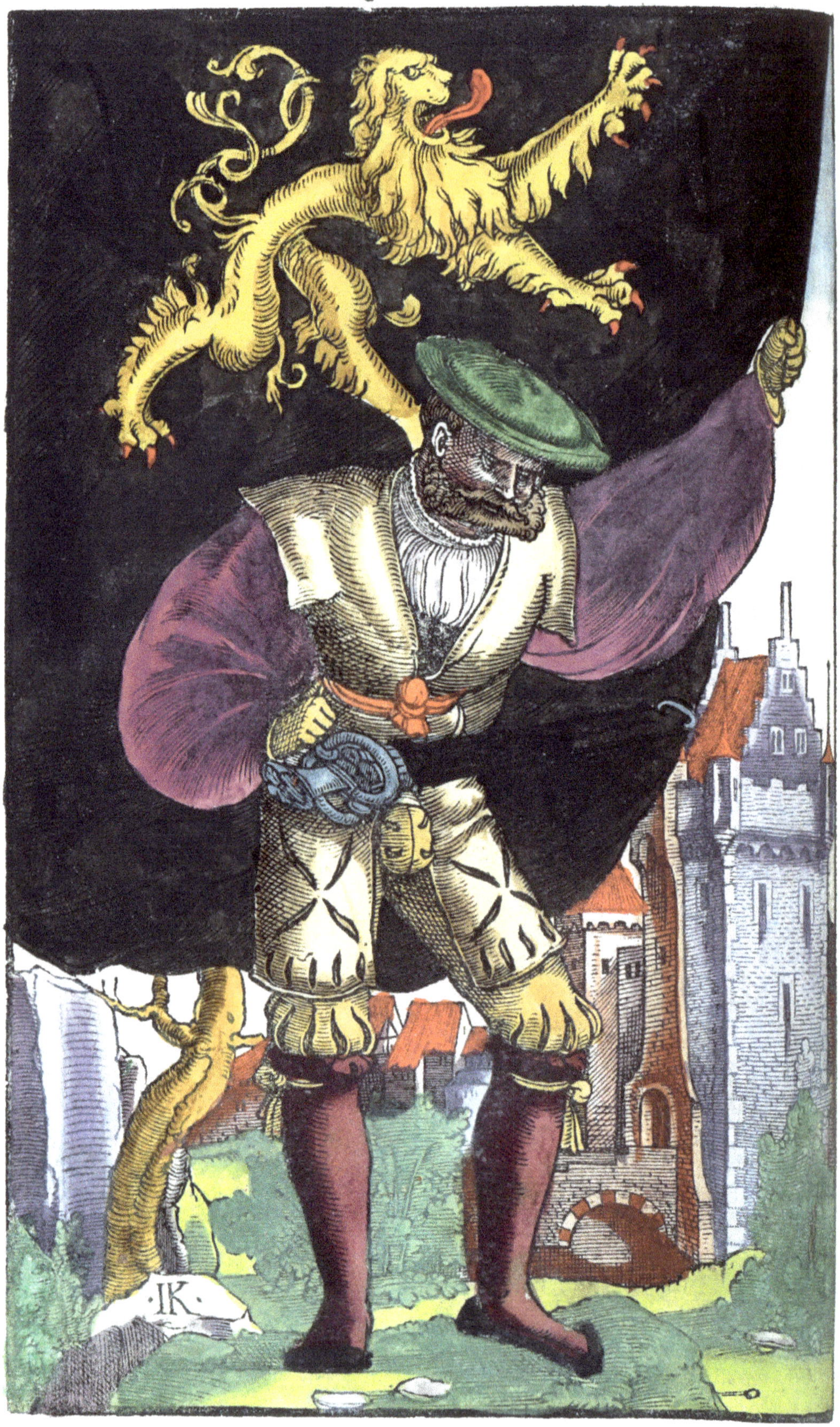

6. Pfalz | Pfaltz The Count Palatine of the Rhine, Imperial Steward

Sachssen.

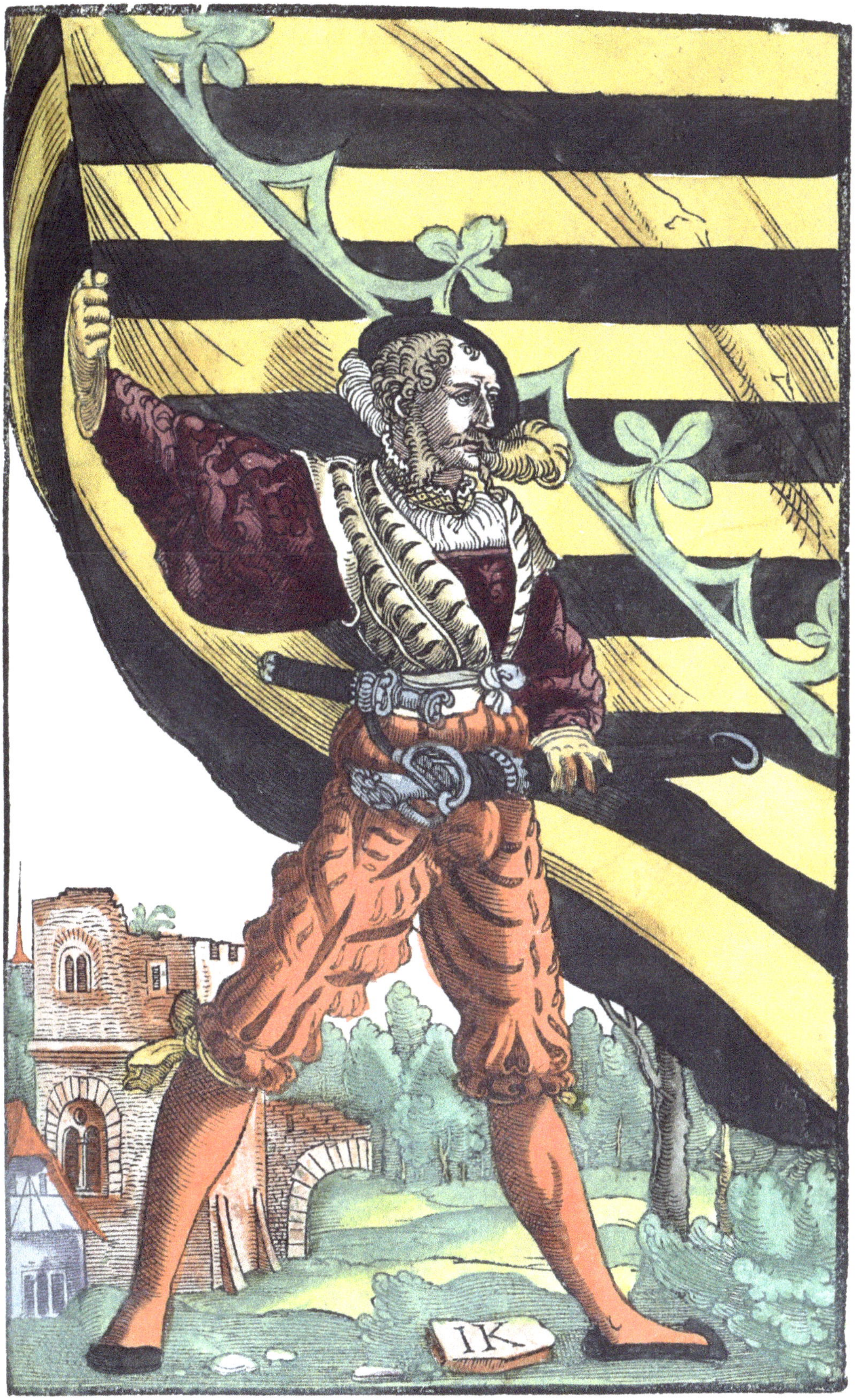

7. Sachssen | Sachsen, The Duchy of Saxony, Imperial Marshal

Brandenburg.

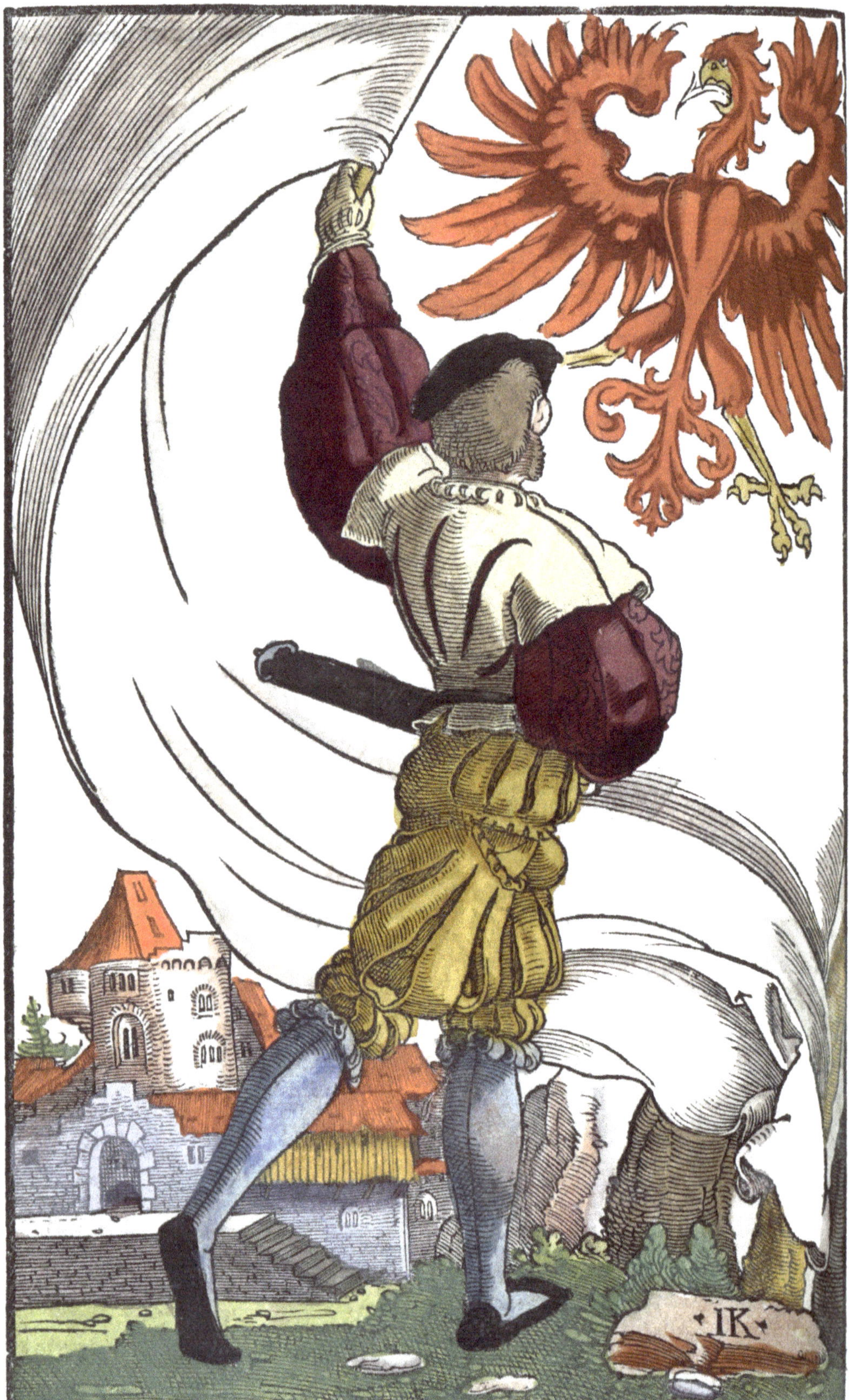

8. Brandenburg | The Margrave of Brandenburg, Imperial Chamberlain

Braunſchwenck.

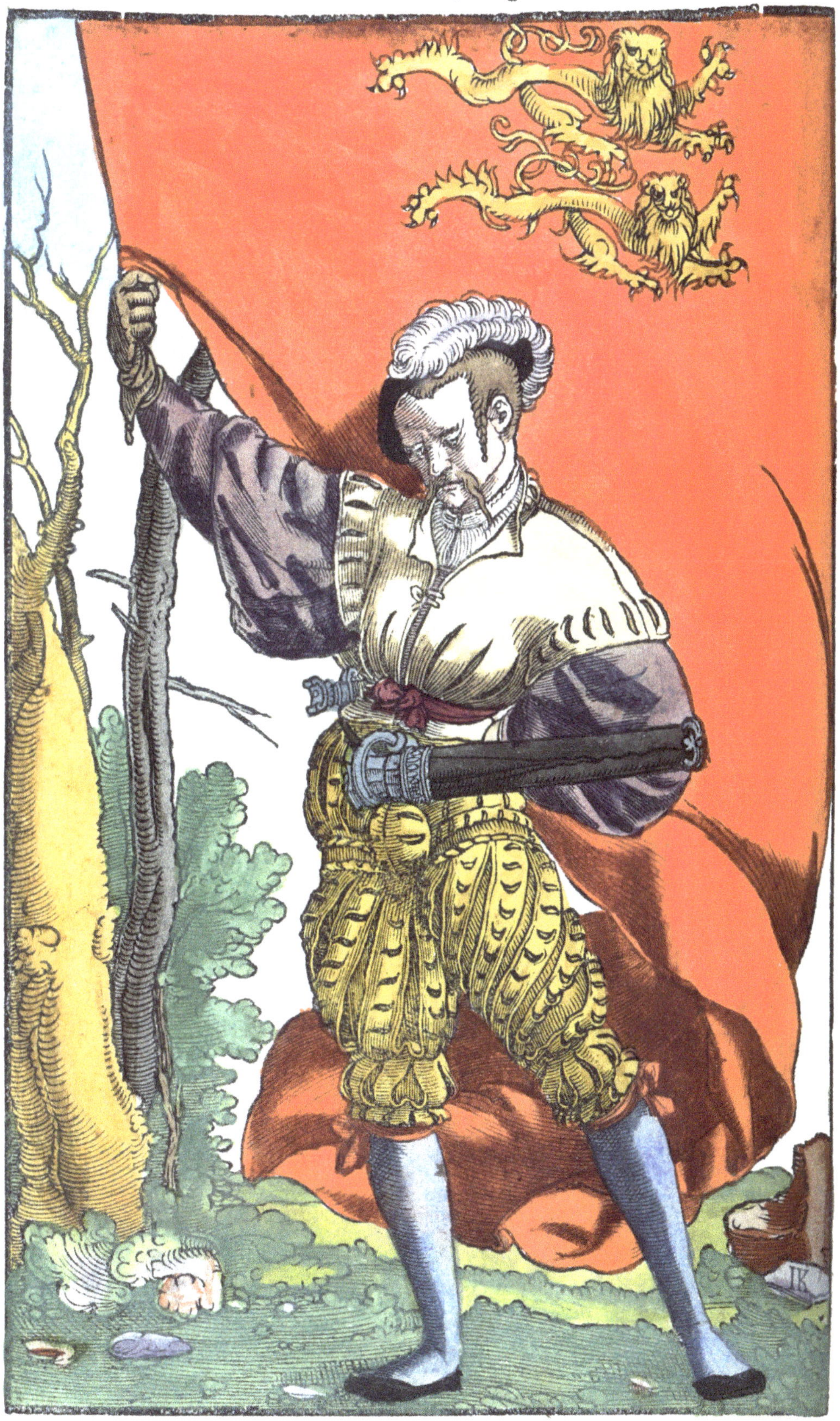

9. Braunschwenck | Braunschweig, Duchy of Brunswick-Lüneburg

Deckendorff.

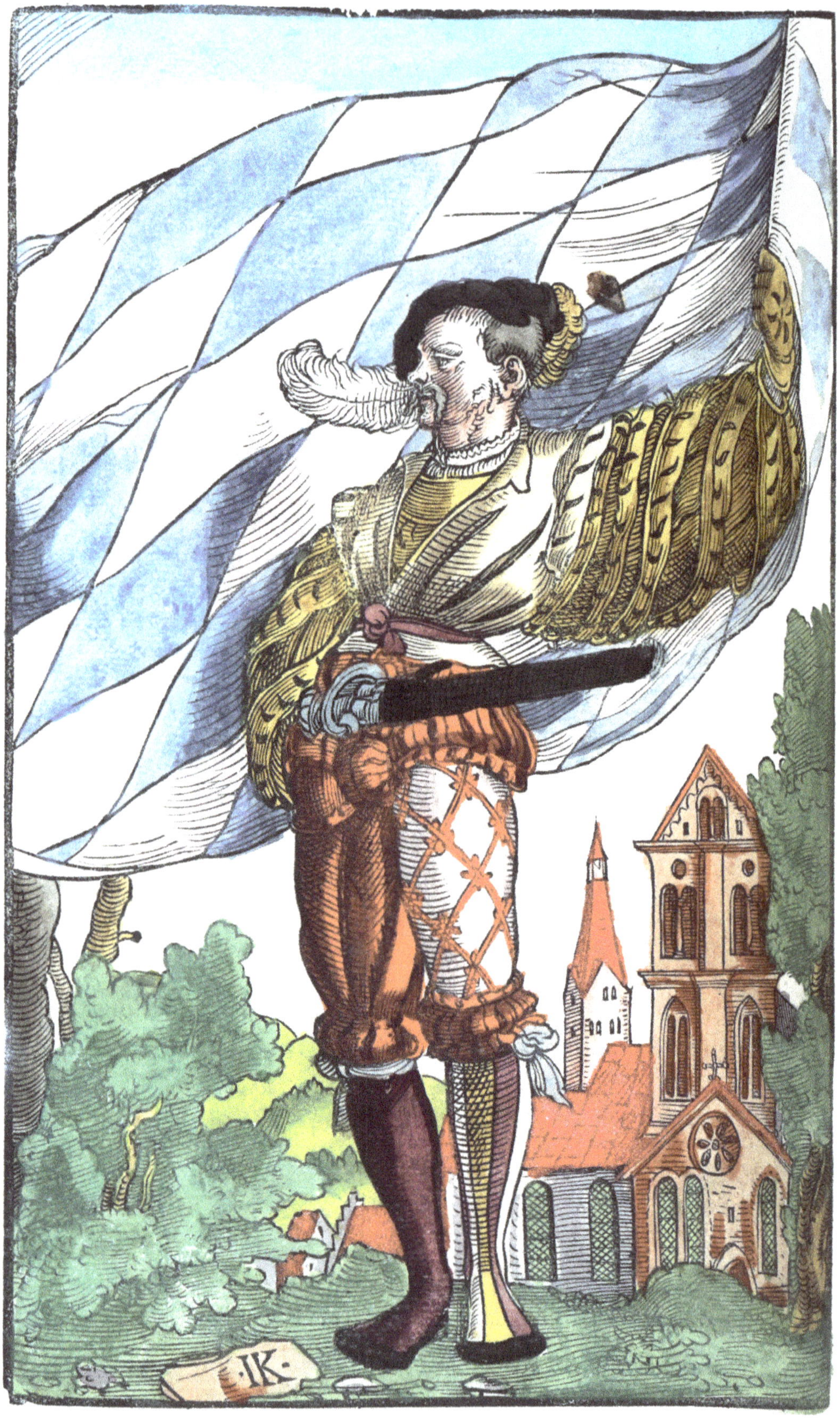

10. Bairn | Bavaria. (Duchy of Bavaria) [Mislabled as Deckendorff

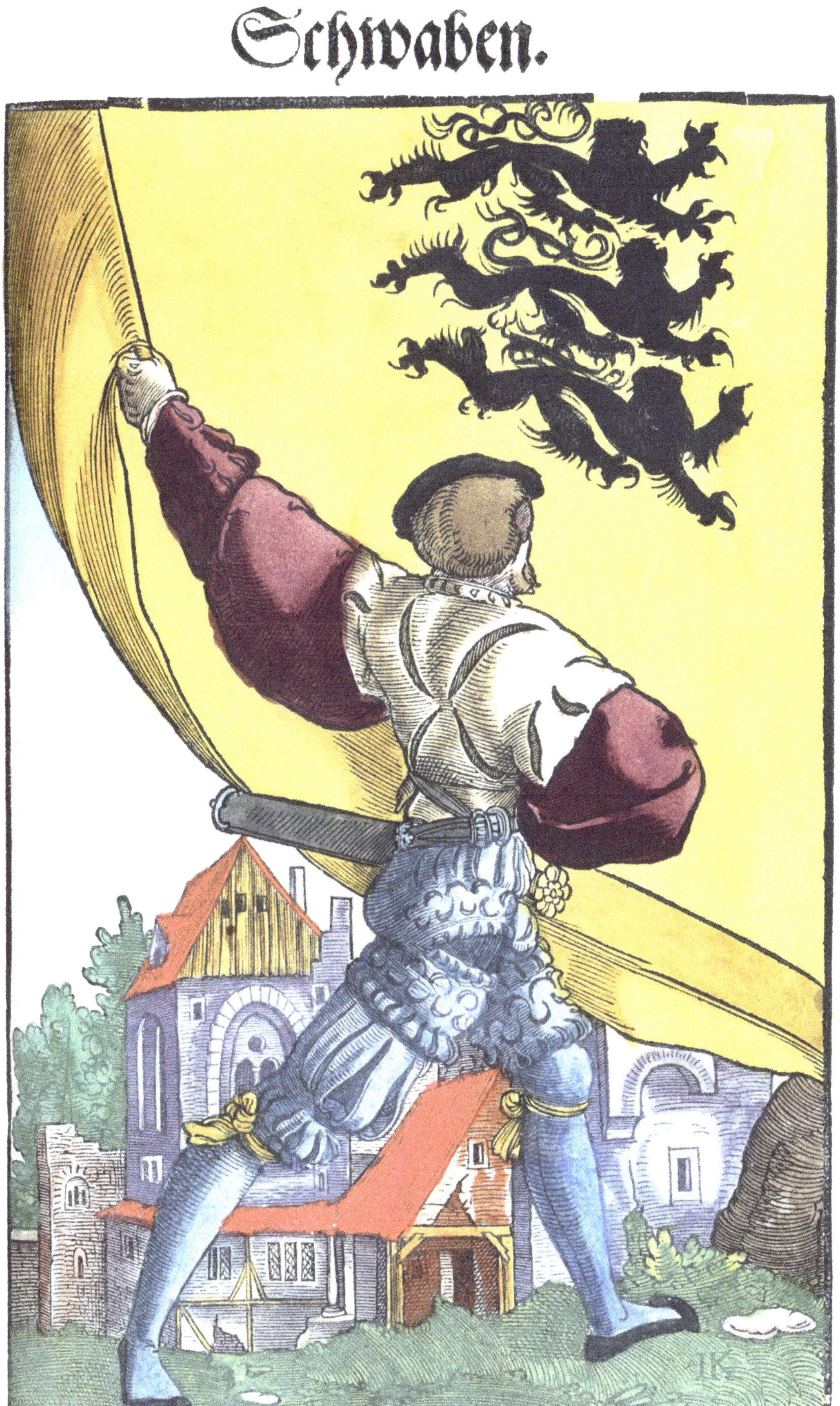

11. Schwaben | Swaben, Duchy of Swabia

Lothringen.

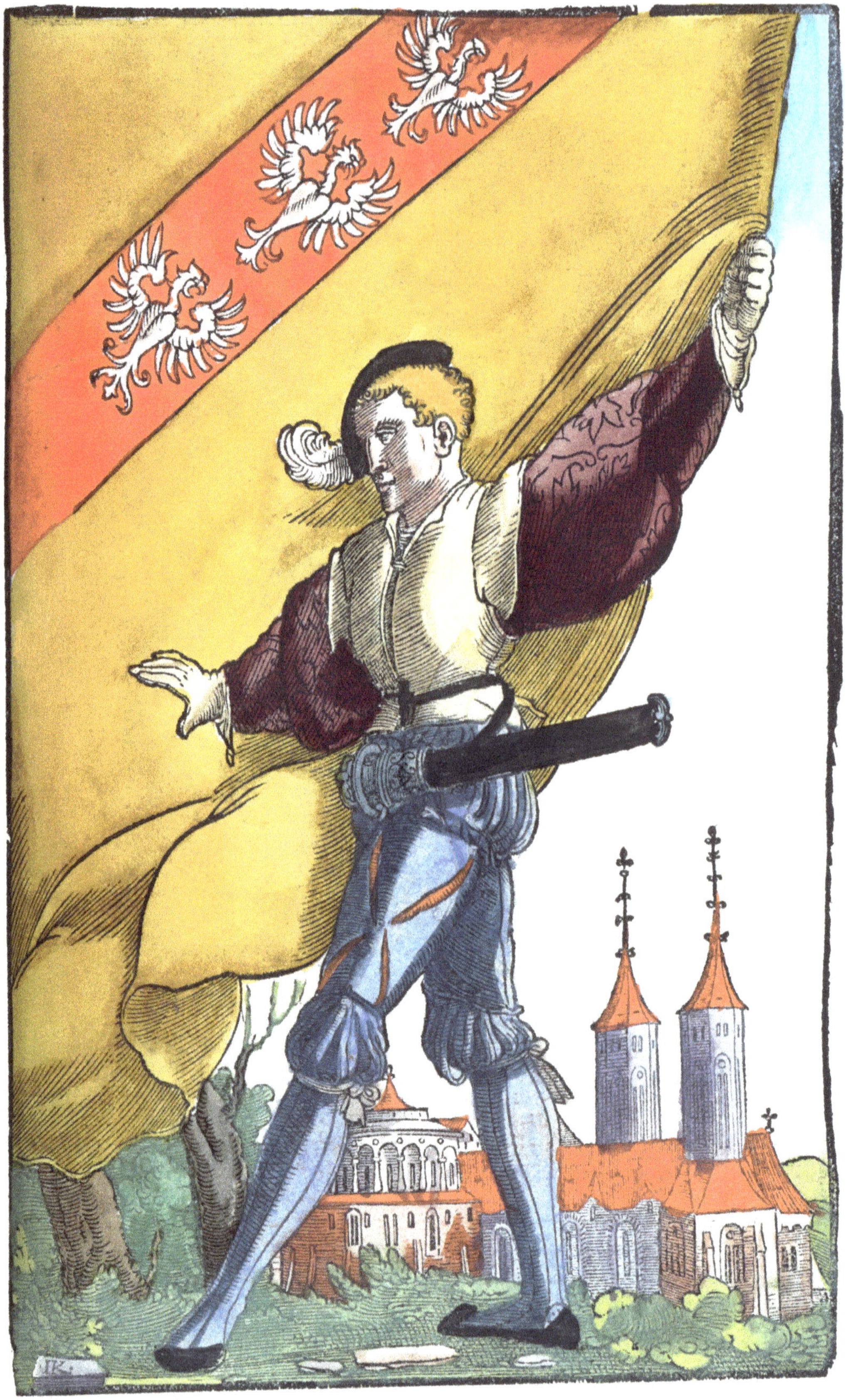

12. Lothringen | Lutring, Duchy of Lorraine

Mörhern.

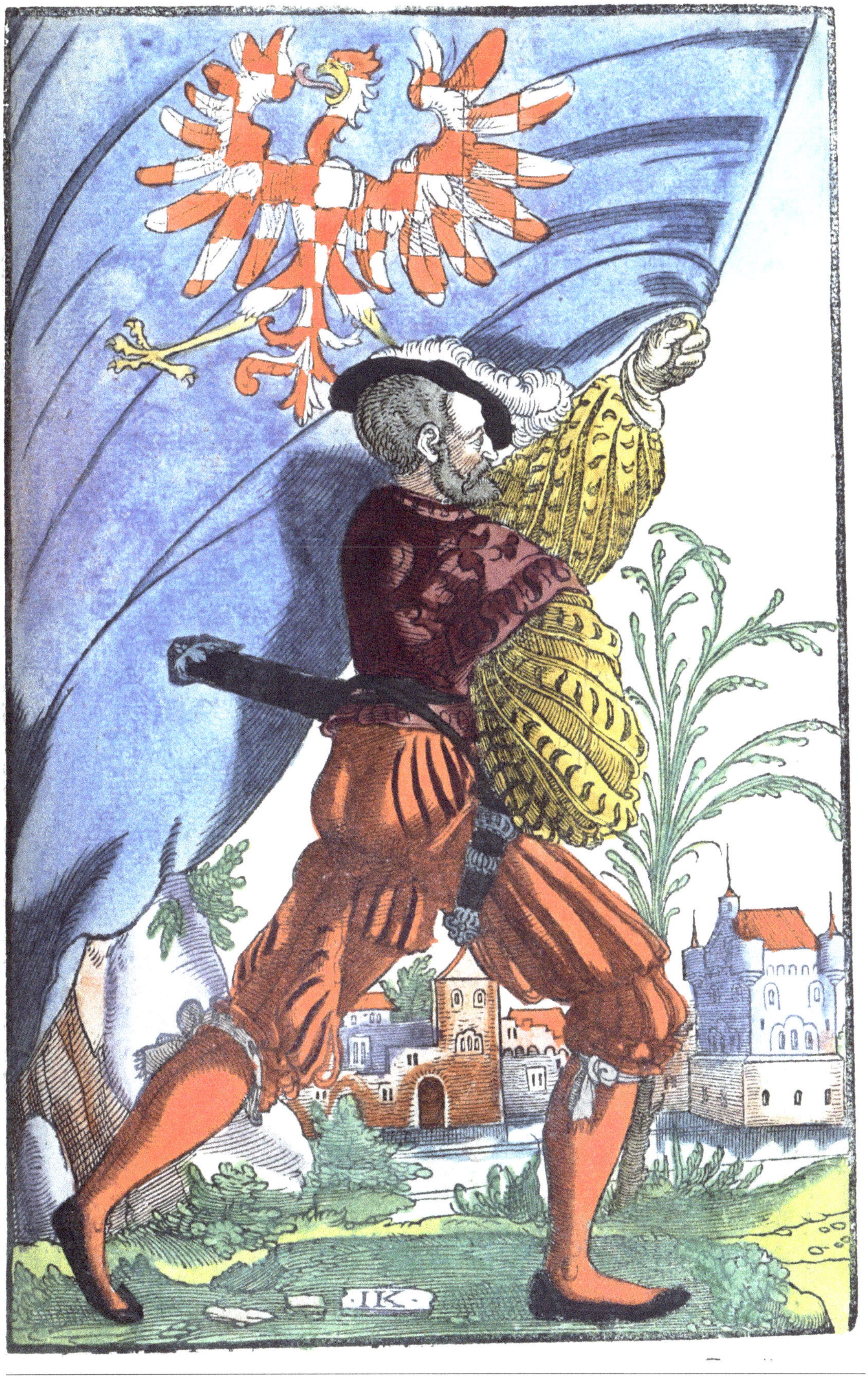

13. Mörhern | Mähren, Moravia

Brandenburg.

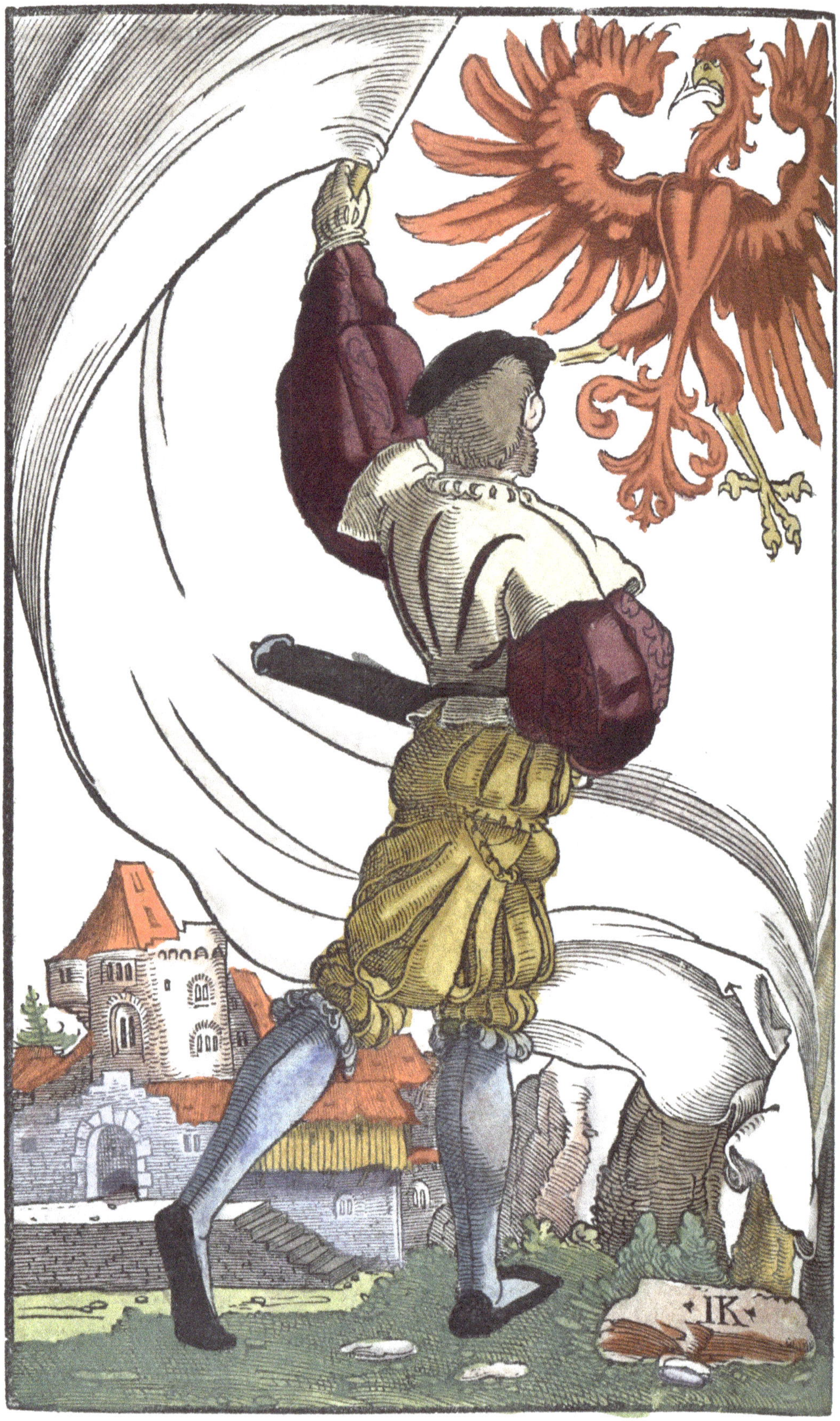

14. Brandenburg | Brandenburg

Landaw.

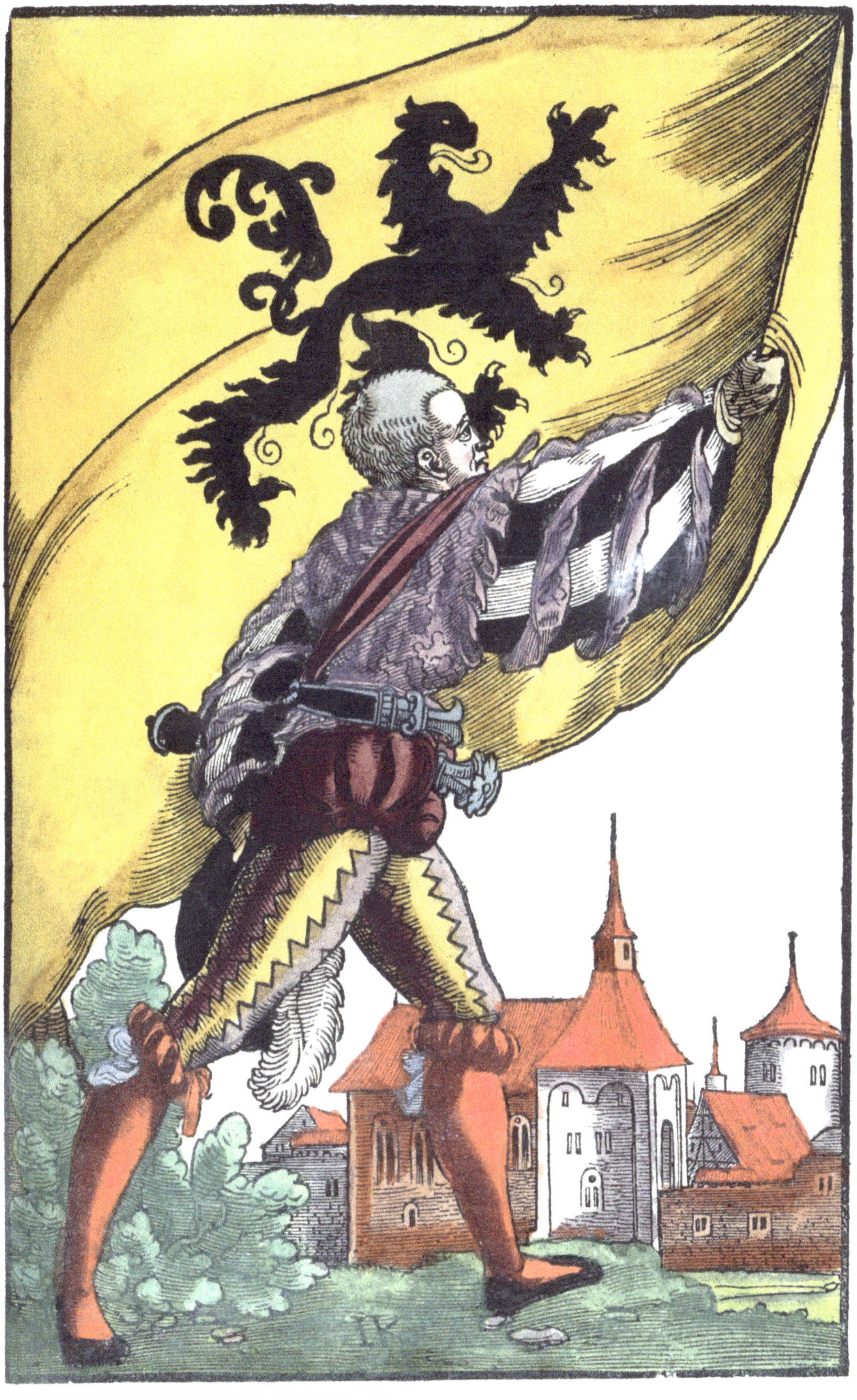

15. Meichssen | Meißen (Mislabeled as Landaw | Landau in der Pfalz)

Baden.

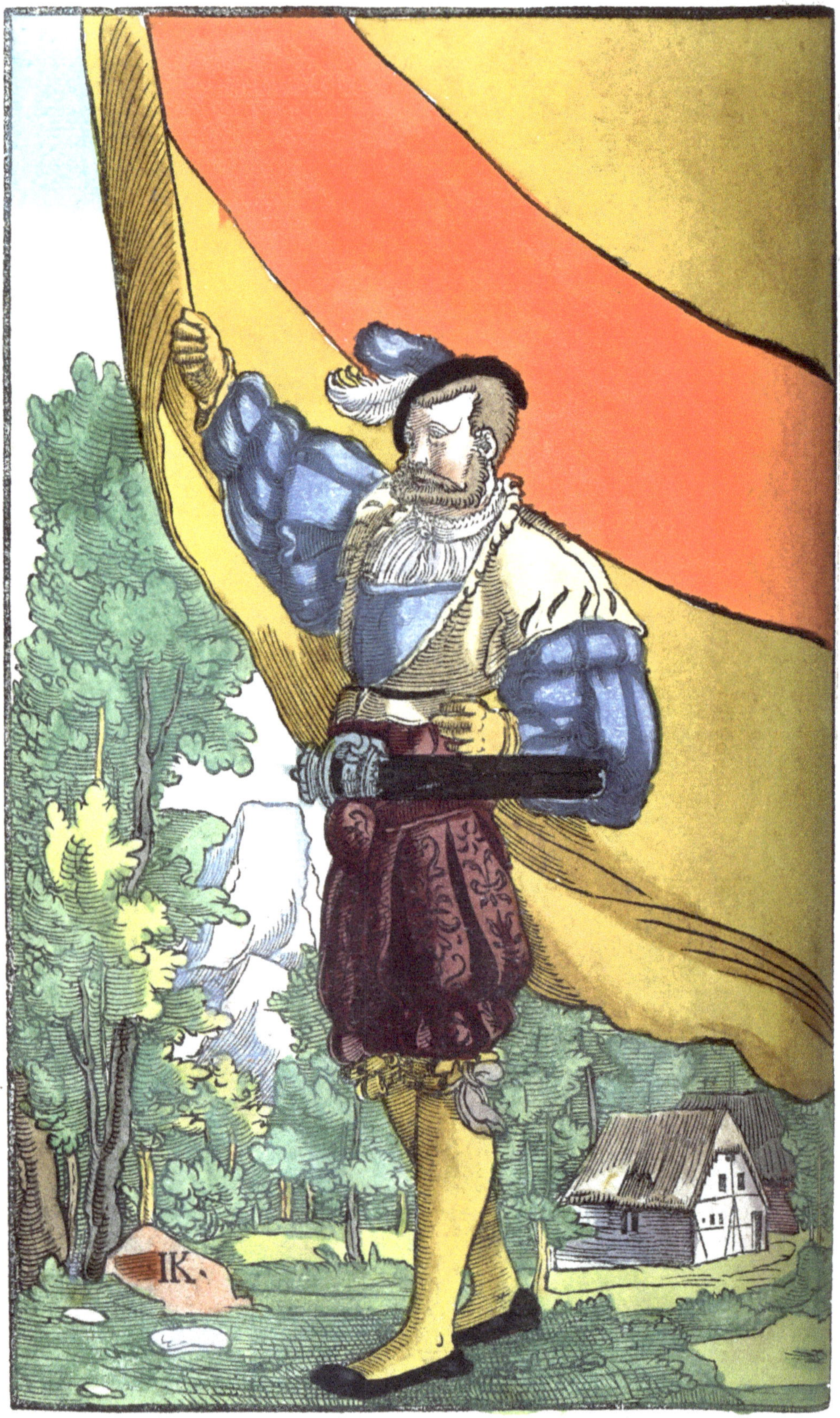

16. Baden | Baden

Düringen.

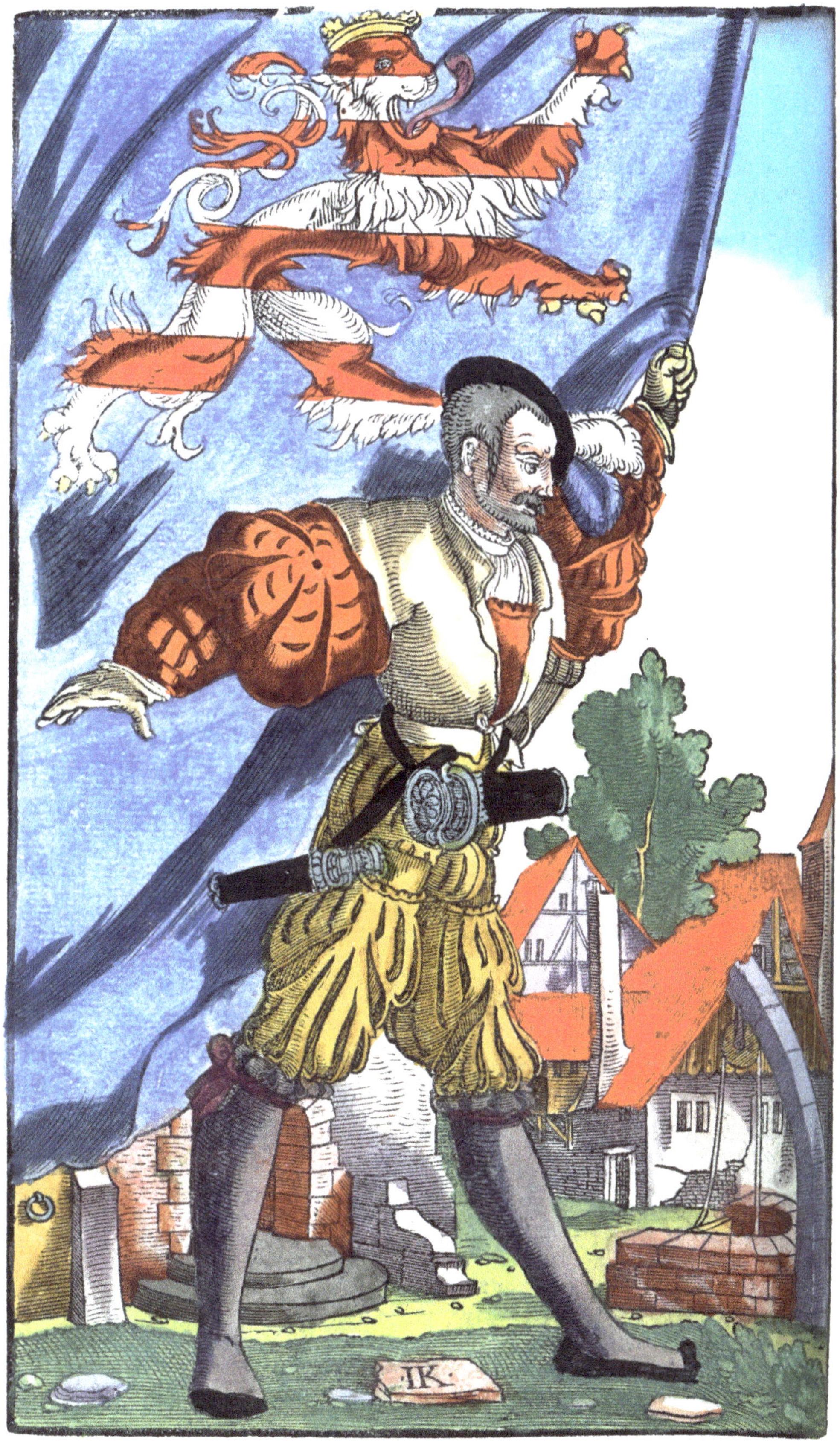

17. Düringen | Düring, Thüringen, Thuringia

Leuchtenburg.

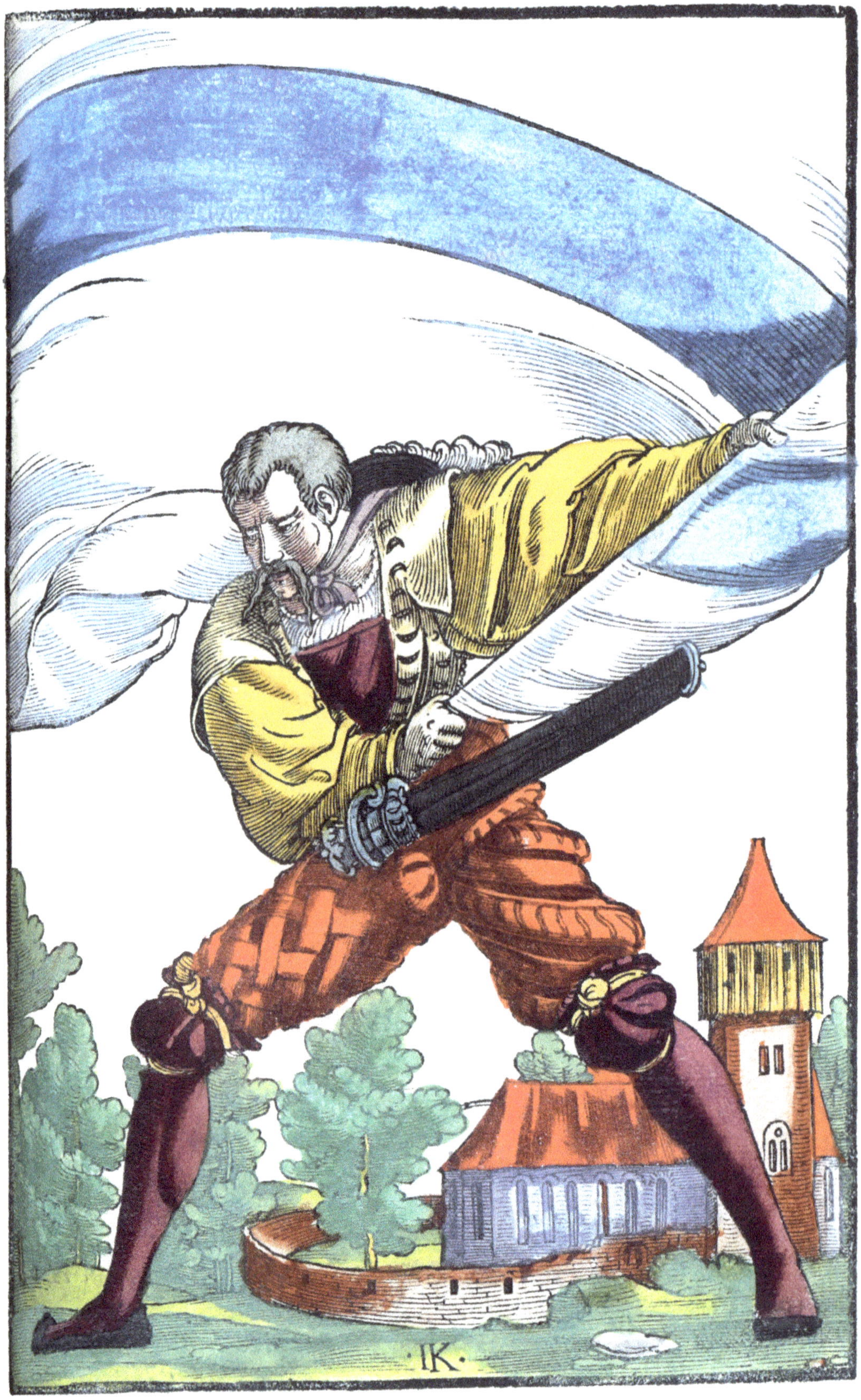

18. Leuchtenburg | Leuchtenberg

Důringen.

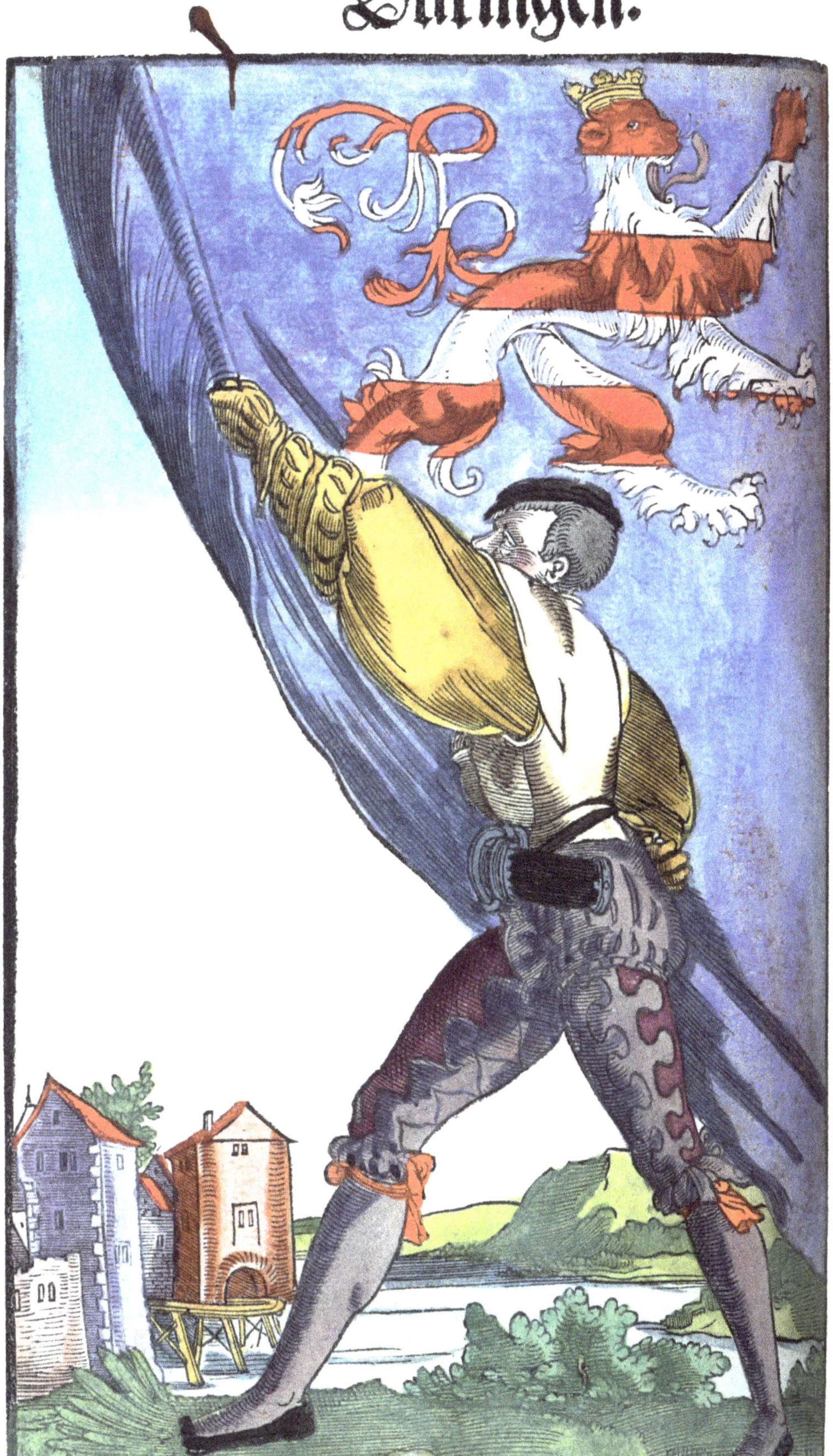

19. Hessen | Hessen, (Mislabeled as Düringen | Thuringia)

Edelsaß.

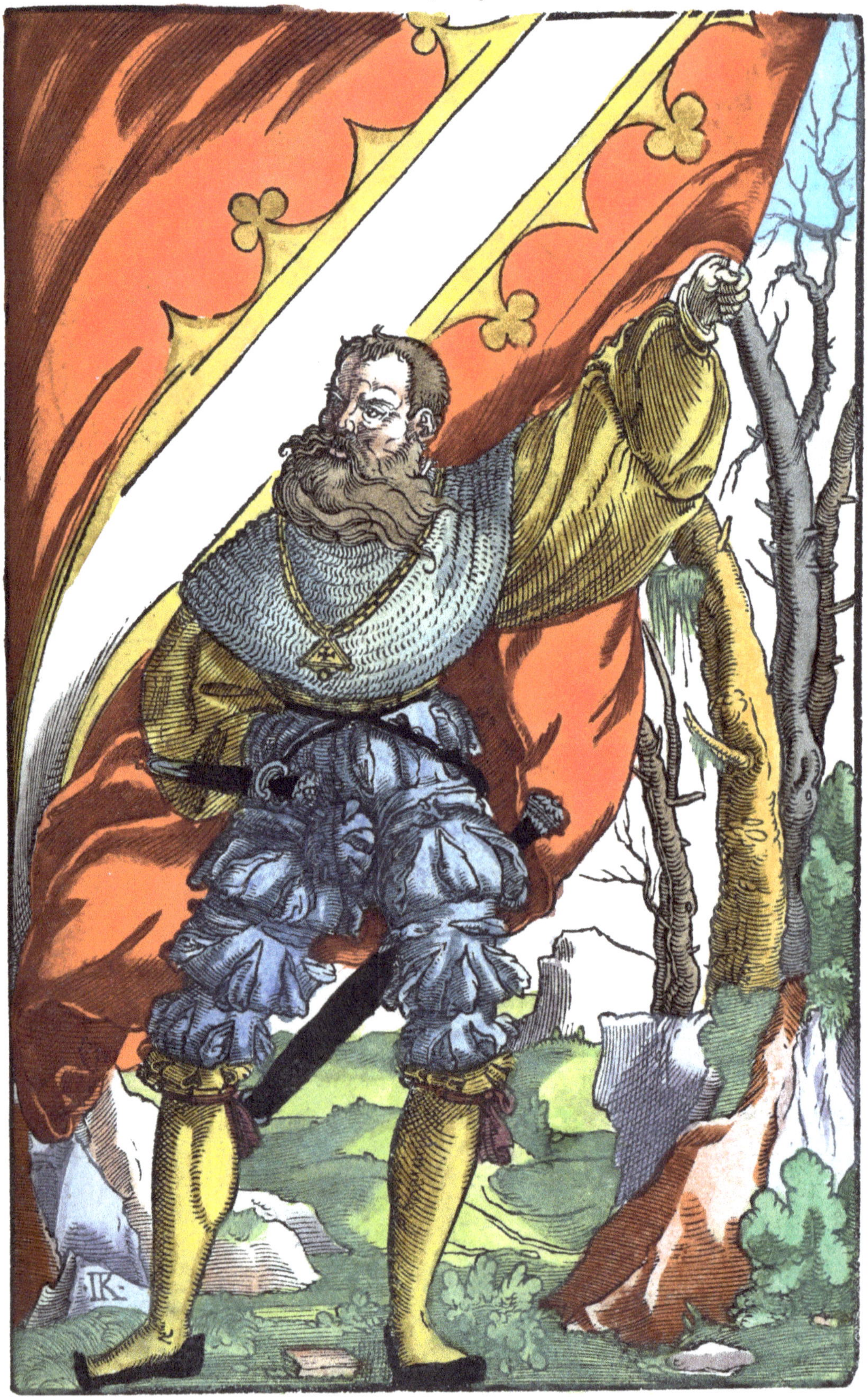

20. Edelsasz | Edelsaß, Landgrafschaft Niederelsass, Bailiffs of Lower Alsace

Nürenberg.

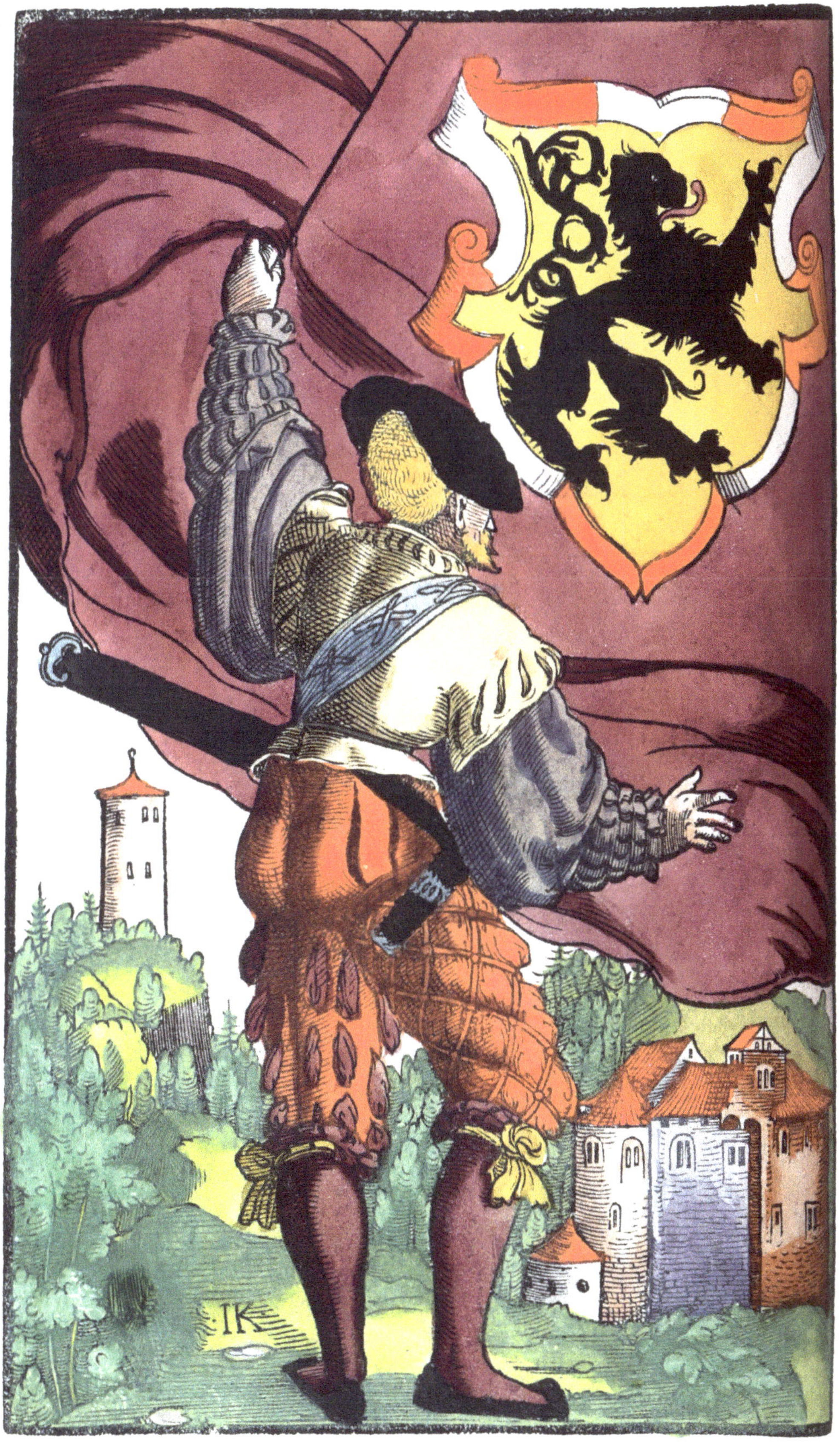

21. Nürenberg Burggraffen | Nürnberg (Nuremberg) Burgraviate Flag

Maidburg.

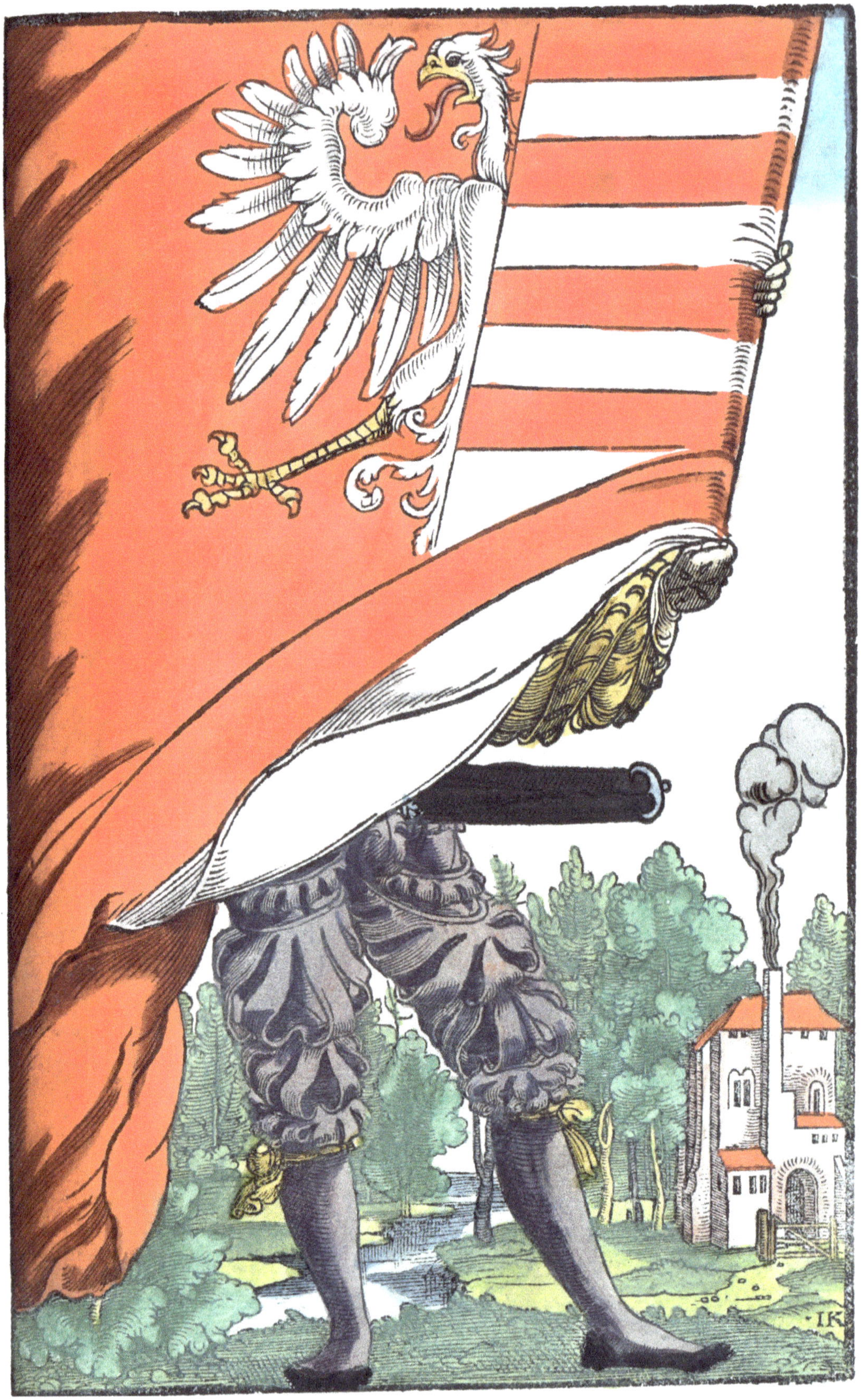

22. Maidburg | Maidburg, Hardegg-Maidburg

Reneck.

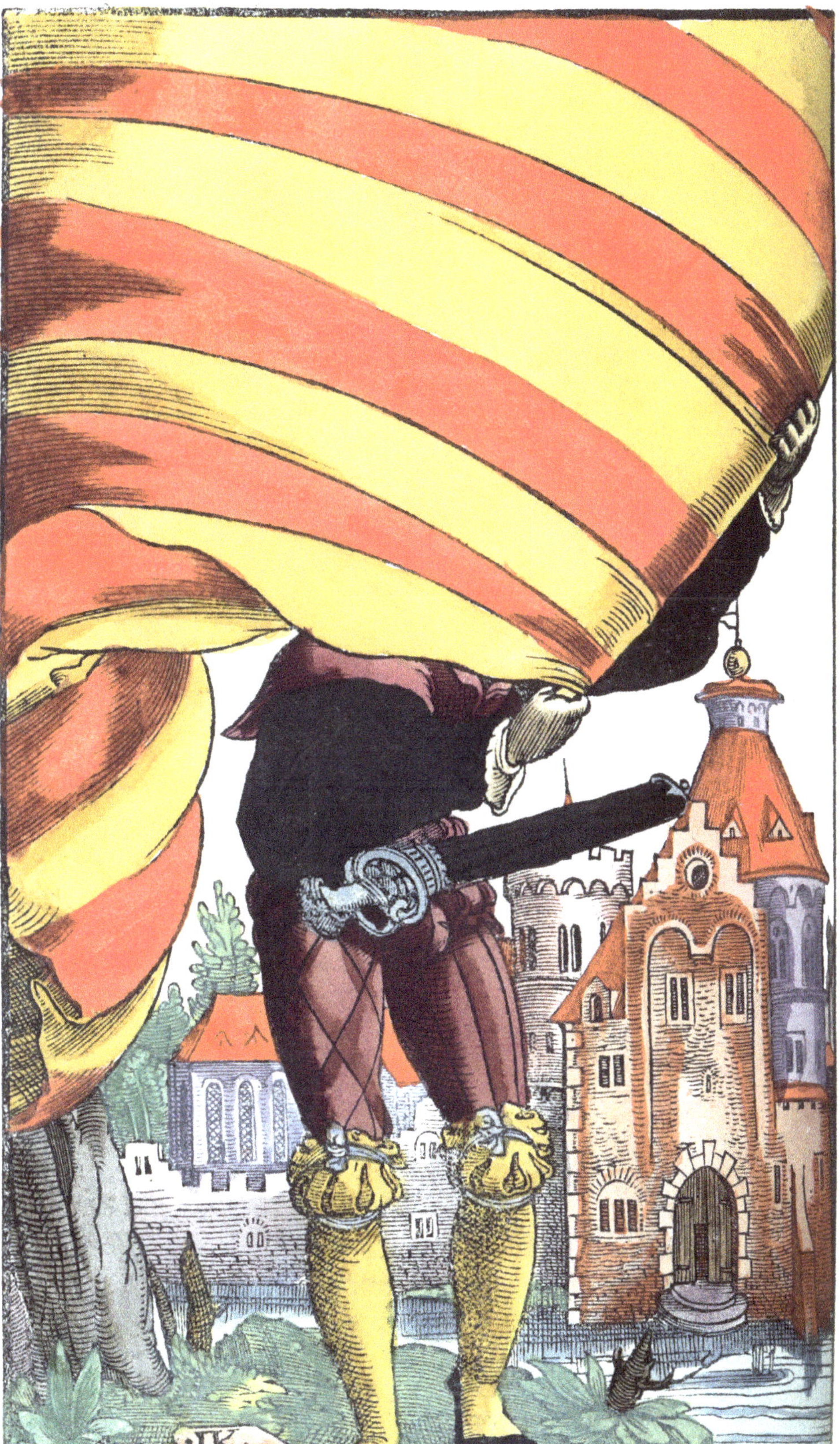

23. Reneck | Reineck

Stramberg.

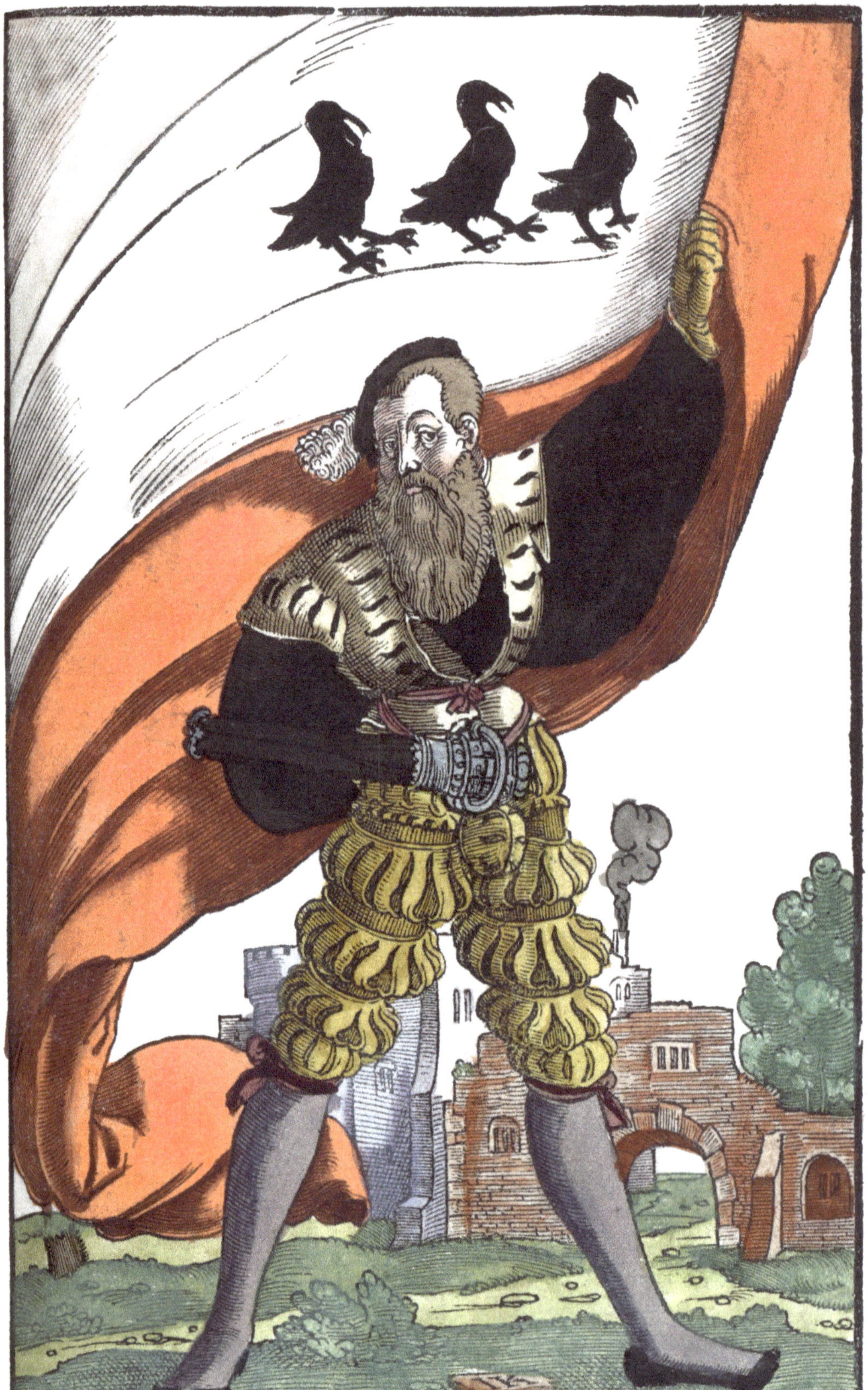

24. Stramberg | Stramberg, Stromberg

Cleue.

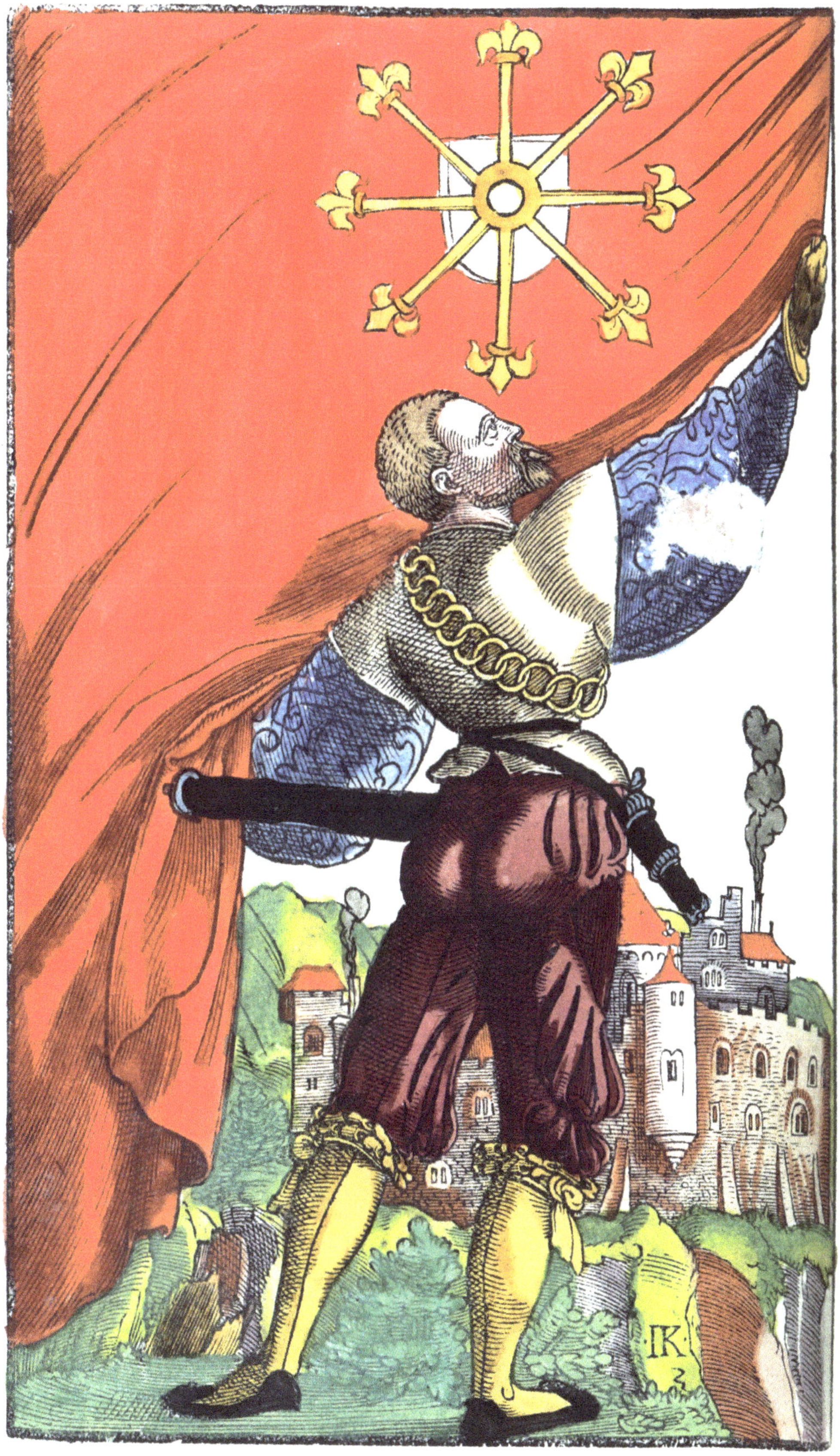

25. Cleue | Grafen von Kleve, Count of Cleves

Saphoy.

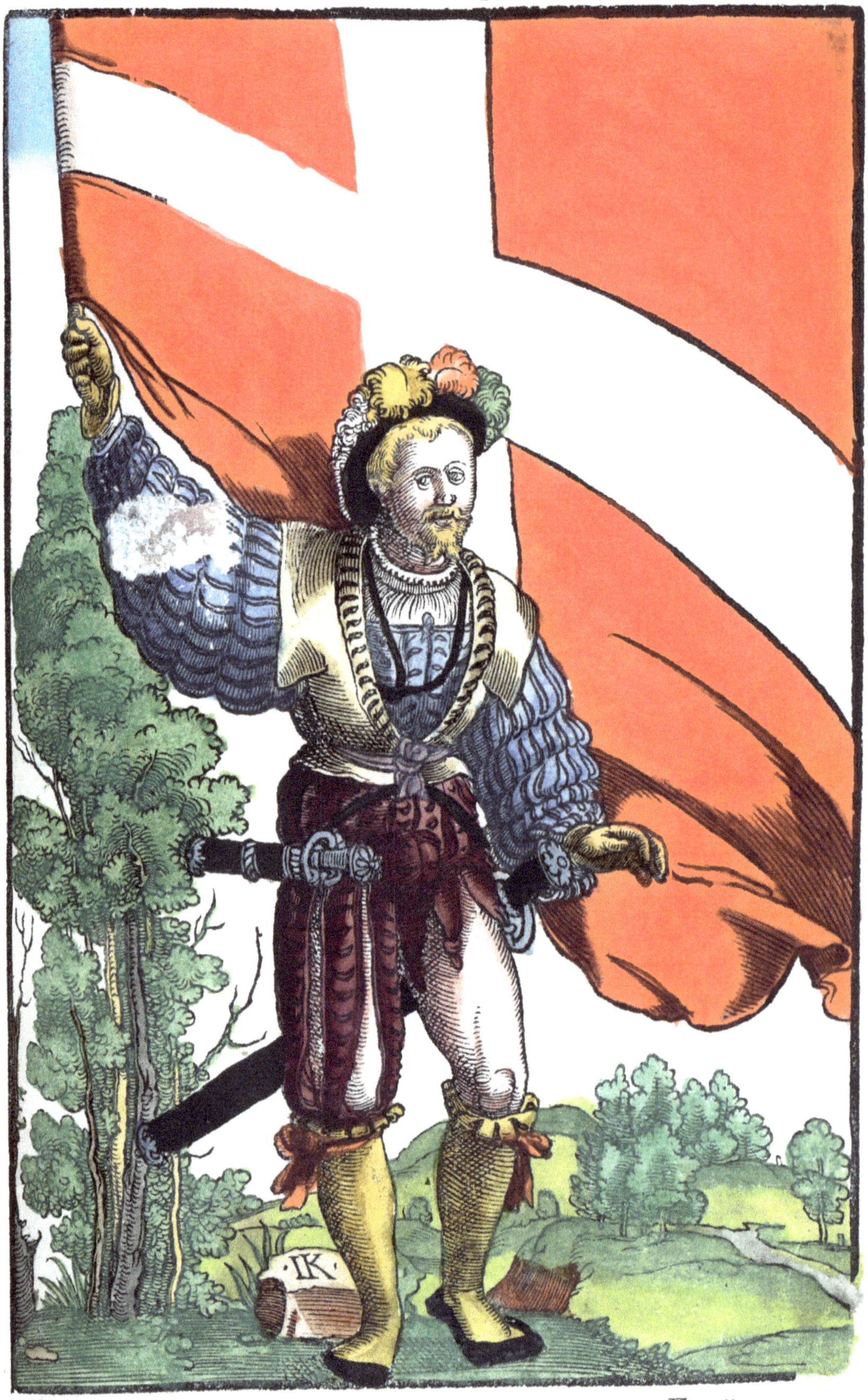

26. Saphoy | Grafen von Savoyen, Count of Savoy

Schwarzburgk.

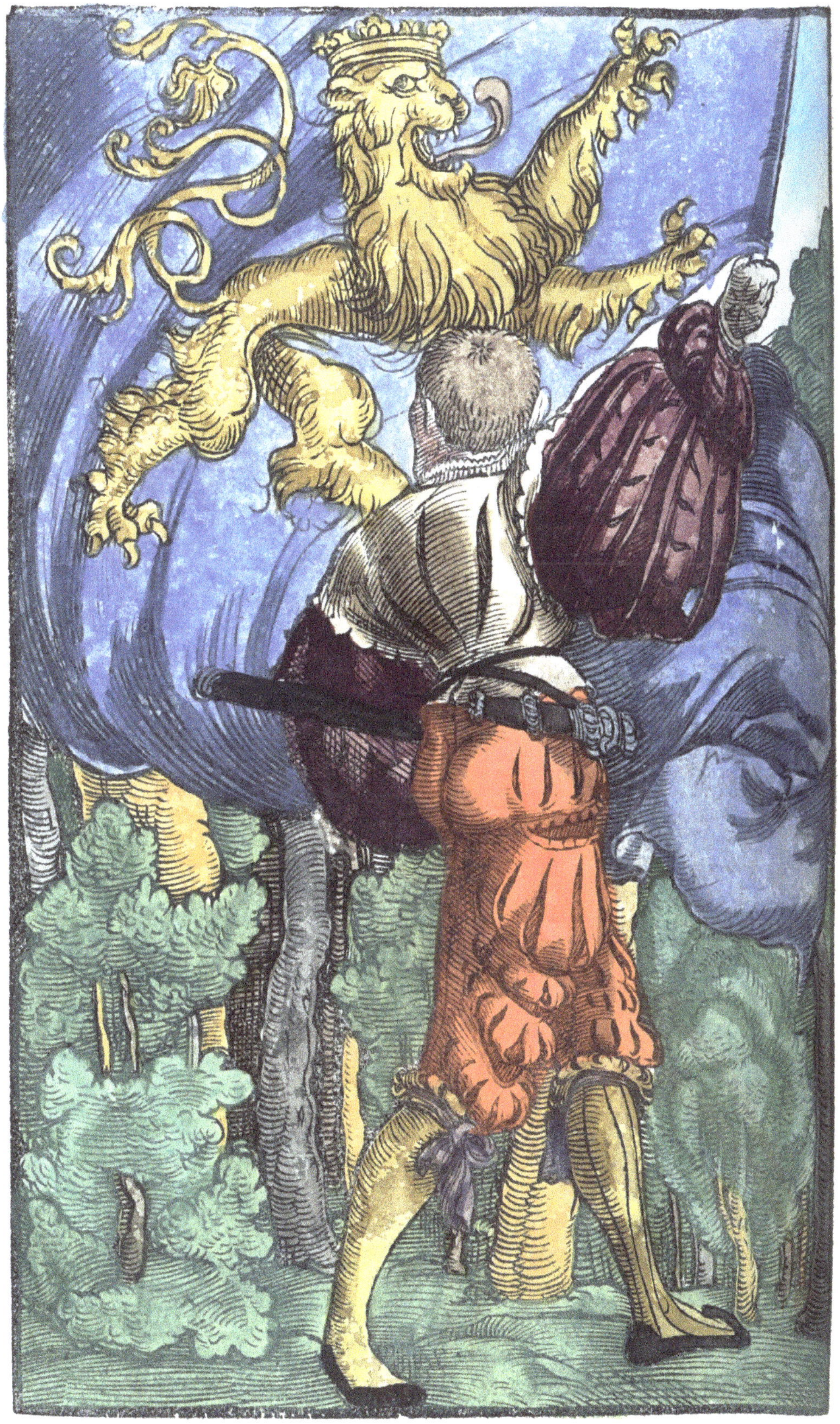

27. Schwanzburgf | Swartzburg, Schwarzburg

Zilli.

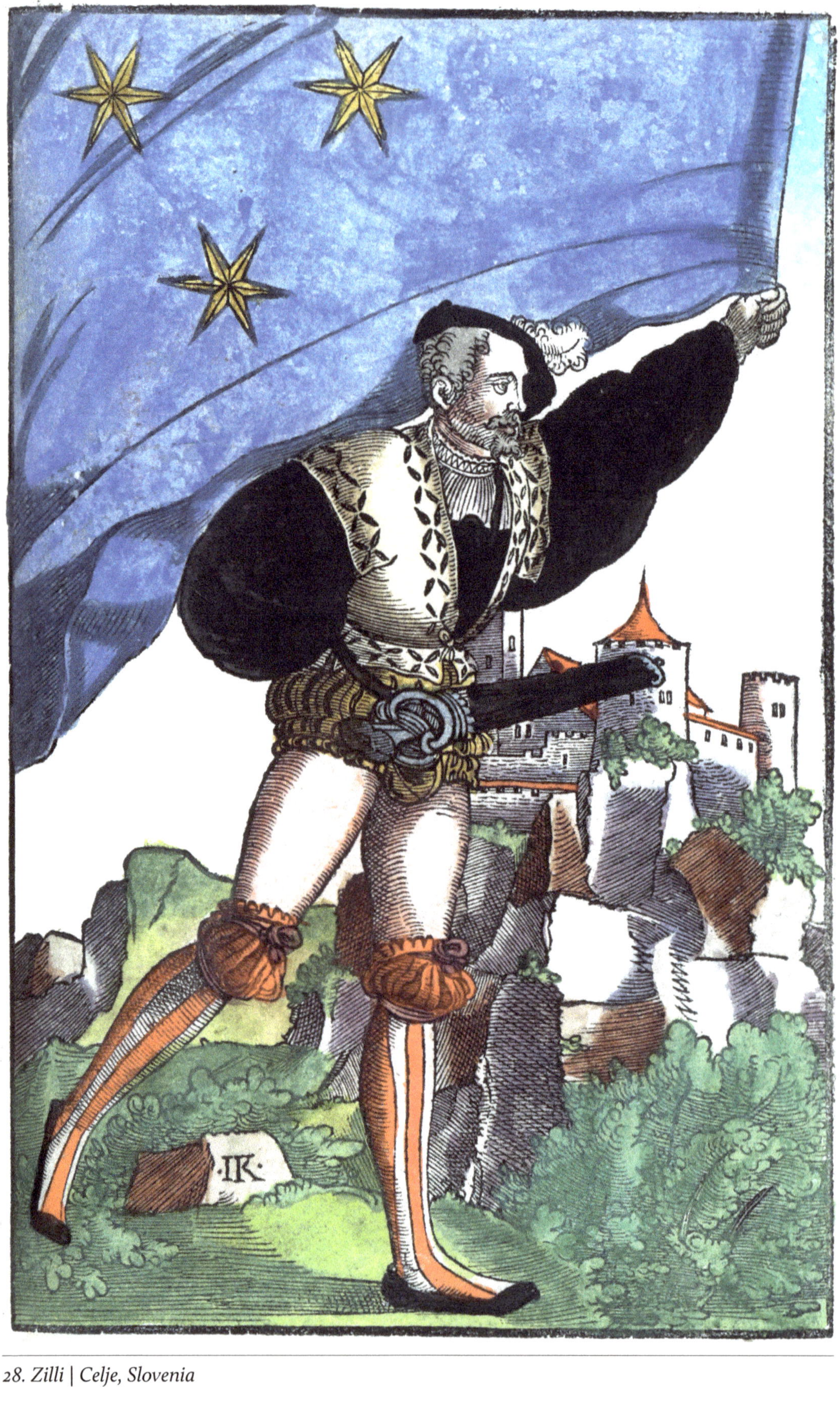

28. Zilli | Celje, Slovenia

Lintpurg.

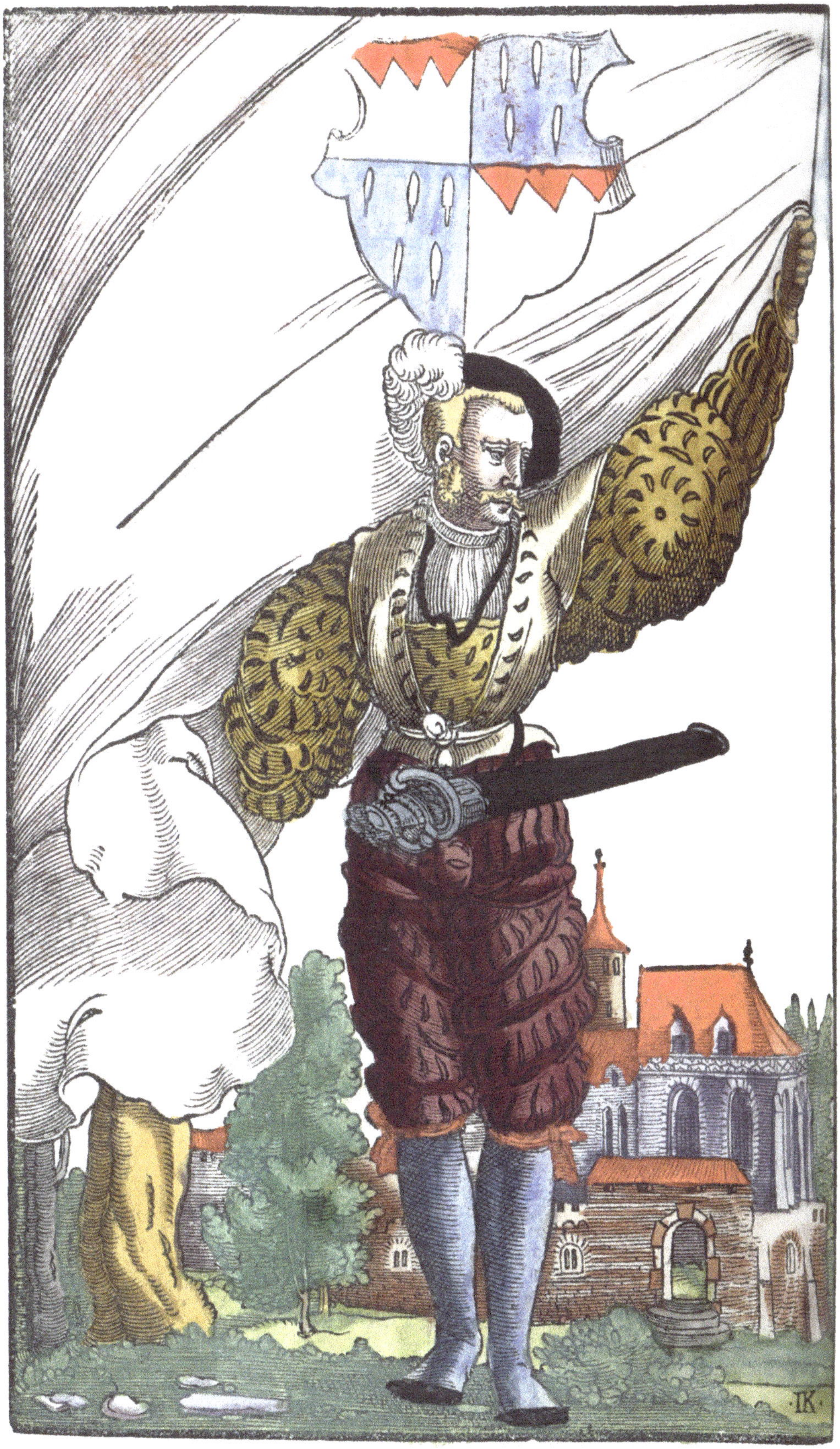

29. Lintpurg | Limpurg

Westerburg.

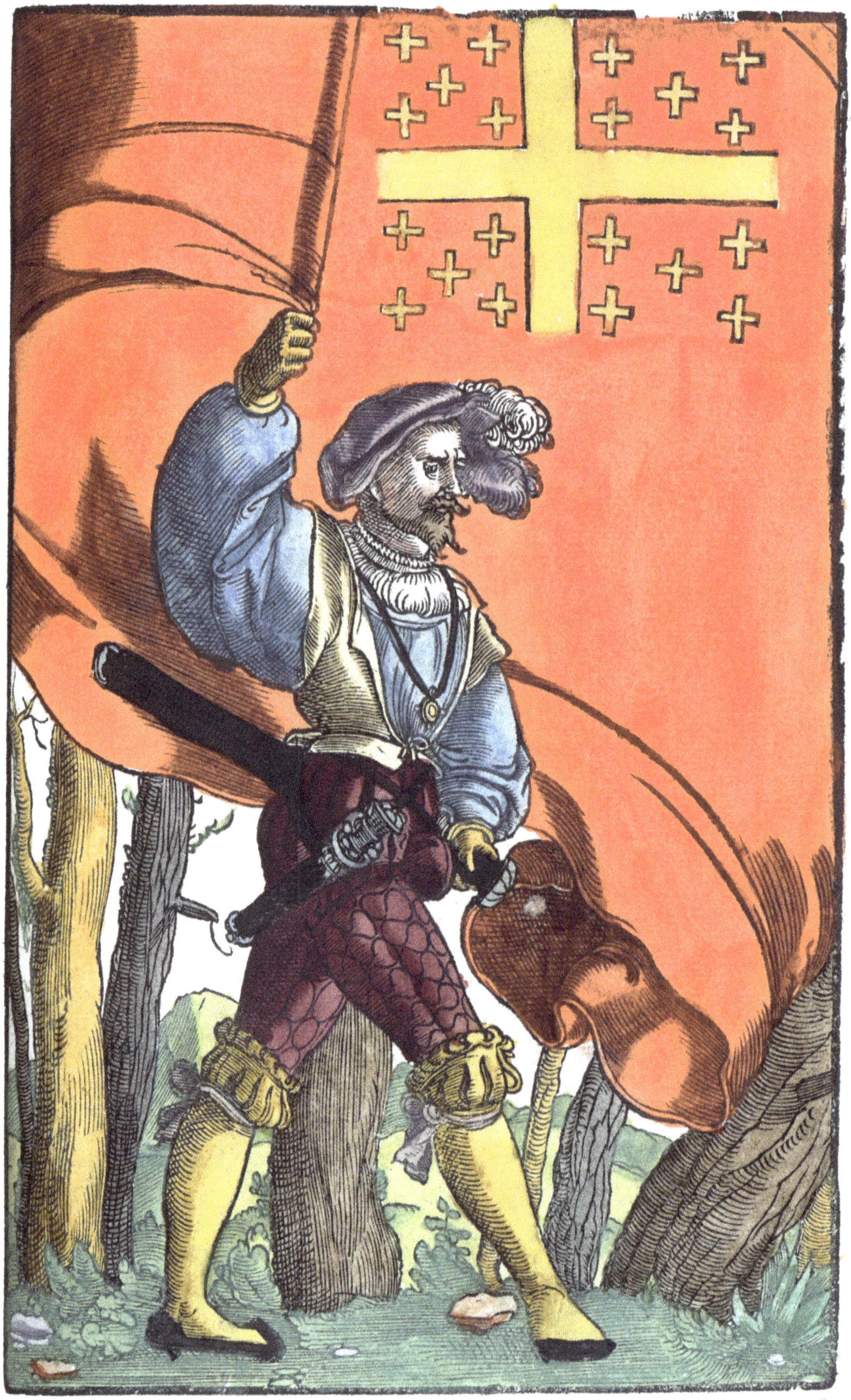

30. Westerburg | Westerburg

Thusis.

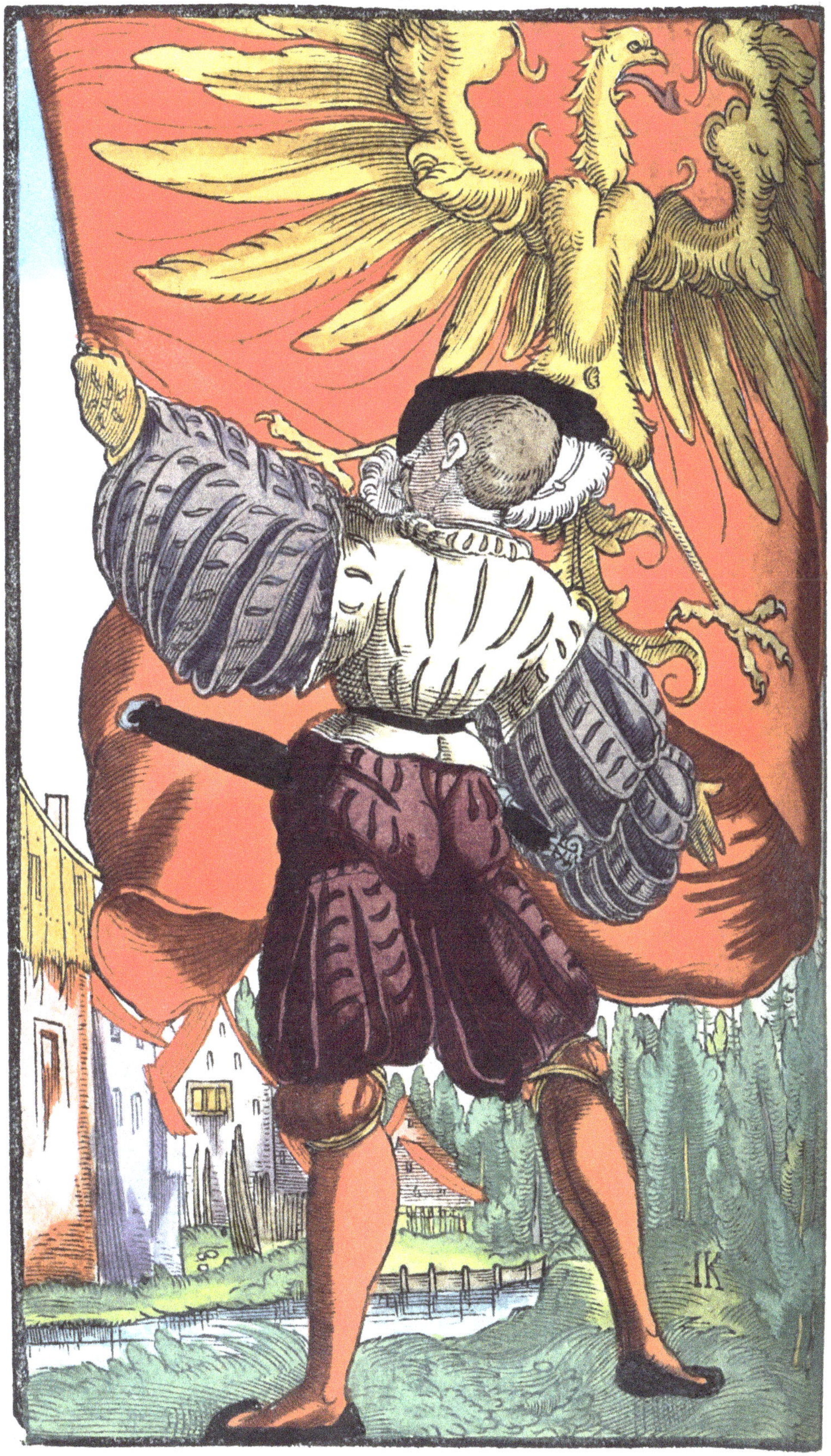

31. Thufis | Thussis, Thusis, Switzerland

Aldenwalden.

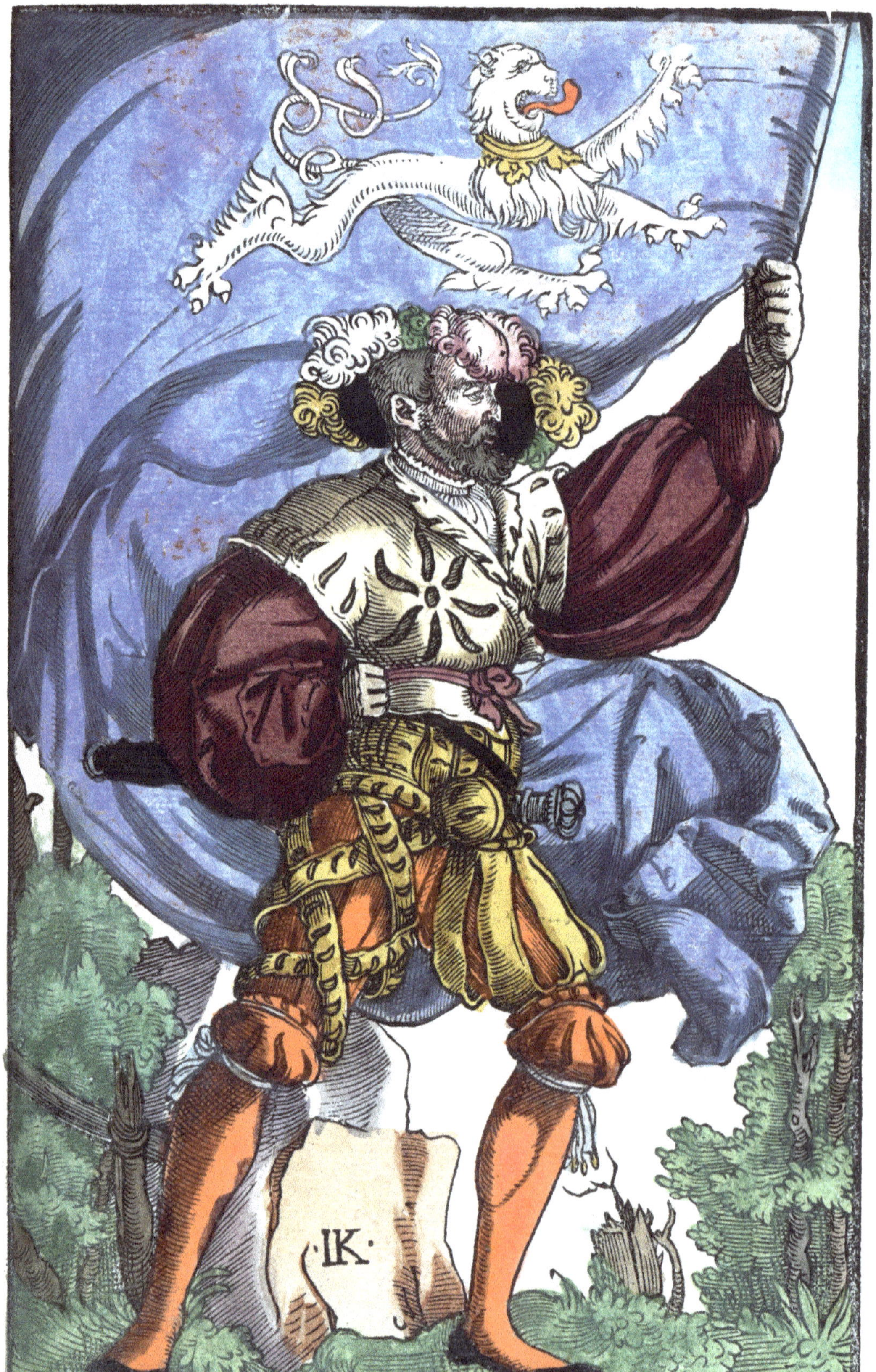

32. Aldenwalden | Alwalden

Andelaw.

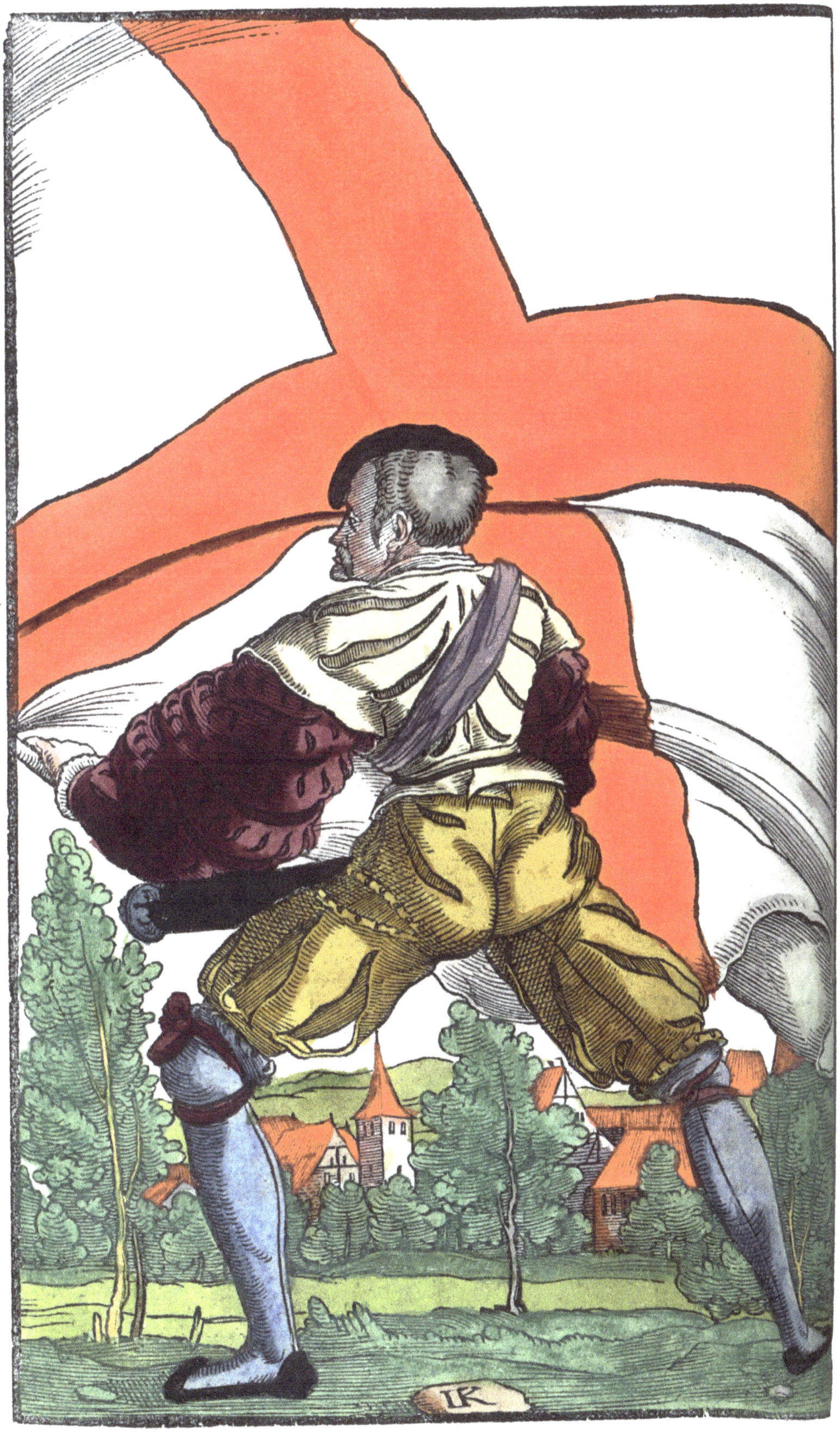

33. Andelaw | Anndelaw (Andlau)

Weissenbach.

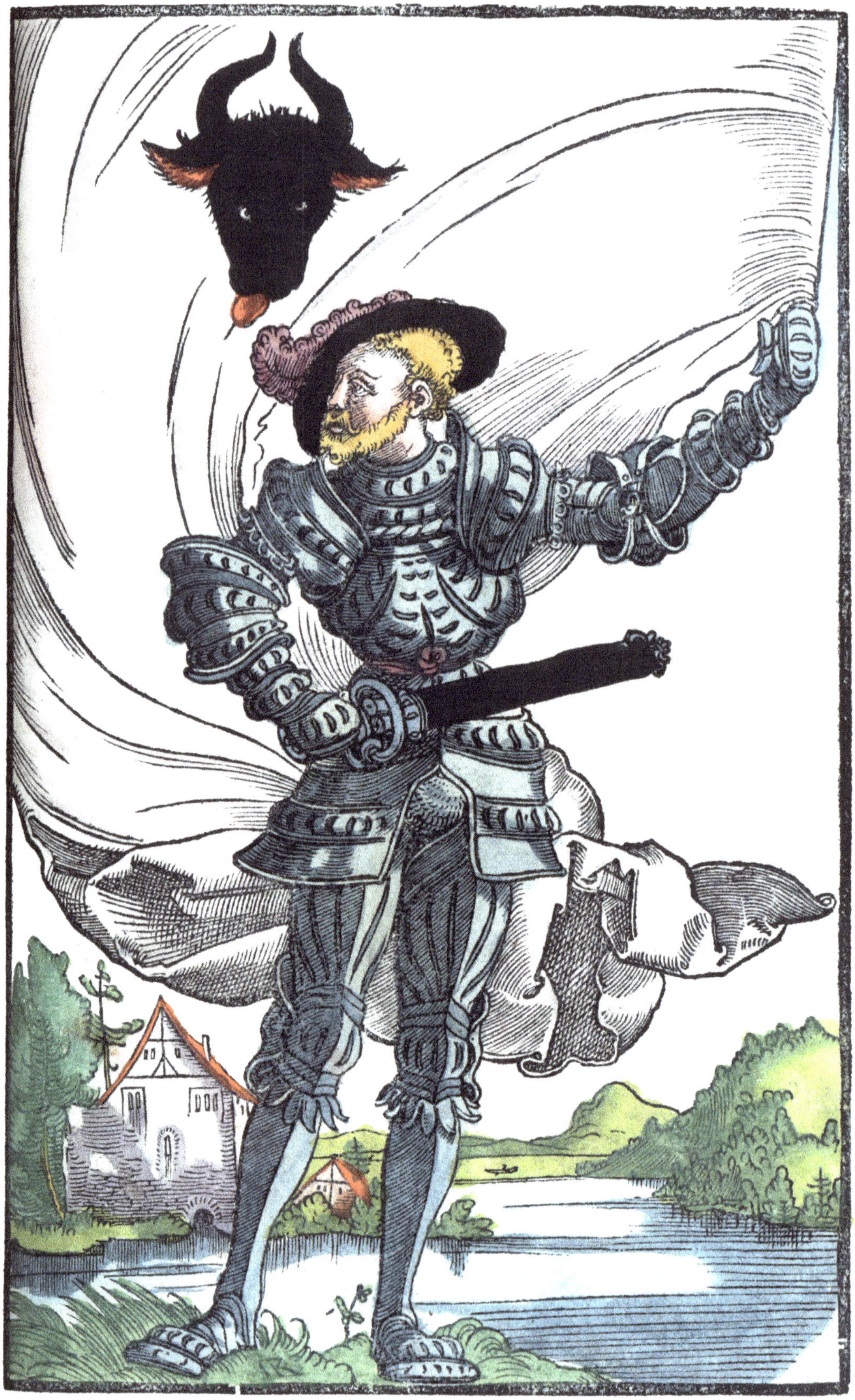

34. Weissenbach | Weissenbach

Frawenberg.

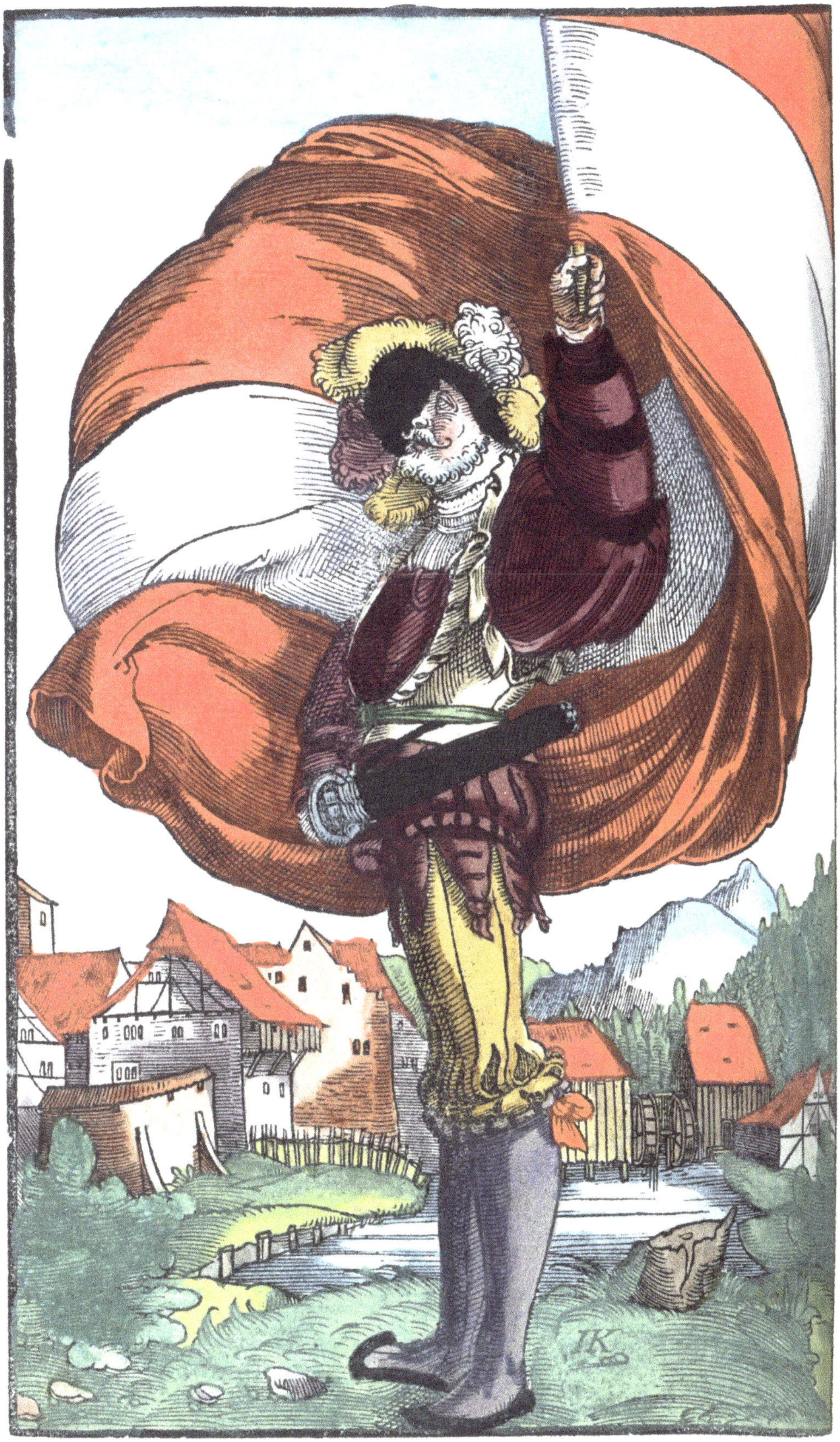

35. Frawenberg | Frawnberg (Fraunberg)

Strundeck.

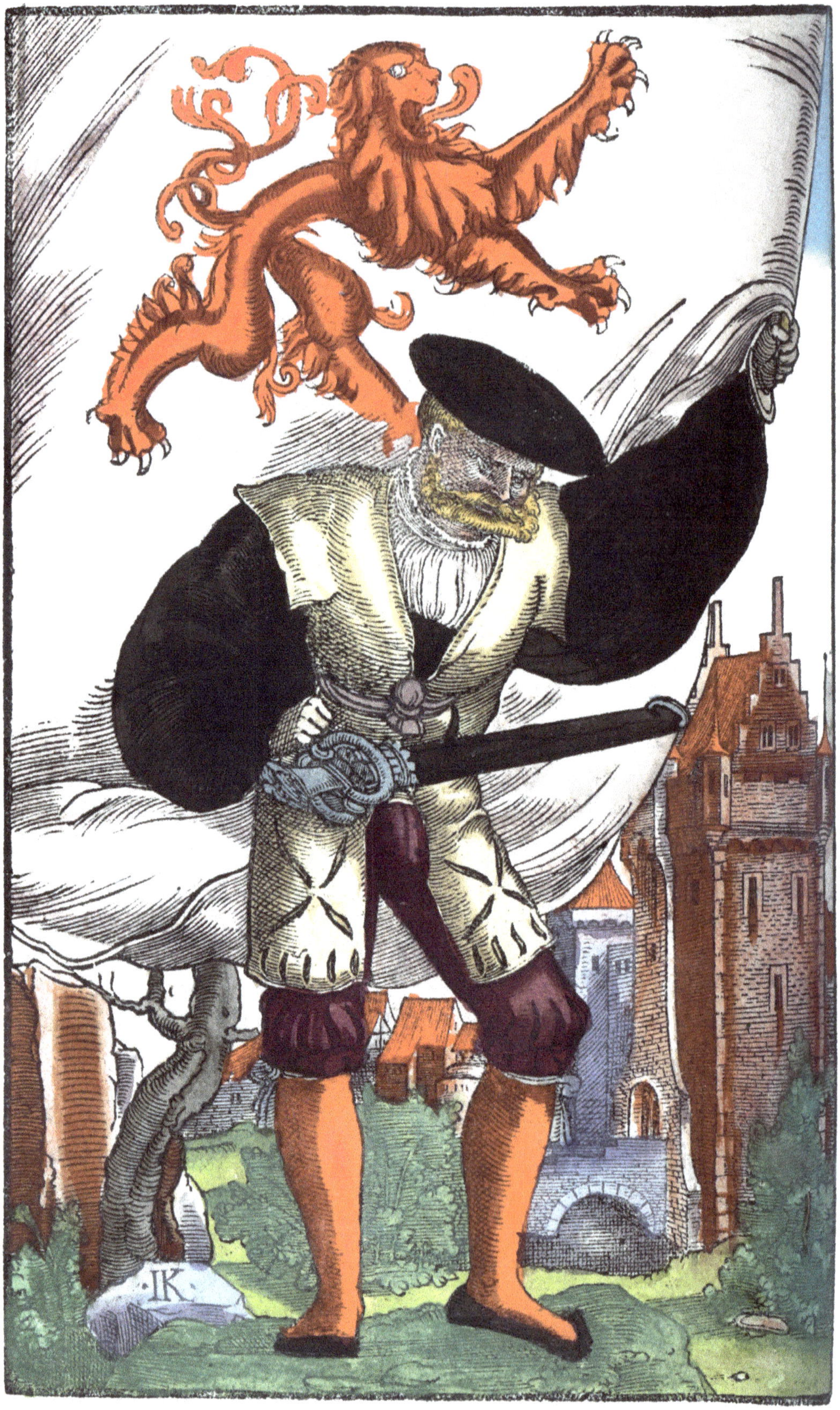

36. Strundeck | Strundeck

Augspurg.

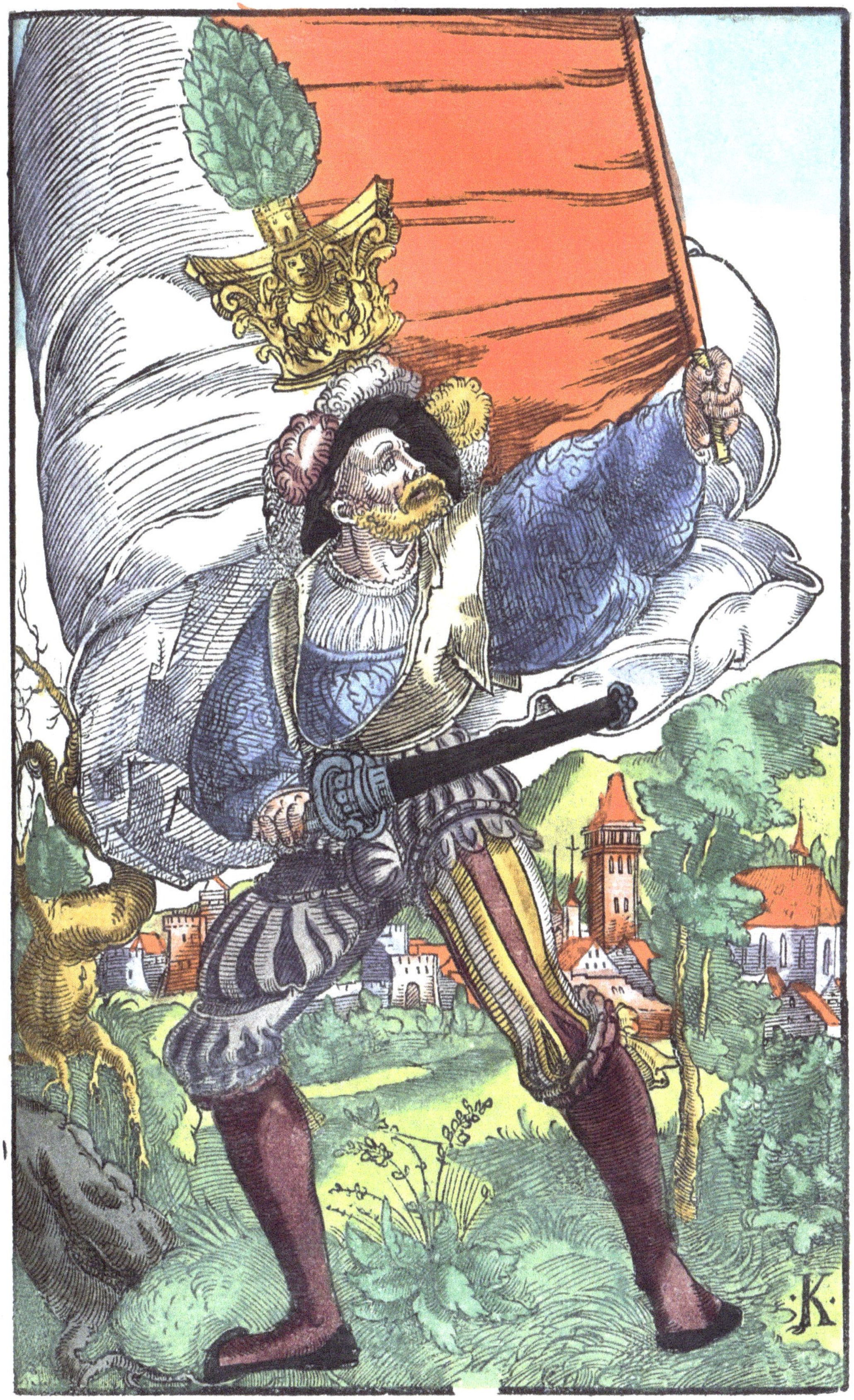

37. *Augspurg | Augsberg*

Mez.

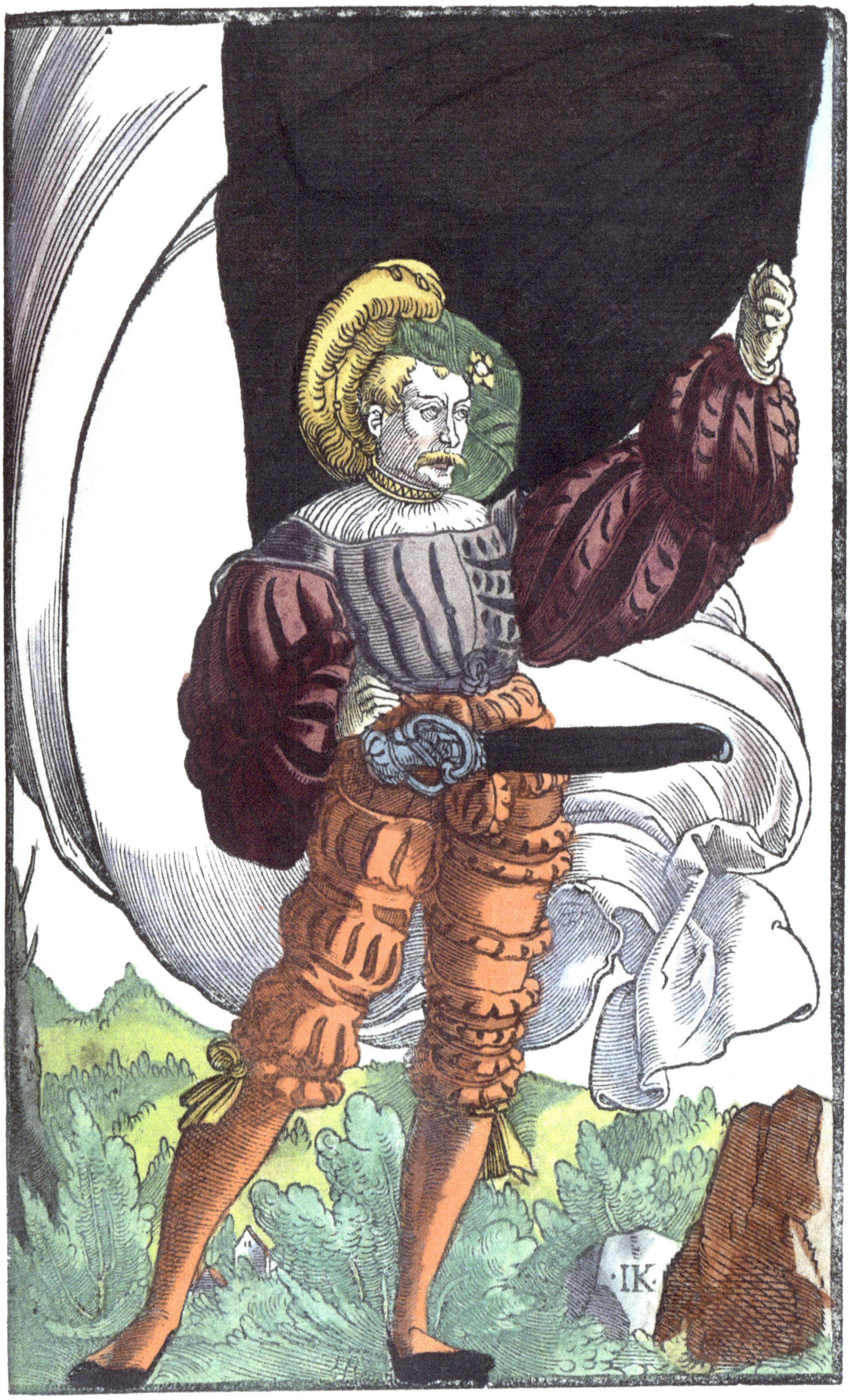

38. Merz | Metz, France

Ach.

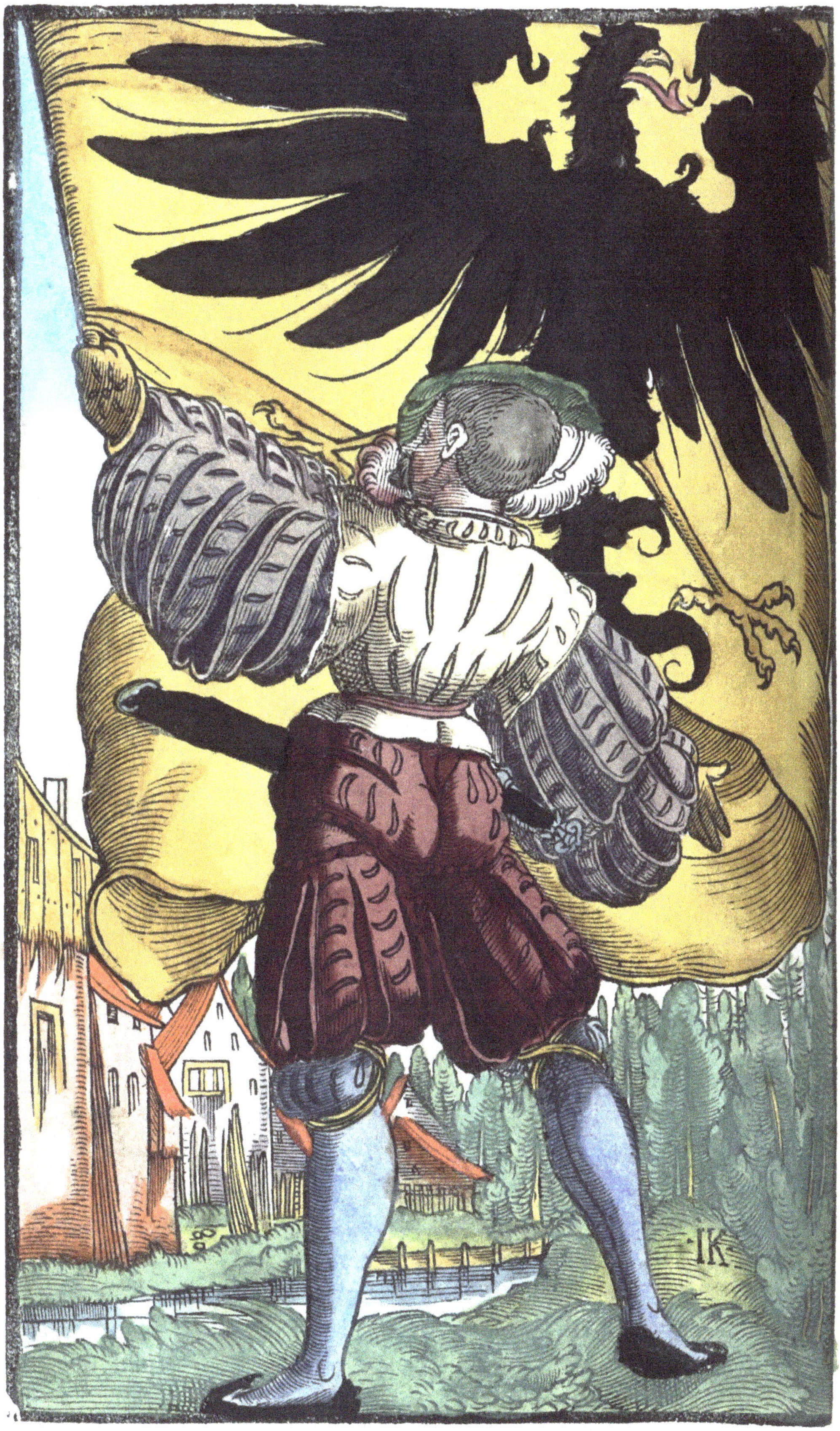

39. Ach | Aachen

Lübeck.

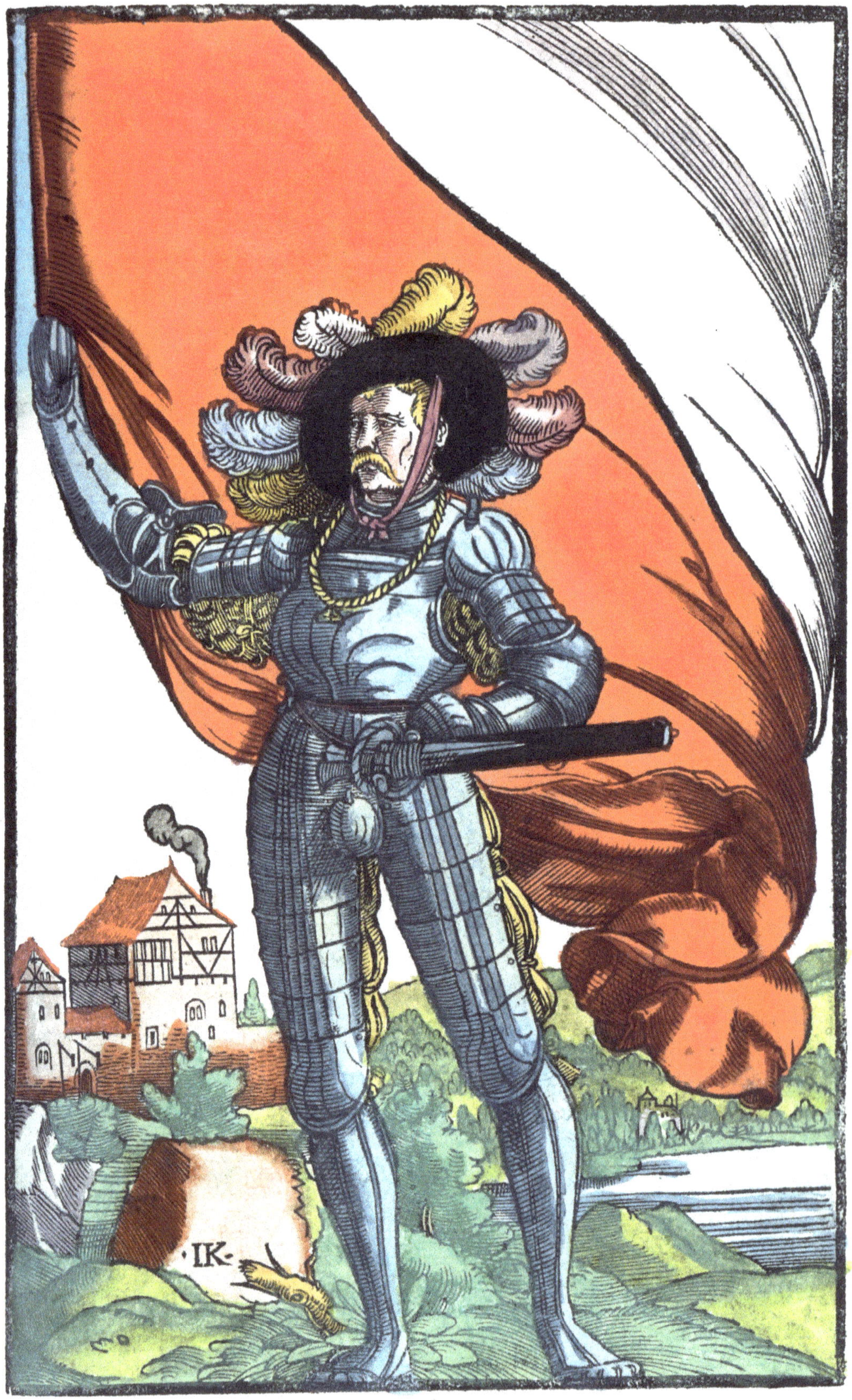

40. Lübeck | Lübeck

Bamberg.

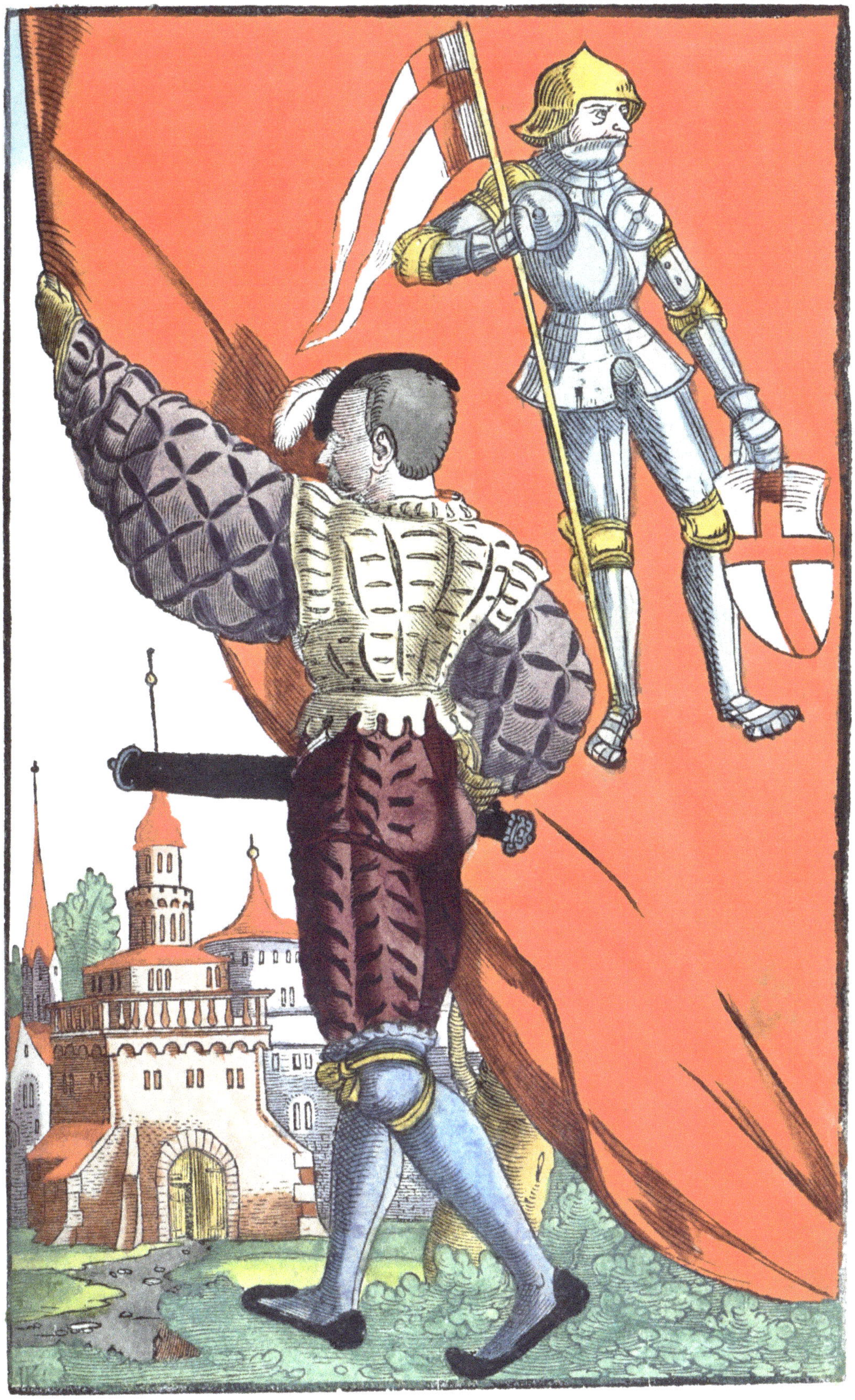

41. Bamberg | Bamberg

Vlm.

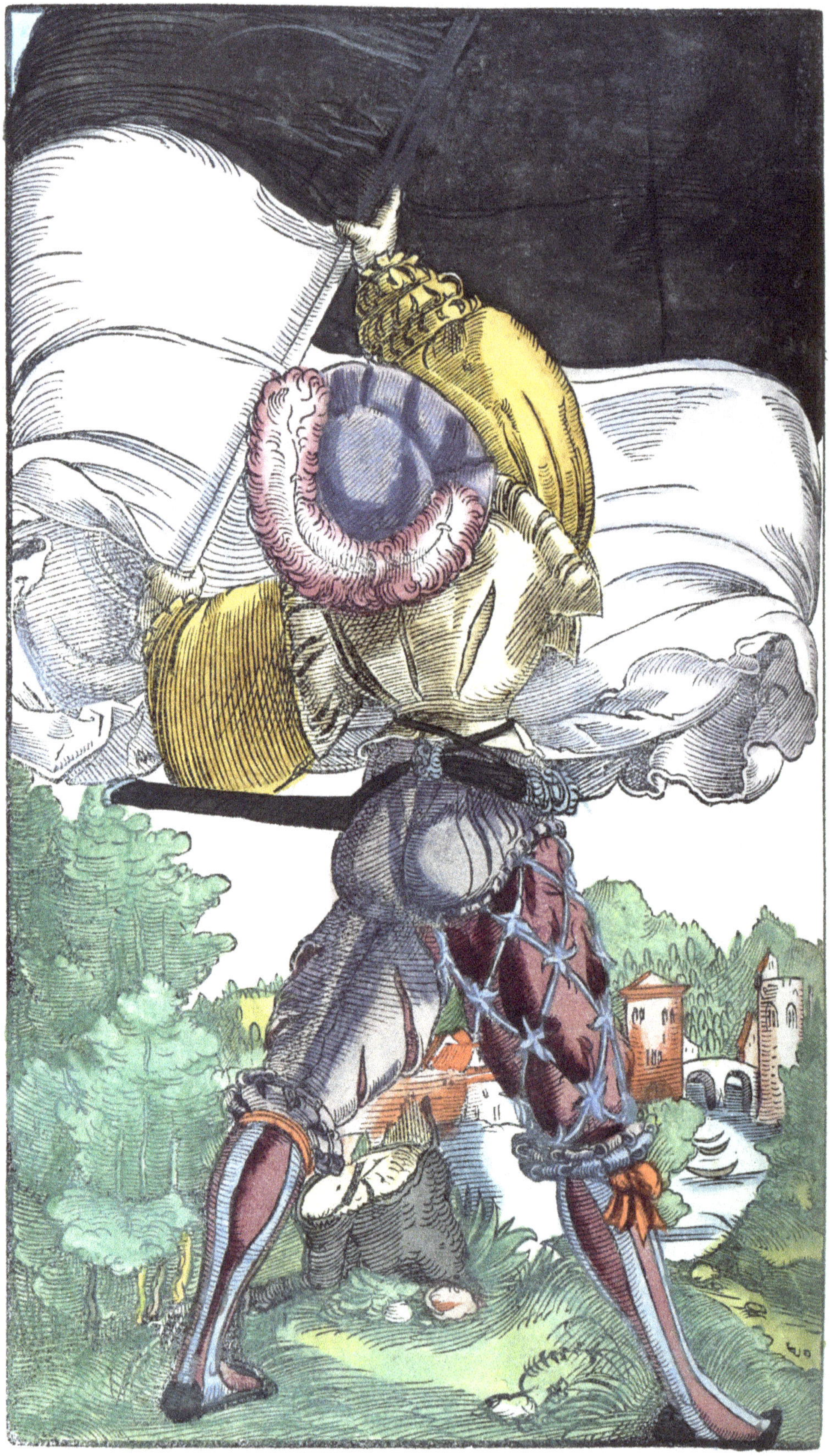

42. *Ulm | Ulm*

Hagnaw.

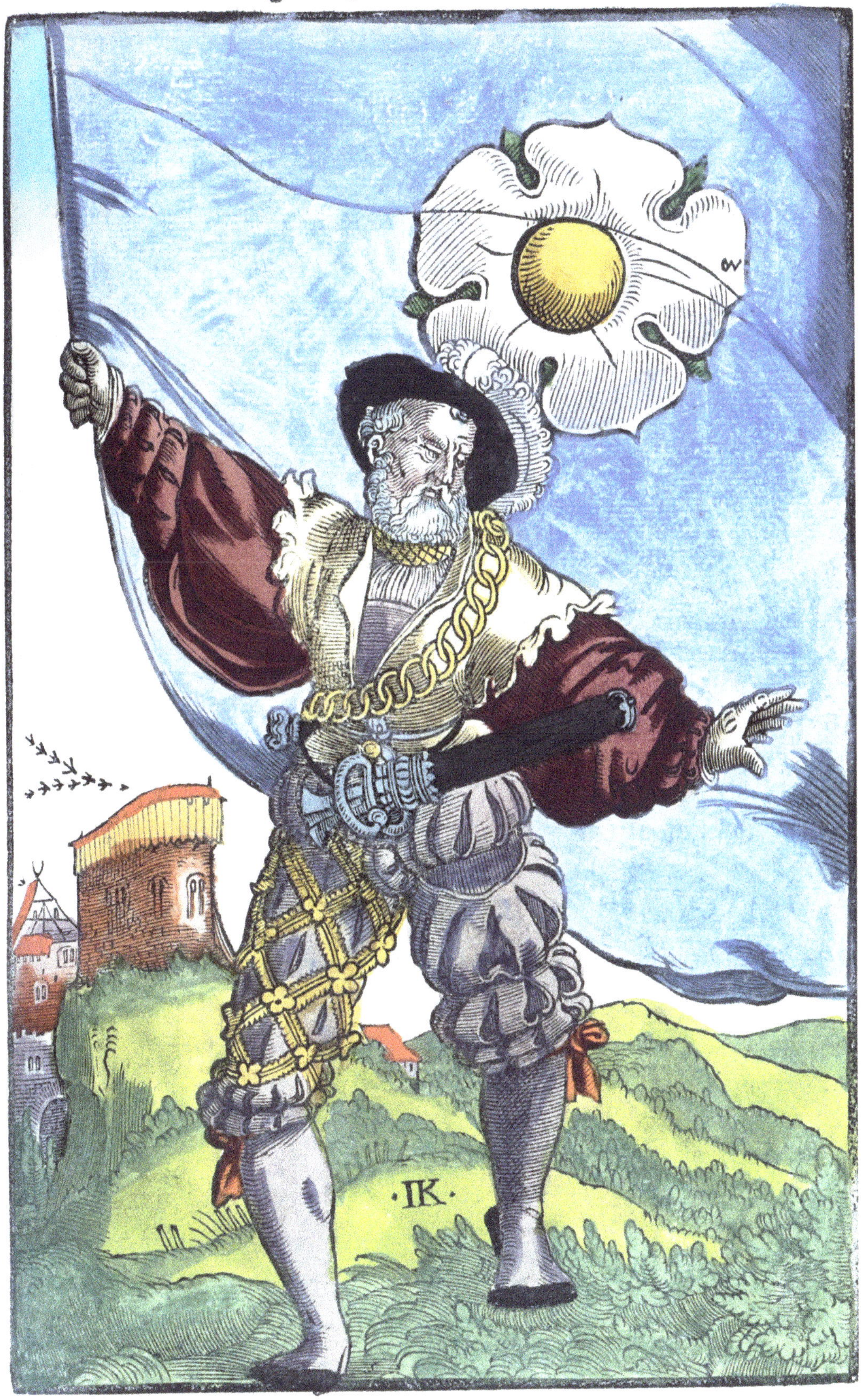

43. Hagnaw | Hagenau

Schlerstat.

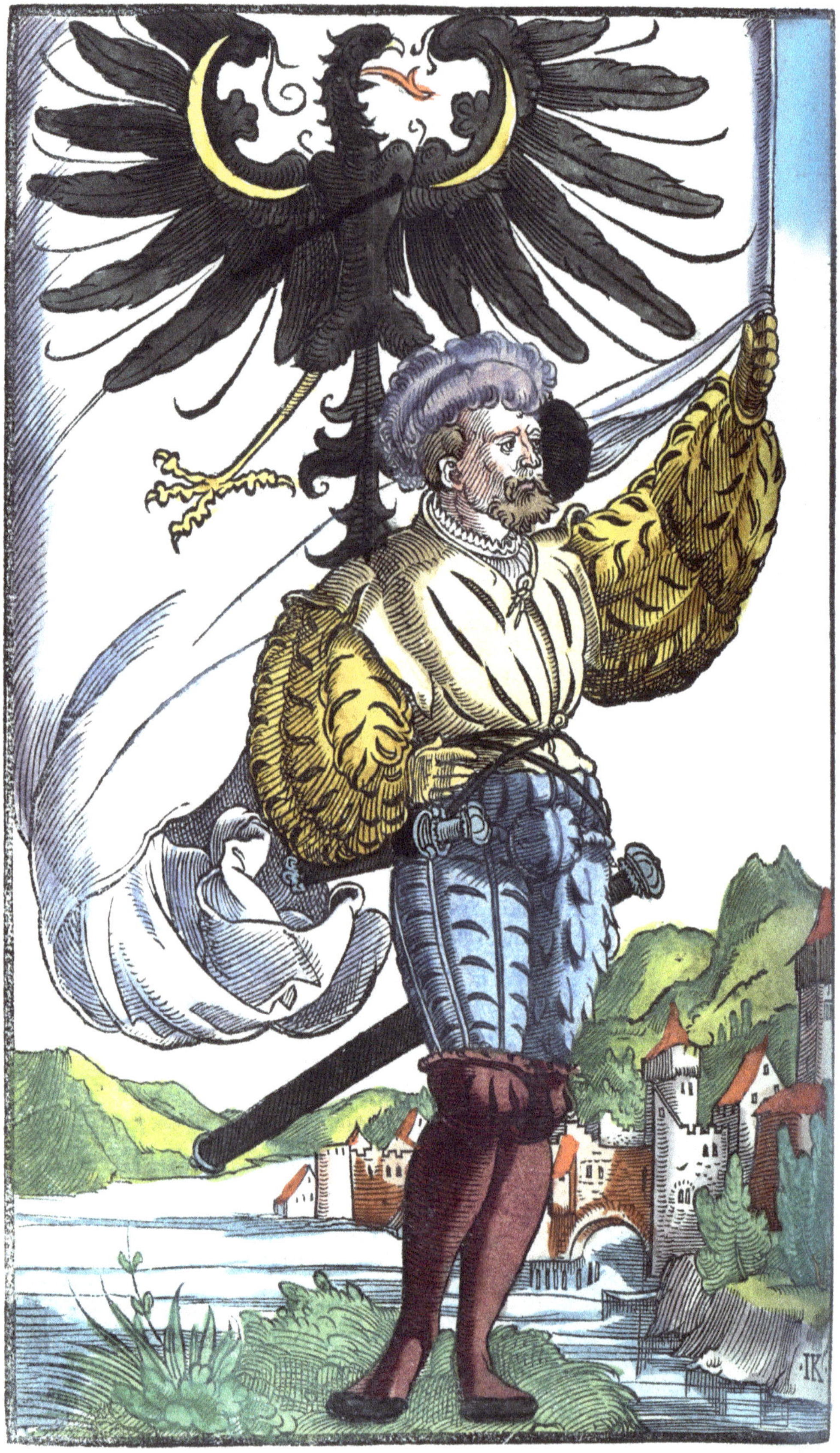

44. Schlerstat | Schlettstatt

Cöln.

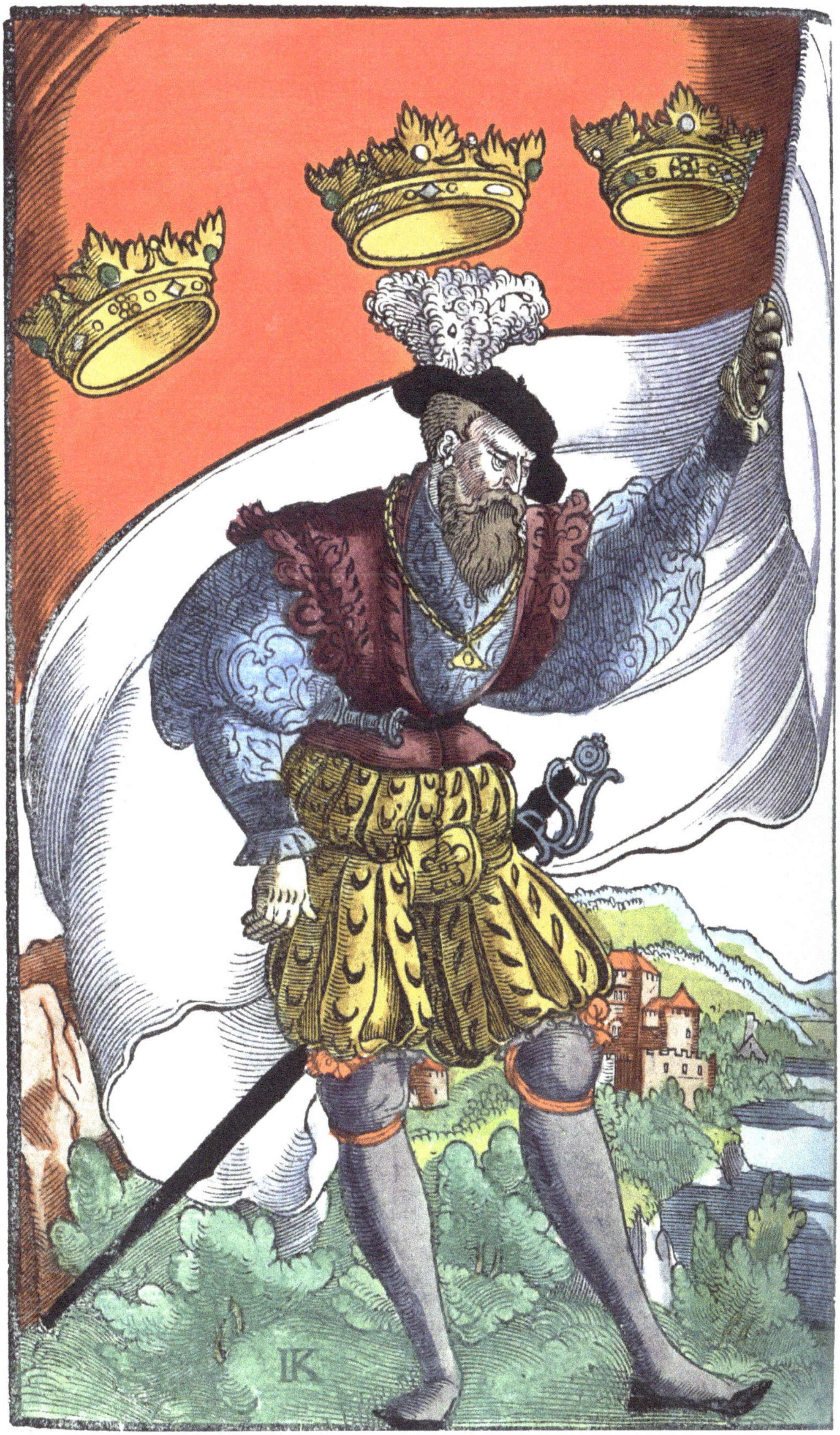

45. Cöln | Koln, Cologne

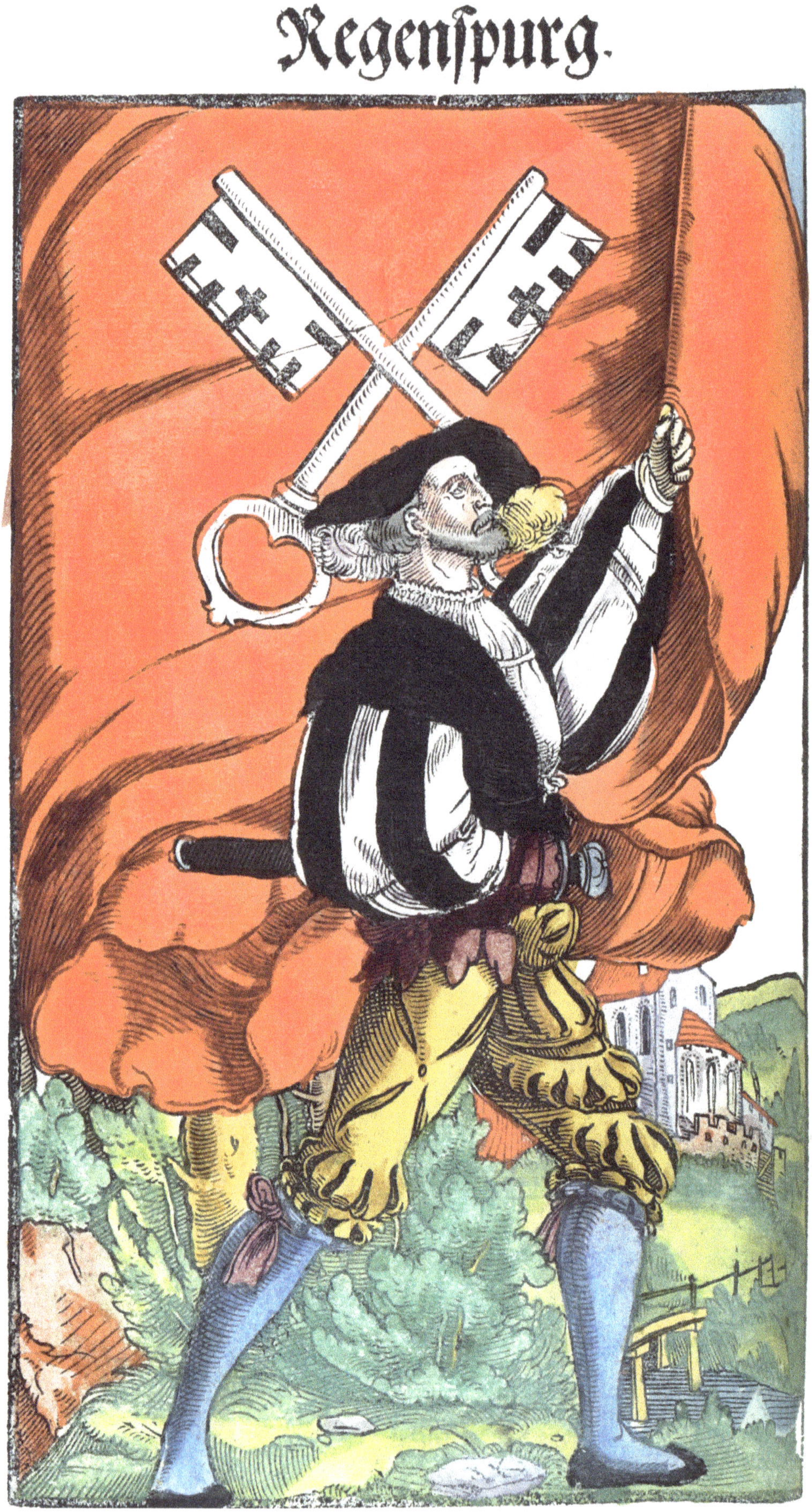

46. Regenspurg | Regensberg

Costniz.

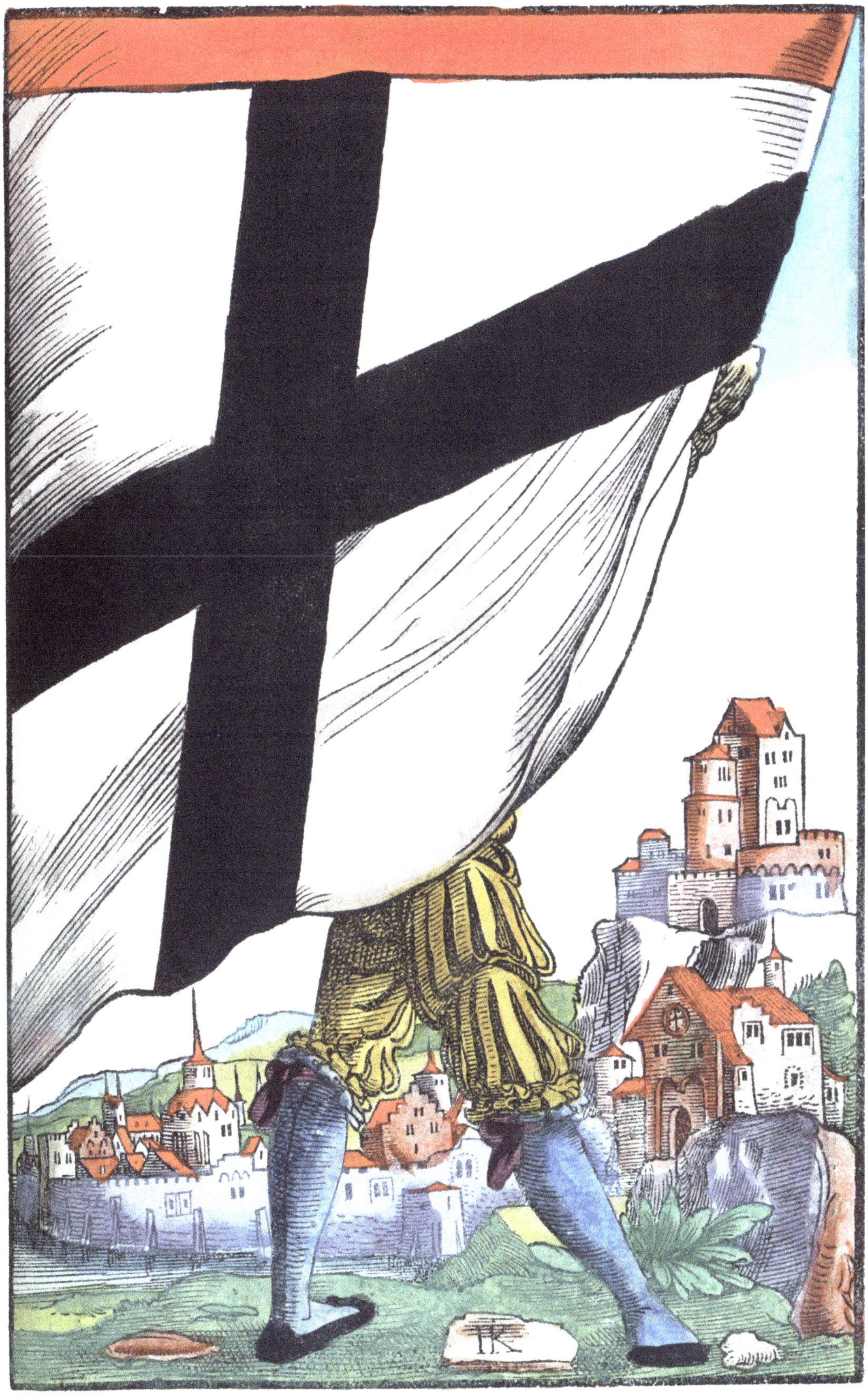

47. Costniz | Konstanz, Constance

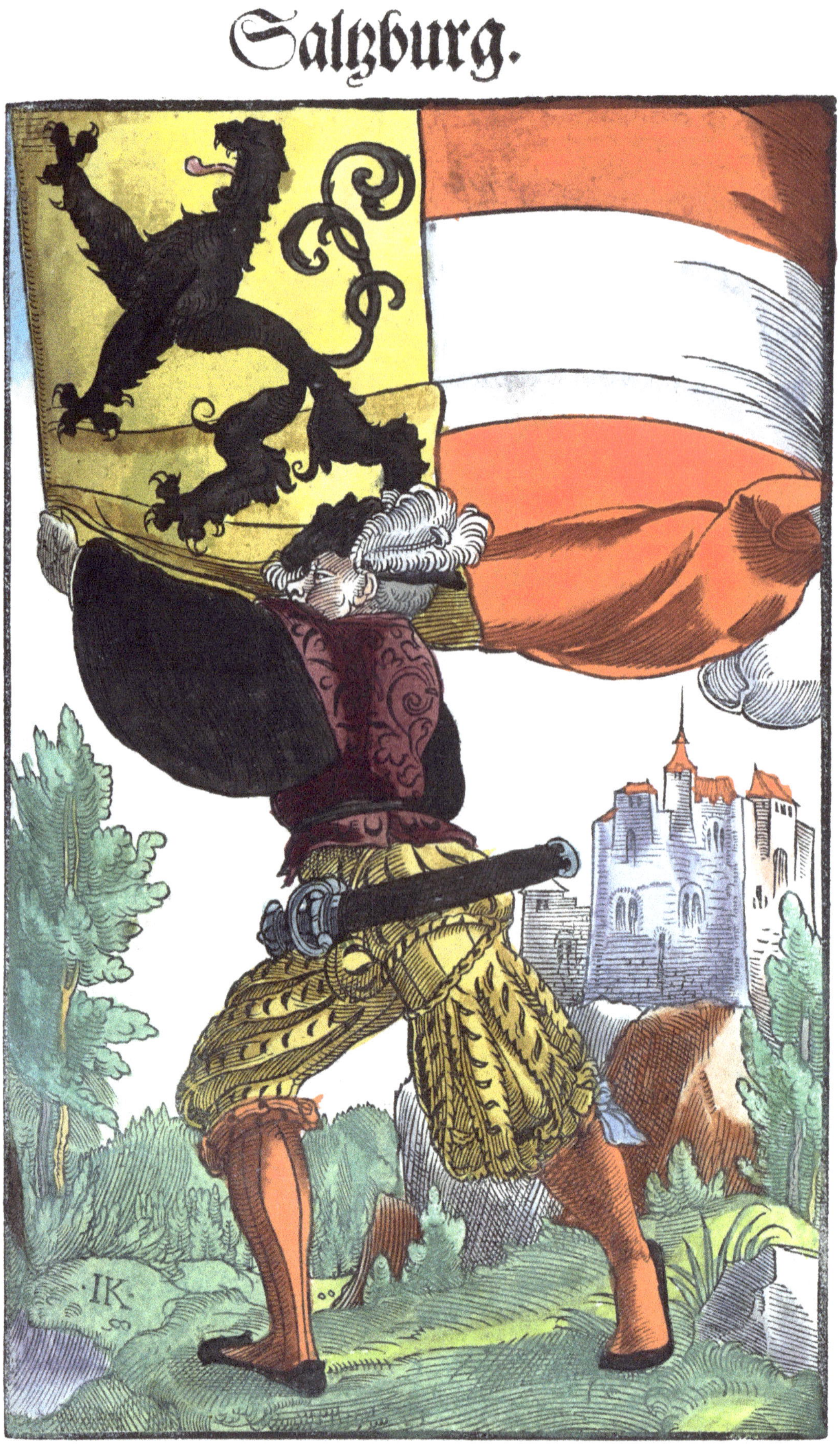

48. Salzburg | Salzburg, Austria

Meydenburg inn sachssen.

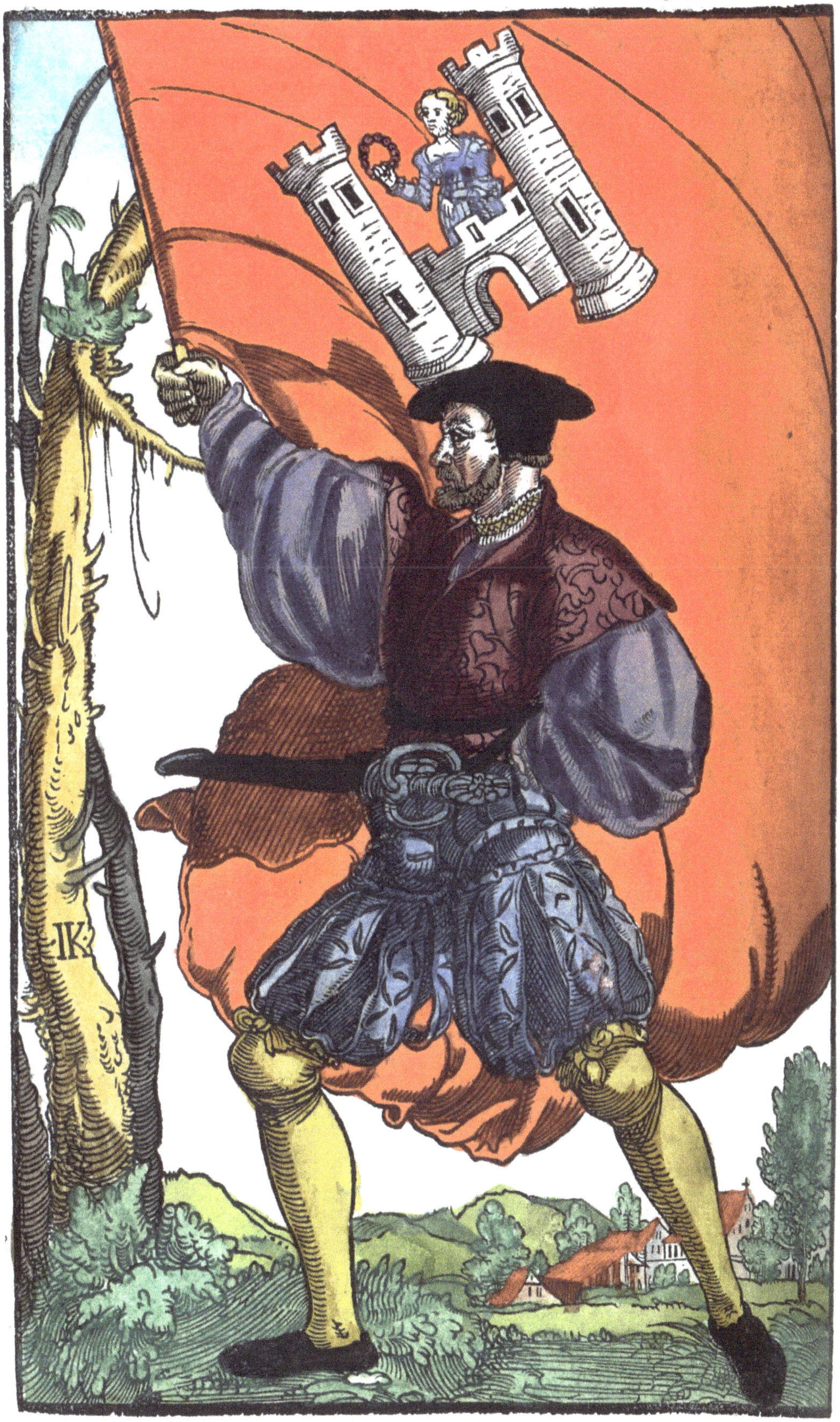

49. Mendenburg inn Sachssen | Meyenburg

Lützelburg.

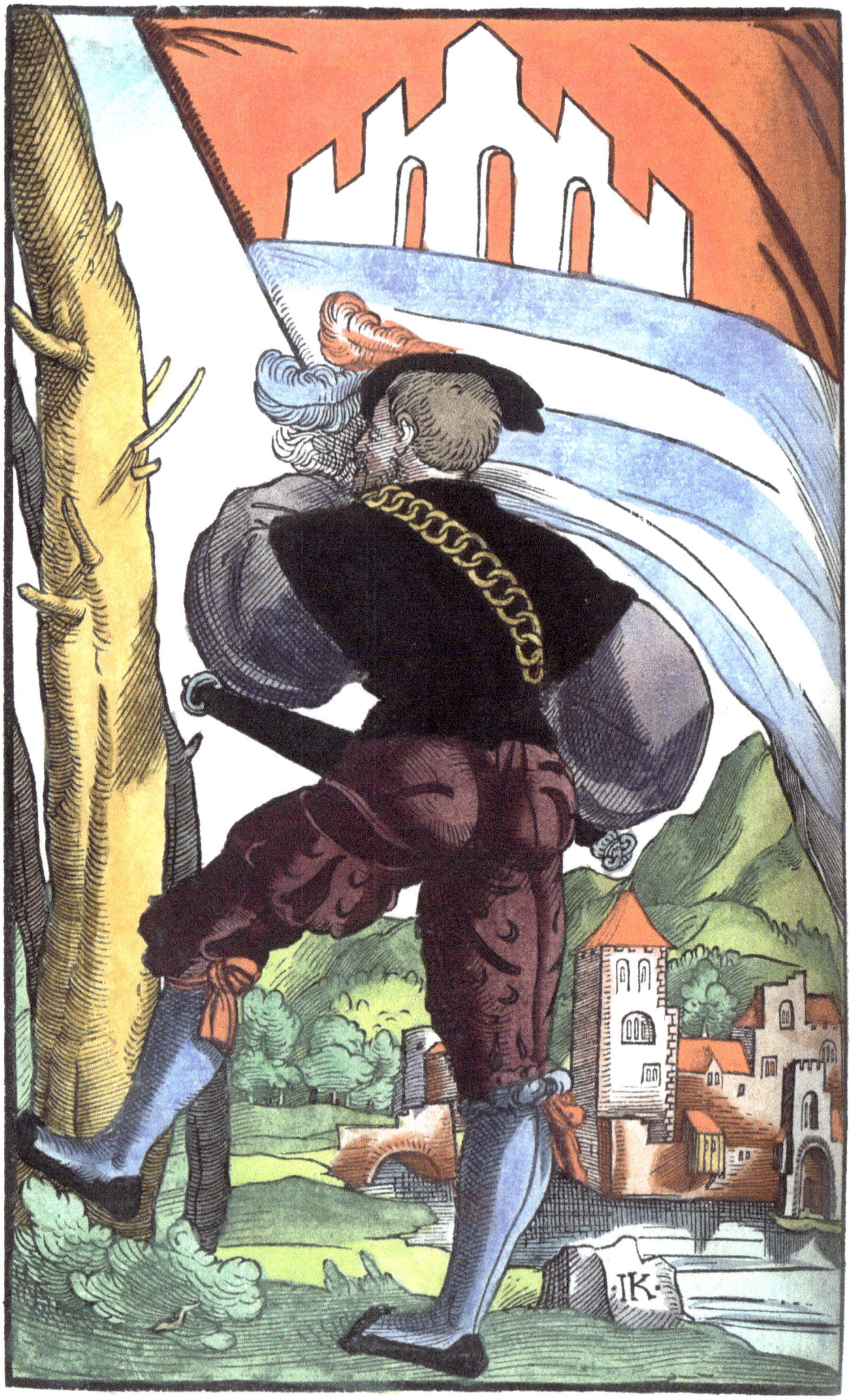

50. Lünzelburg | Luxemburg

Rottenburg am Necker.

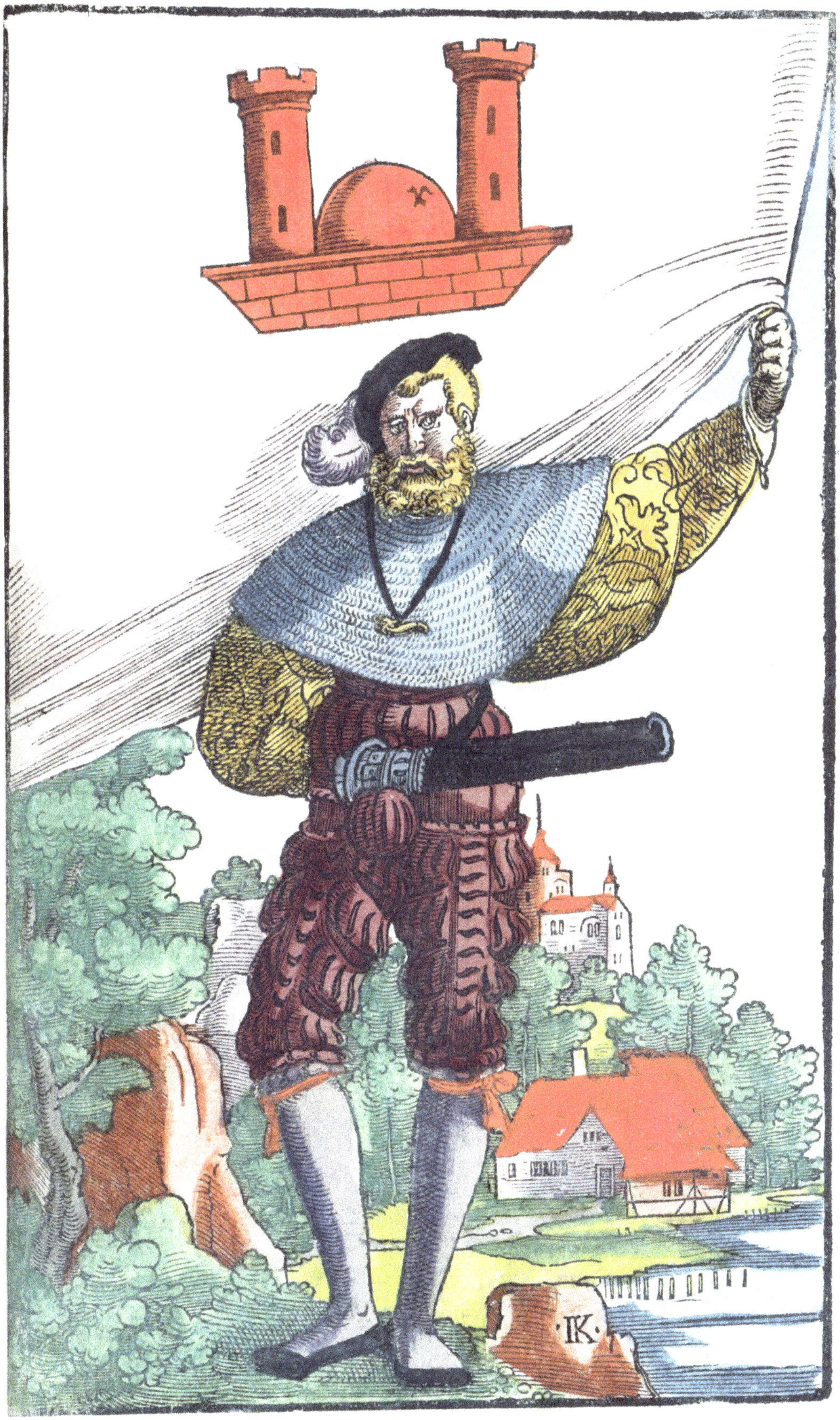

51. Rottenburg am Necker | Rottenburg am Neckar

Aldenburg.

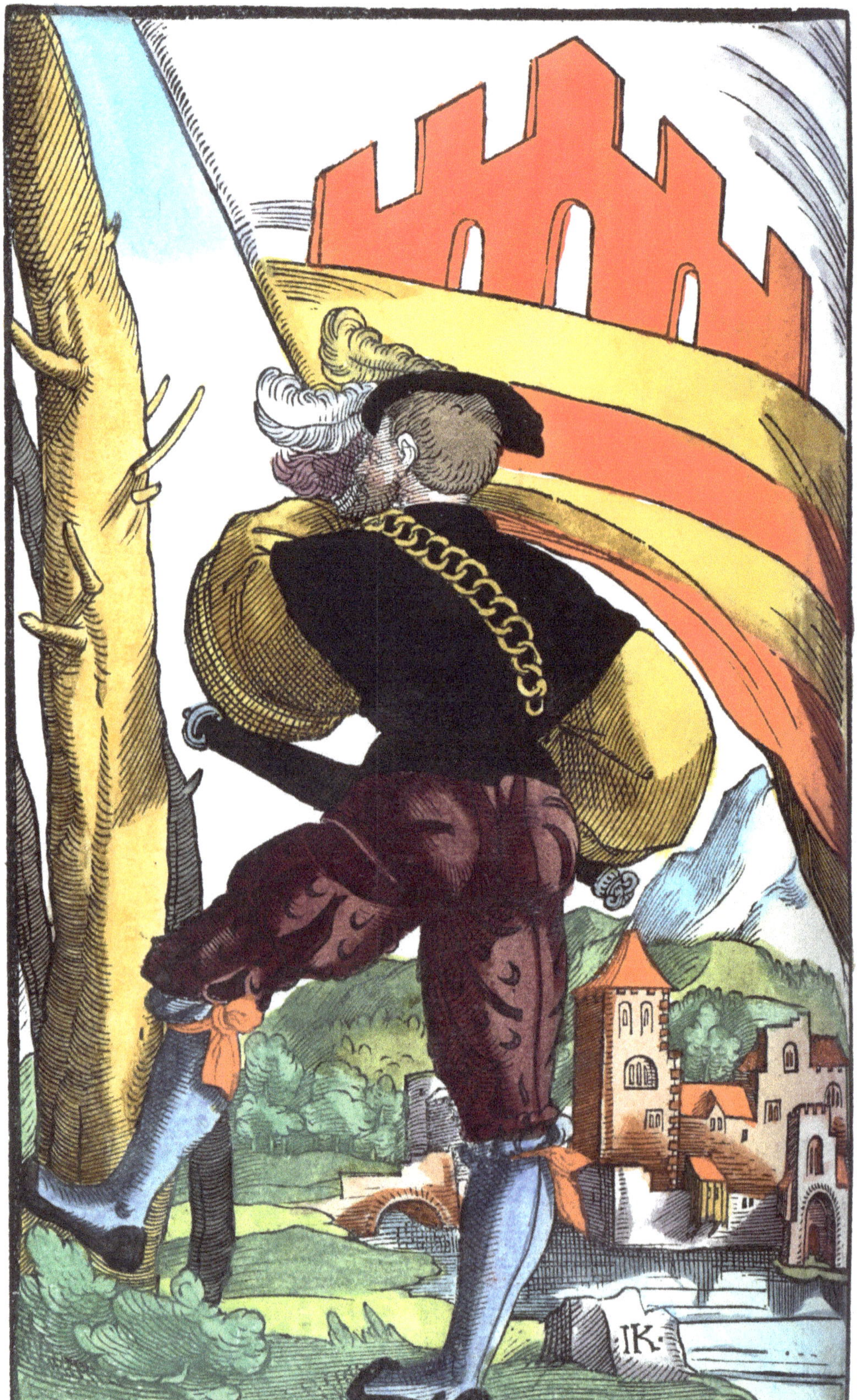

52. *Aldenburg | Altenburg*

Brabandt.

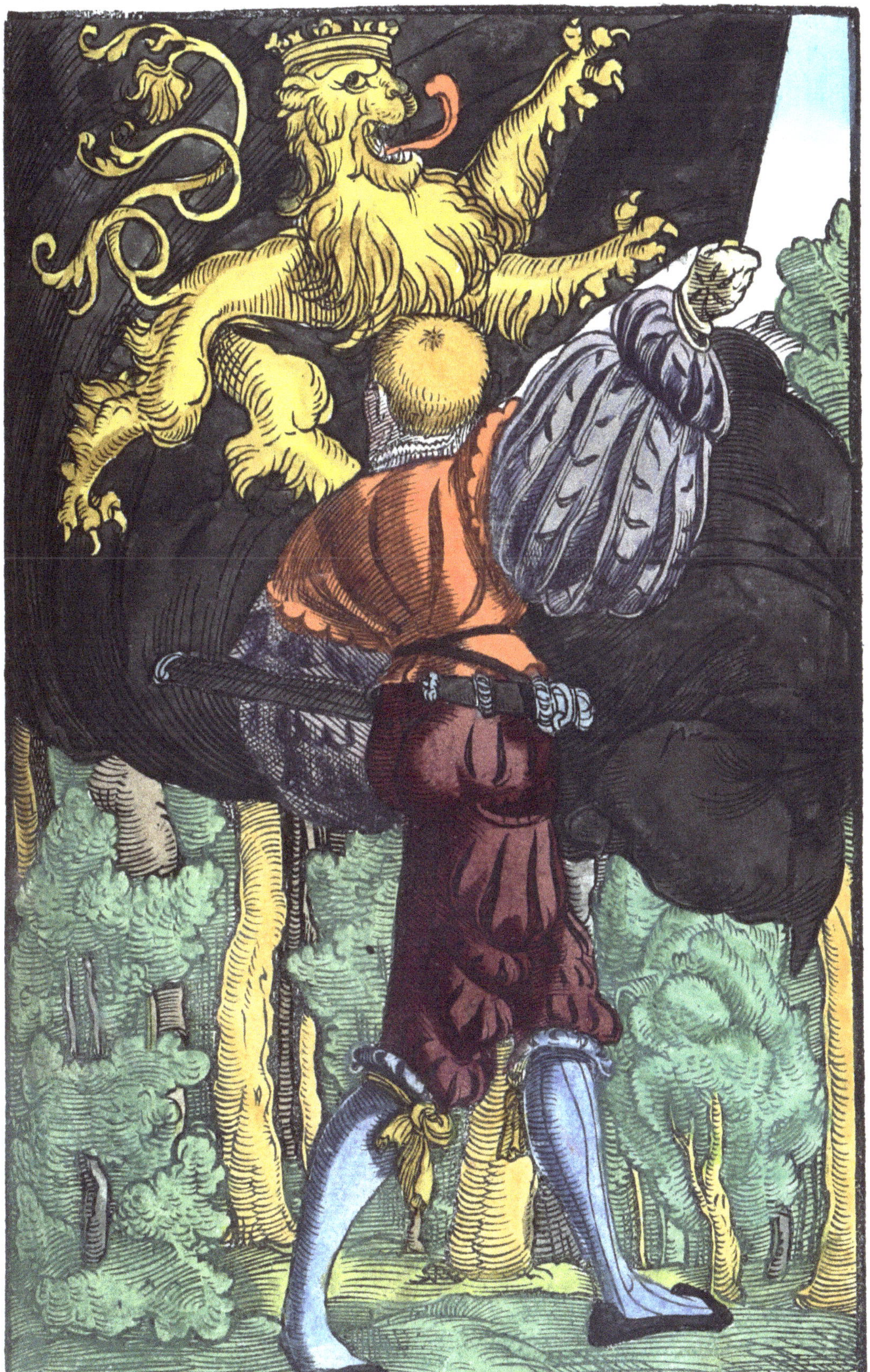

53. Brabandt | Duchy of Brabant

Nider Sachssen.

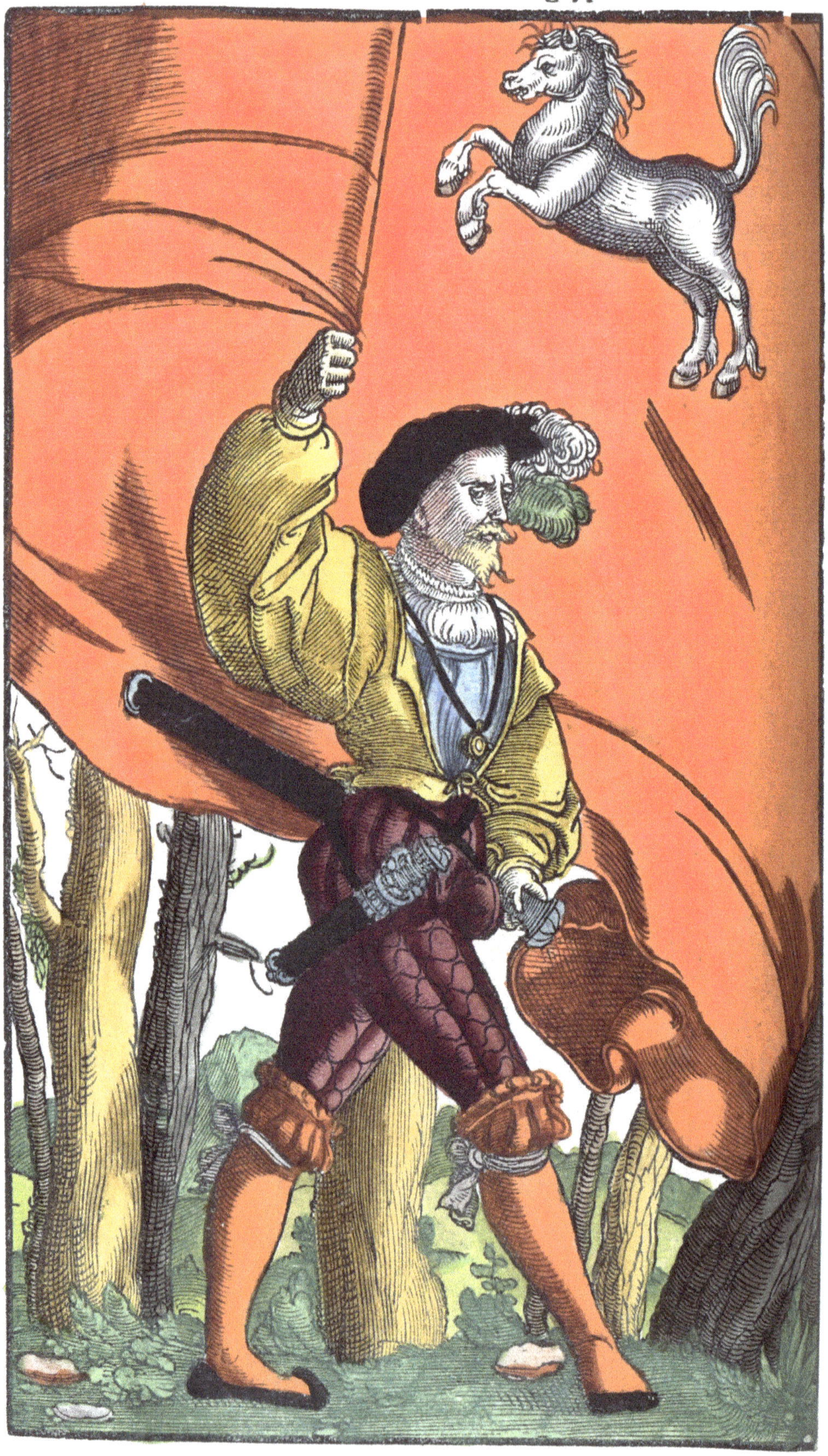

54. *Nider Sachssen* | *Niedersachen, Lower Saxony*

Westerreich.

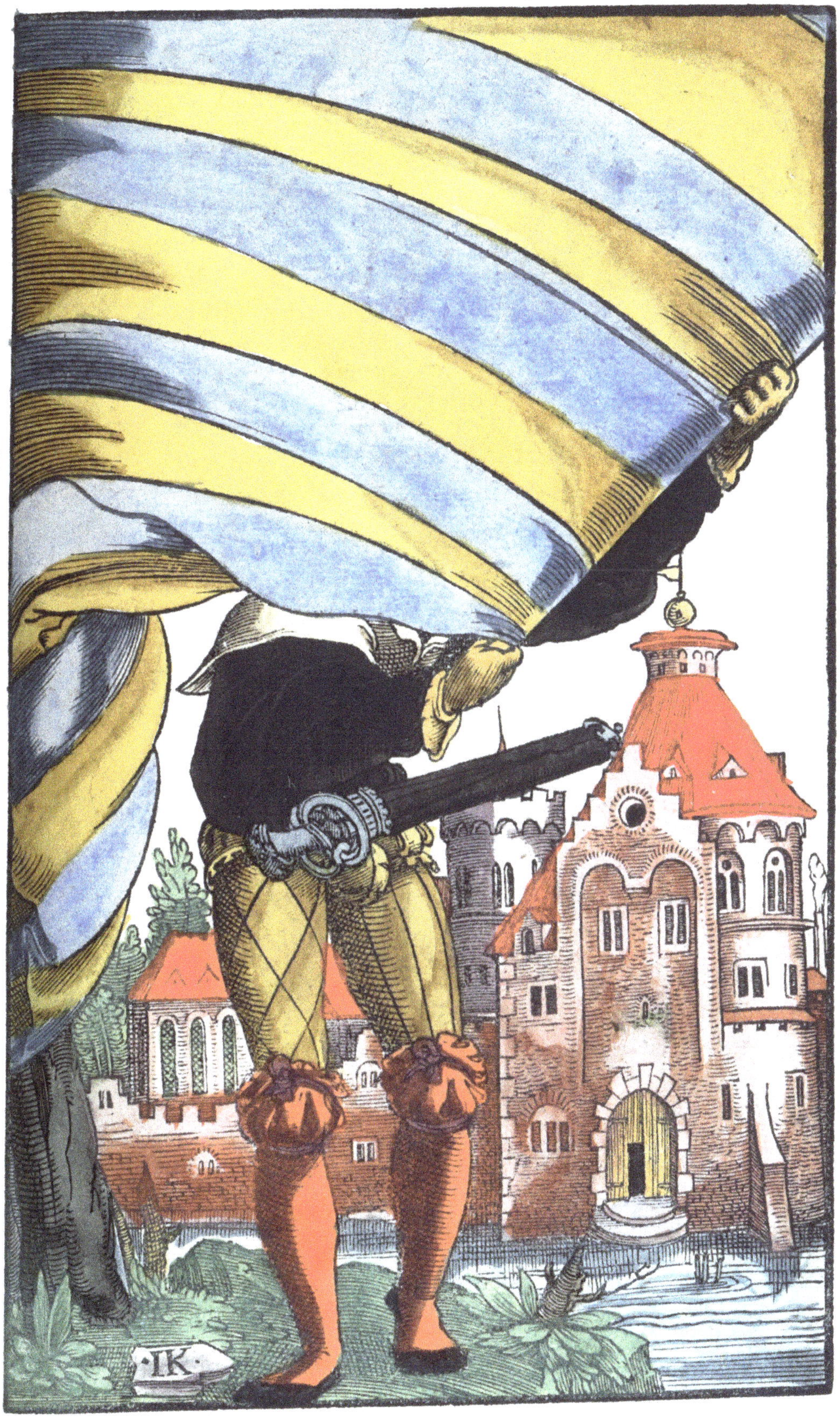

55. Westerreich | Westrich

Schlessi.

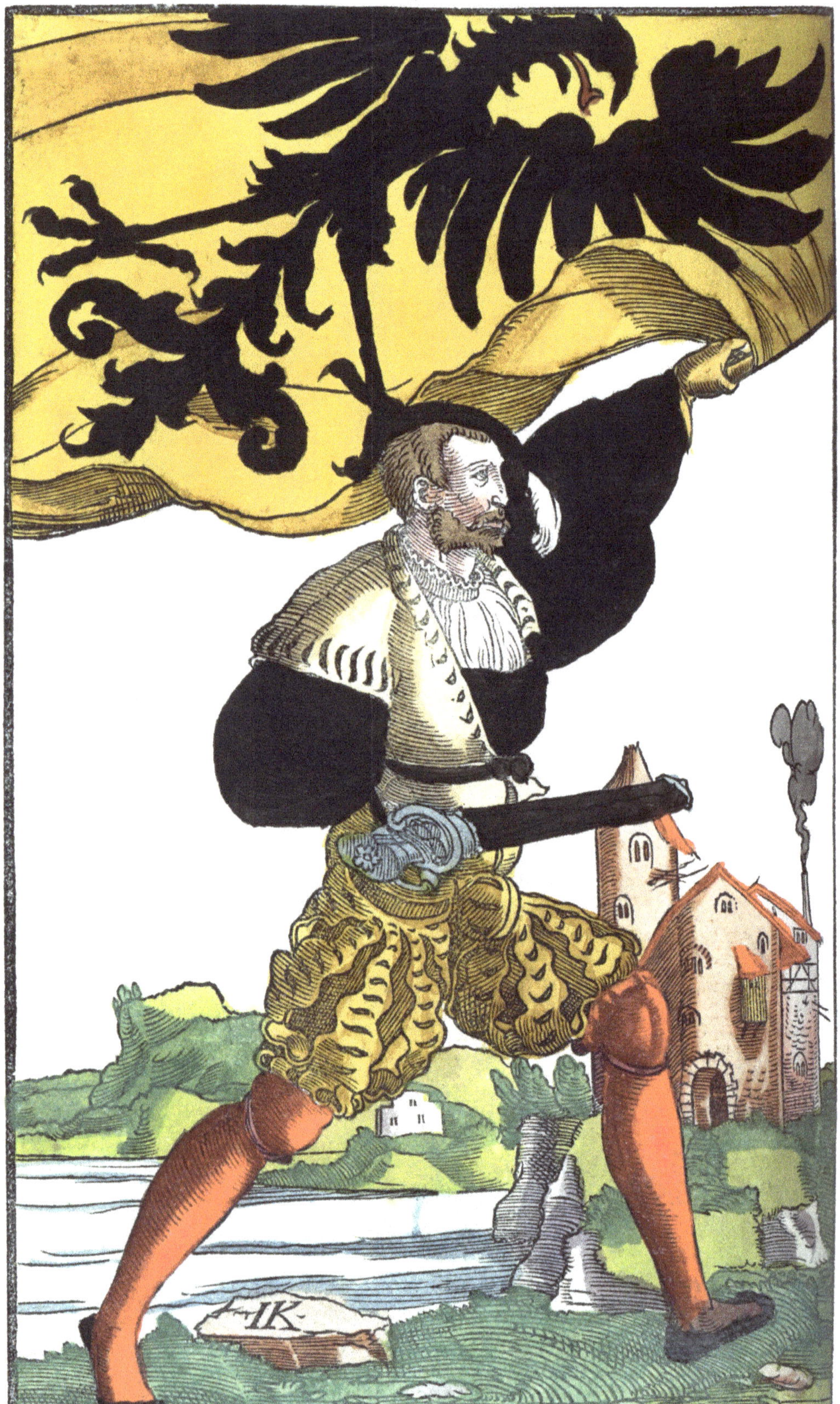

56. Schlessi | Schlesien, Silesia

Offenburg.

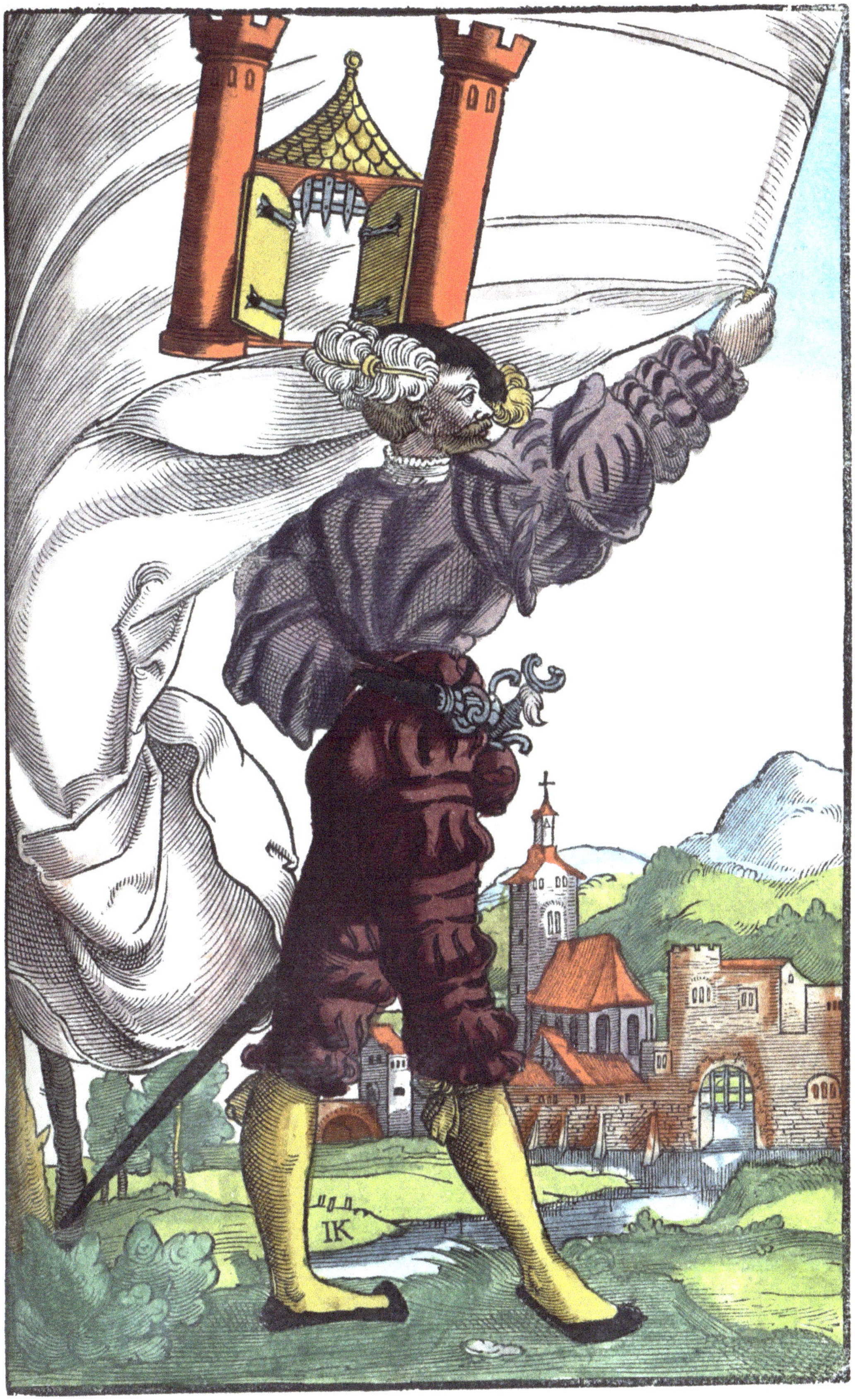

57. Offenburg | Offenburg

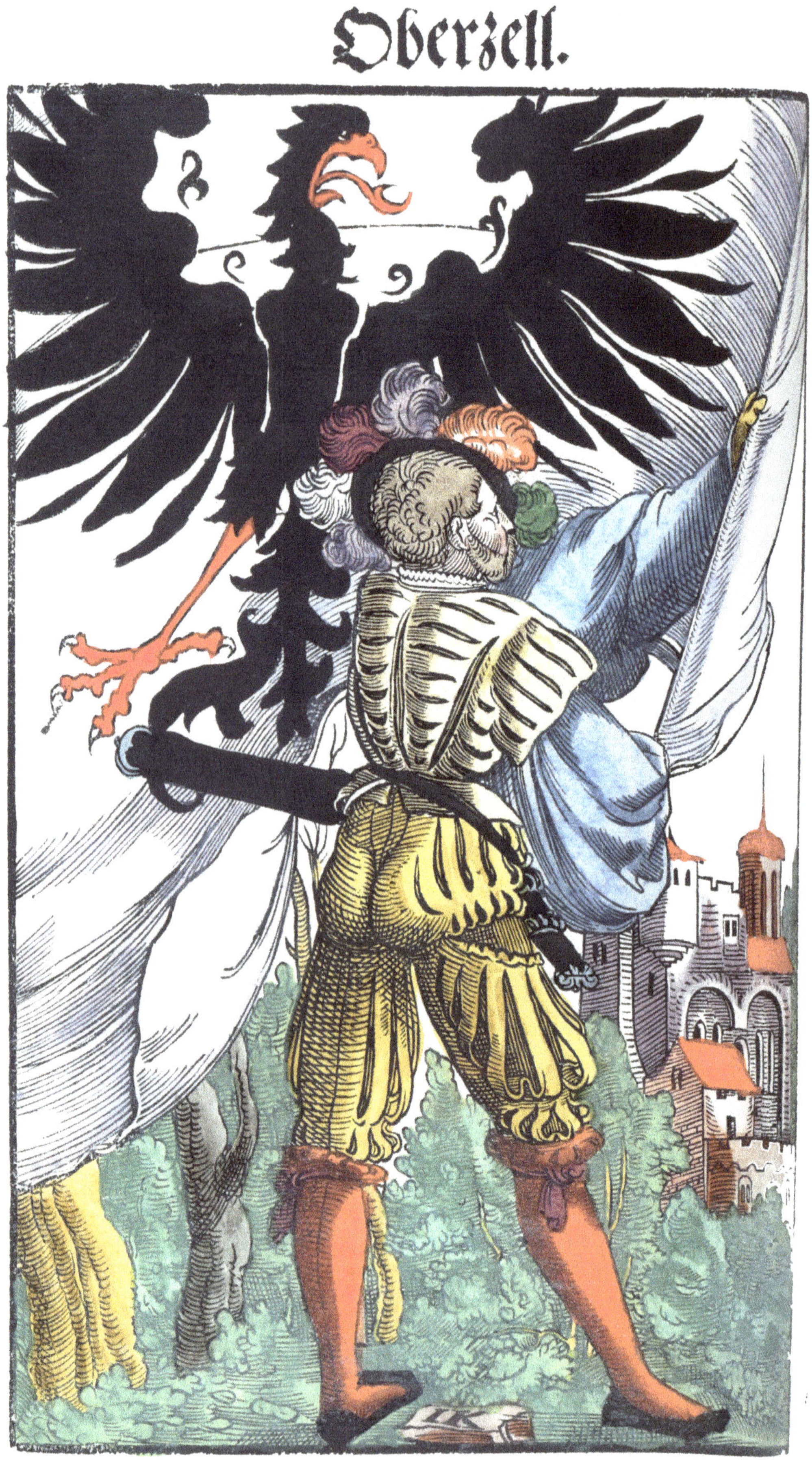

58. Oberzell | Zell am Harmersbach

Nerlingen.

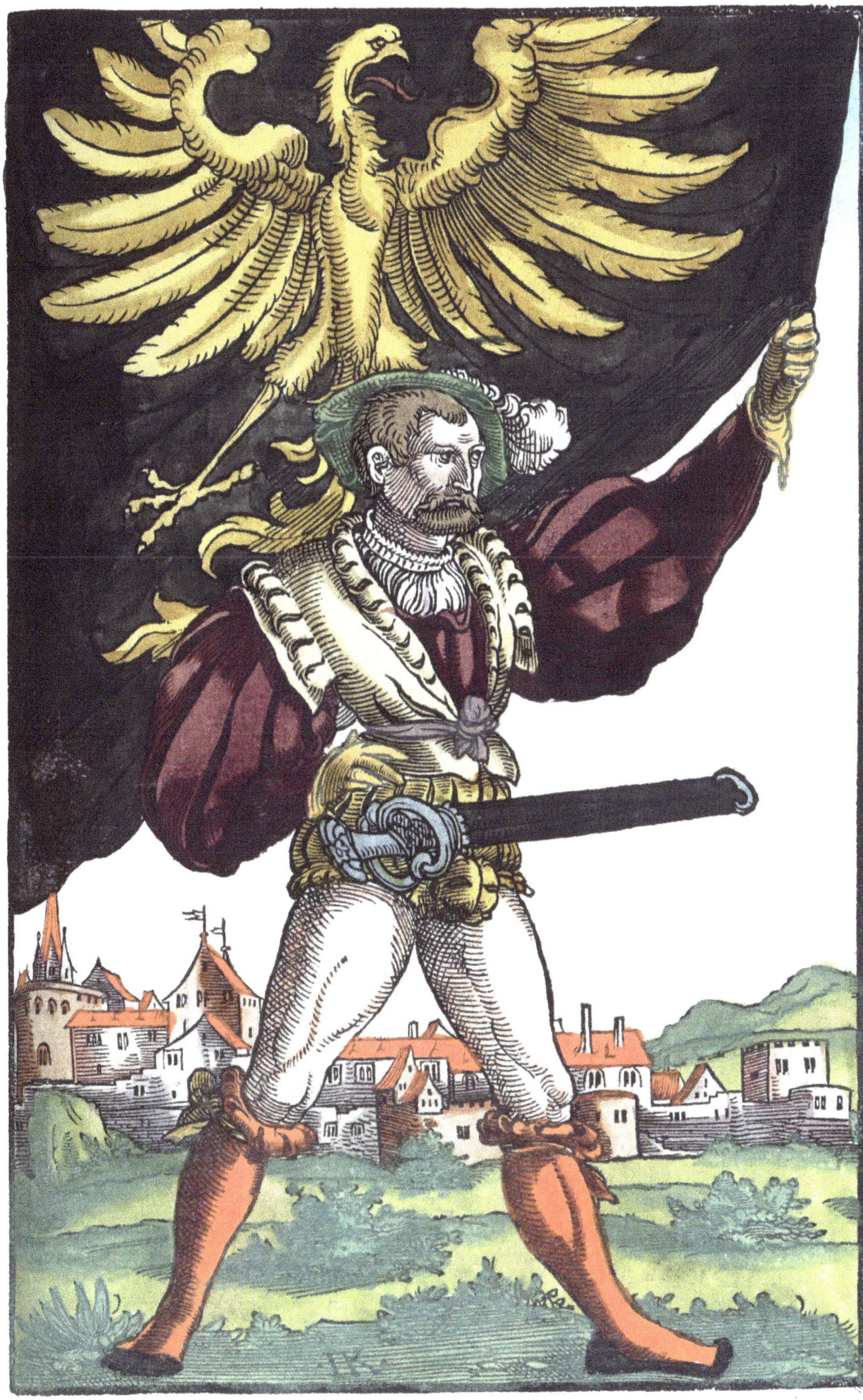

59. Nerlingen | Nördlingen

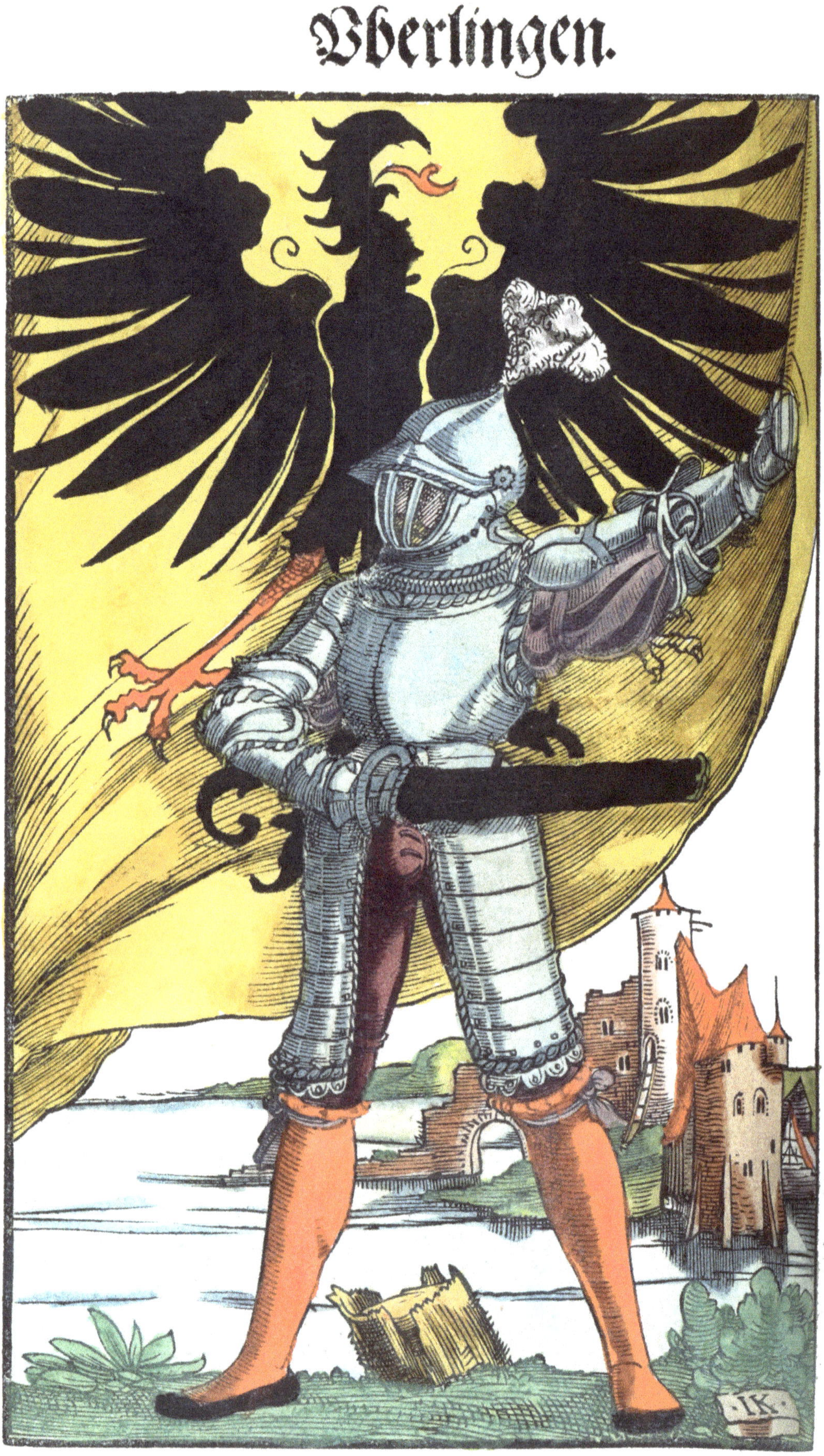

60. *Uberlingen | Überlingen*

Rosenheym.

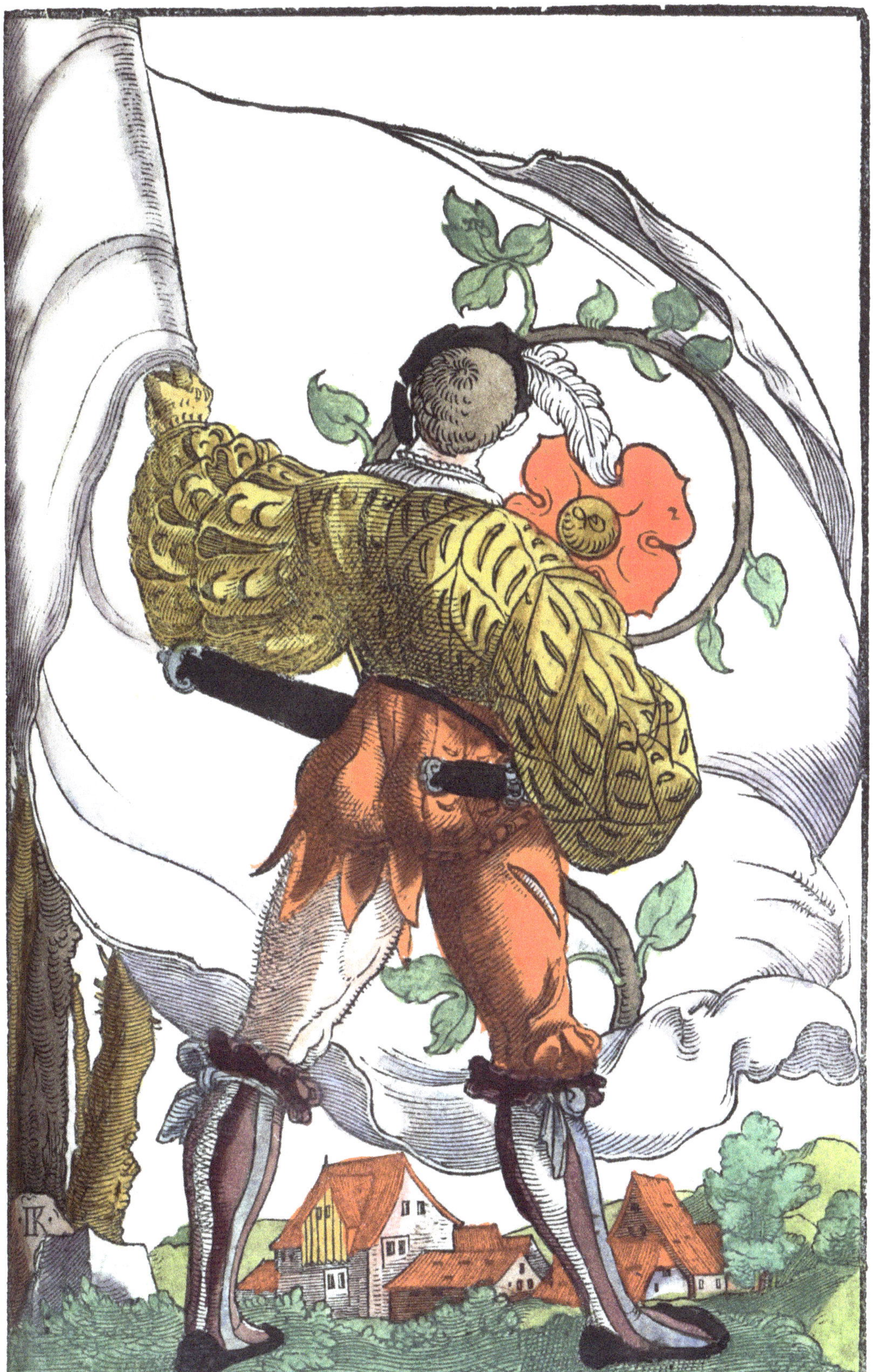

61. Rosenhenm | Rosenheim

Wimpfem.

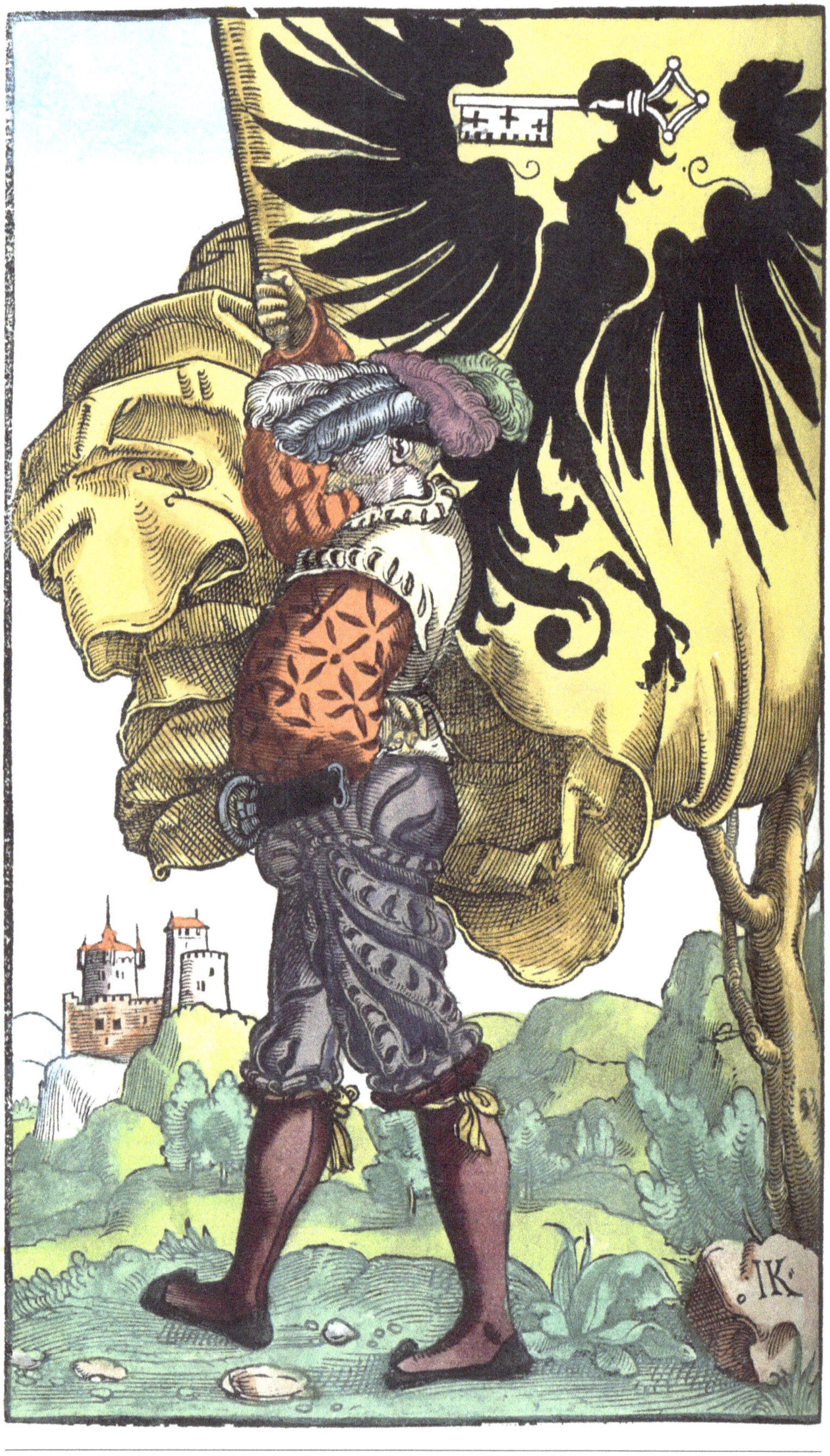

62. *Wimpfem | Bad Wimpfen*

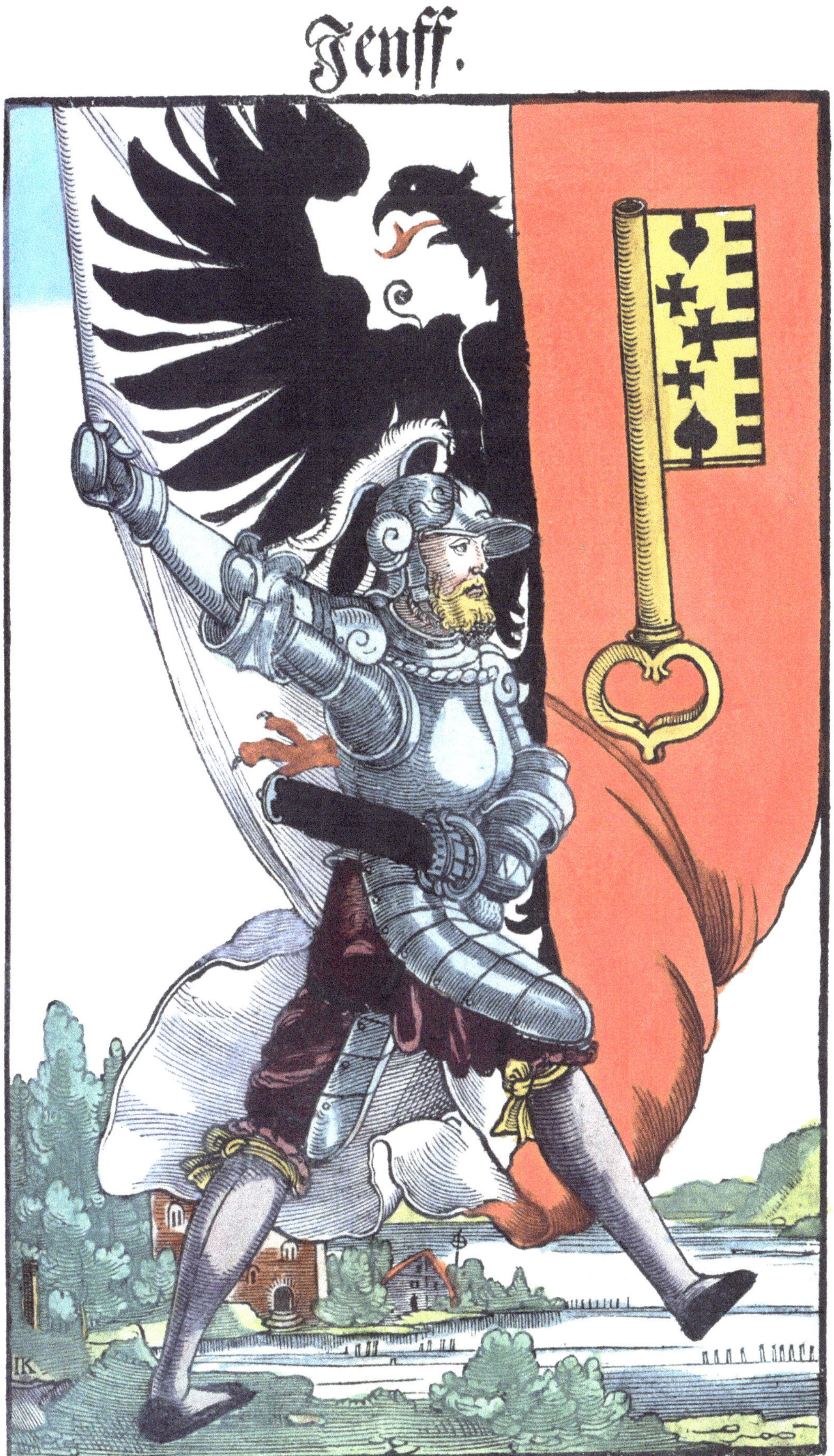

63. Jenff | Genf, Geneva Switzerland

Toll.

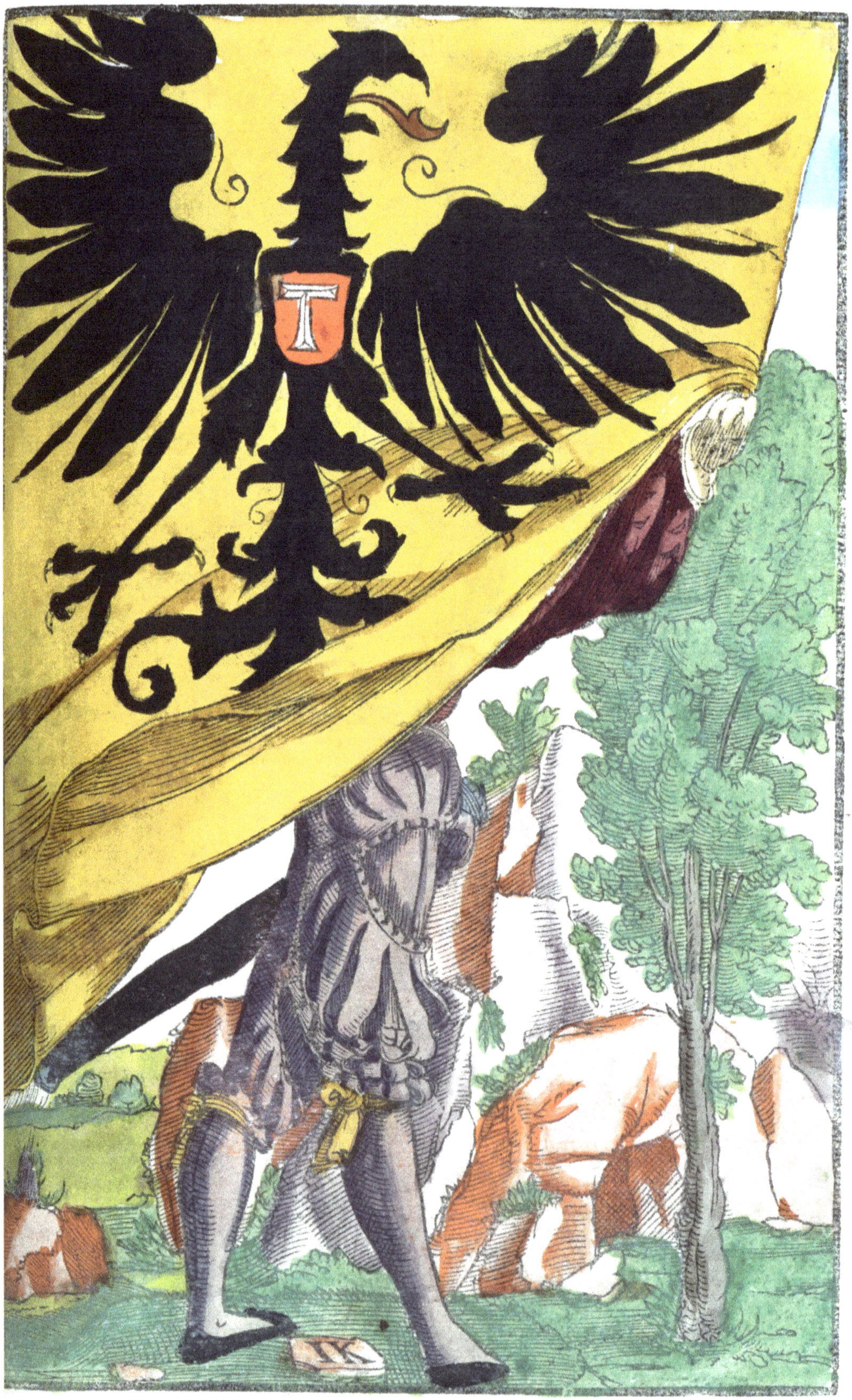

64. Toll | Toul, France

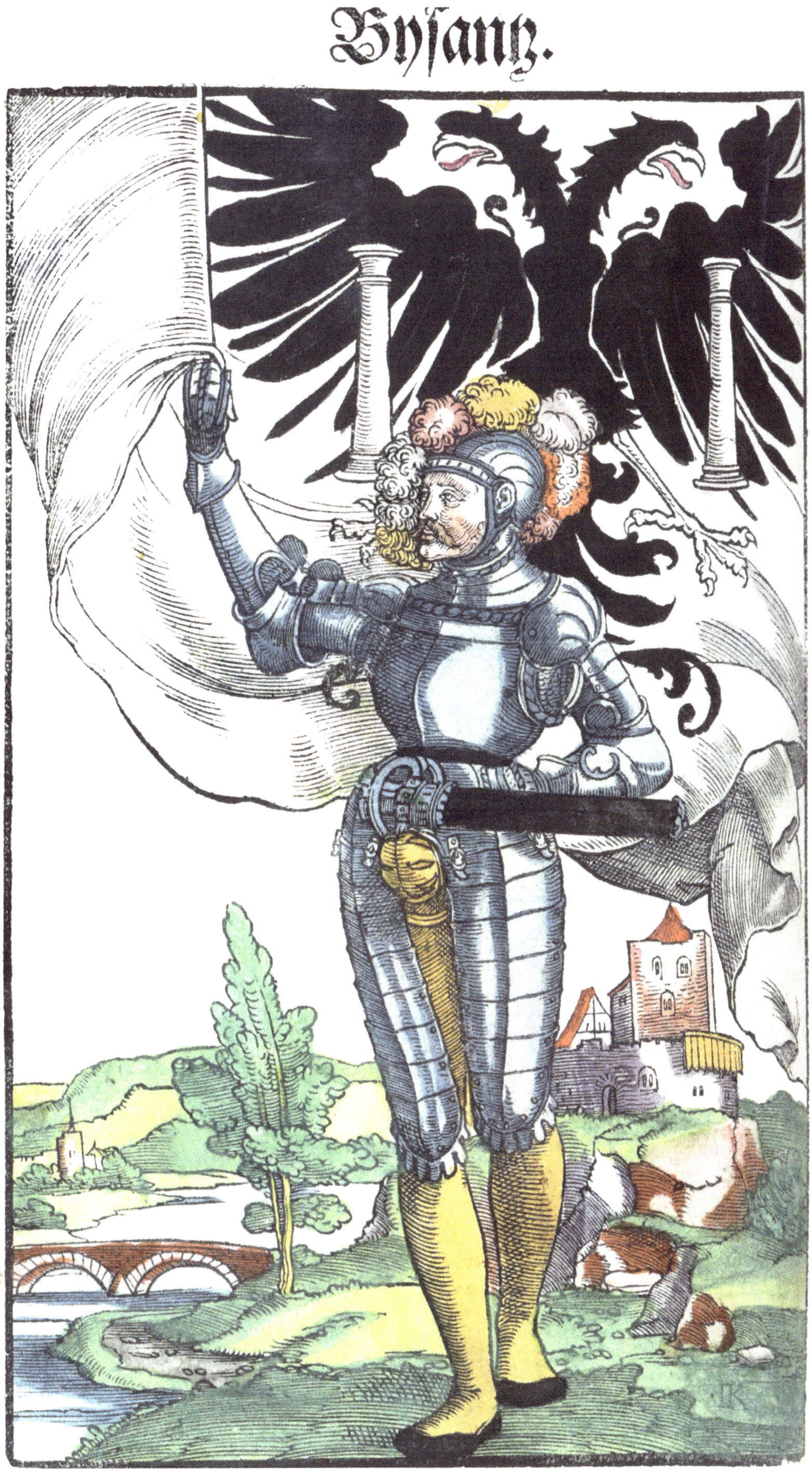

65. *Bisfantz* | *Besançon, France*

Villingen.

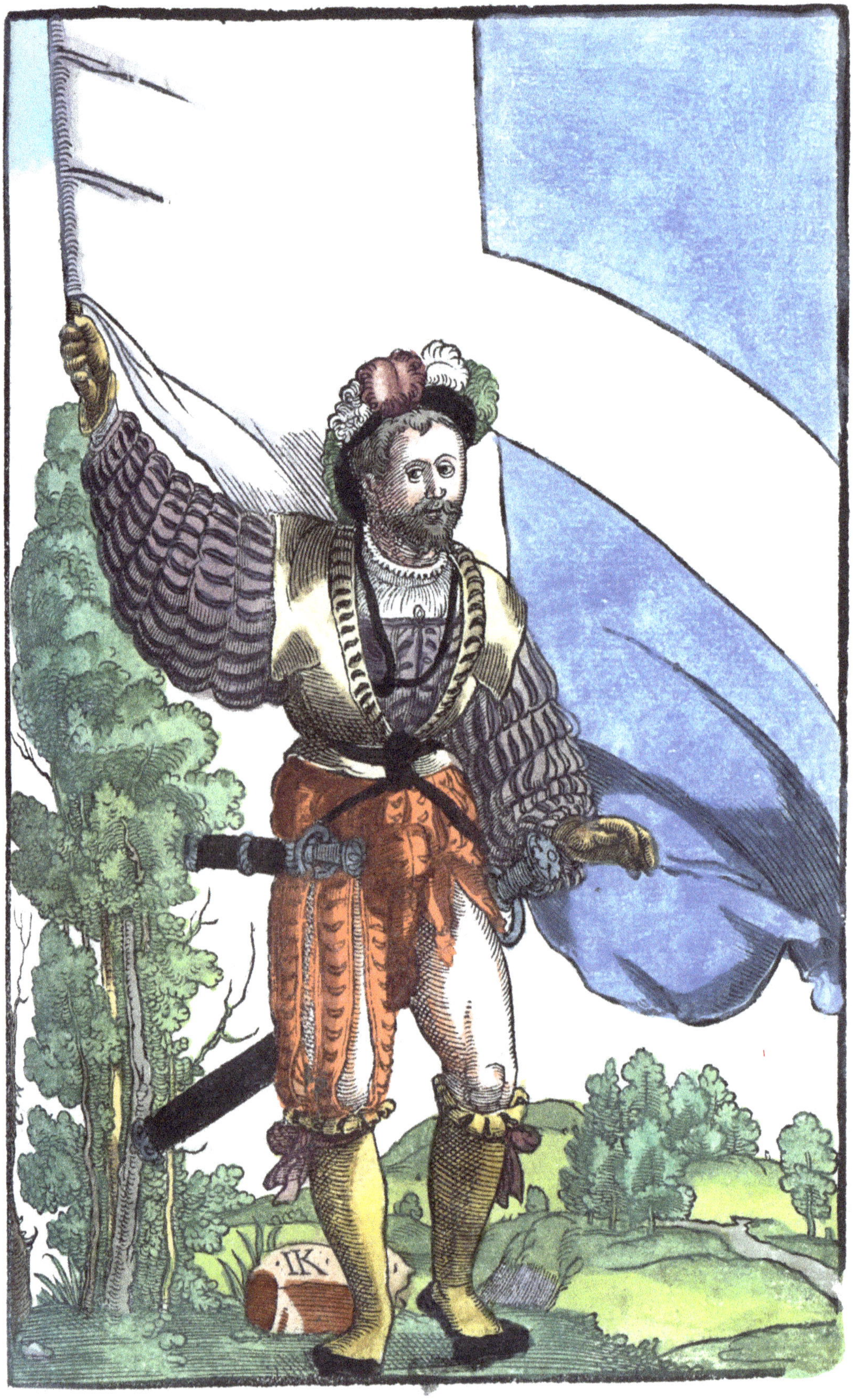

66. Villingen | Villingen

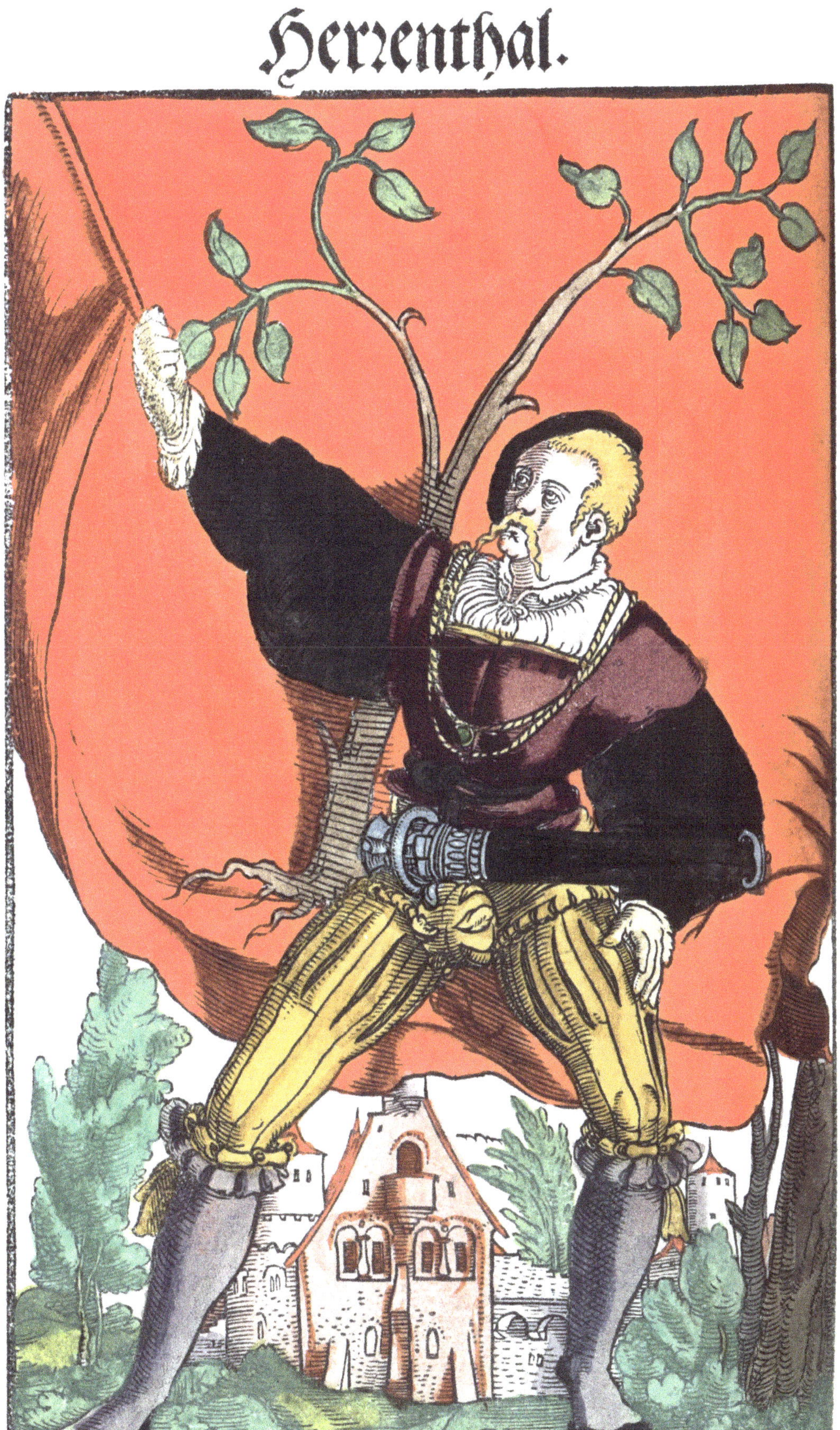

67. *Herzenthal* | *Herentals, Belgium*

Benfelt.

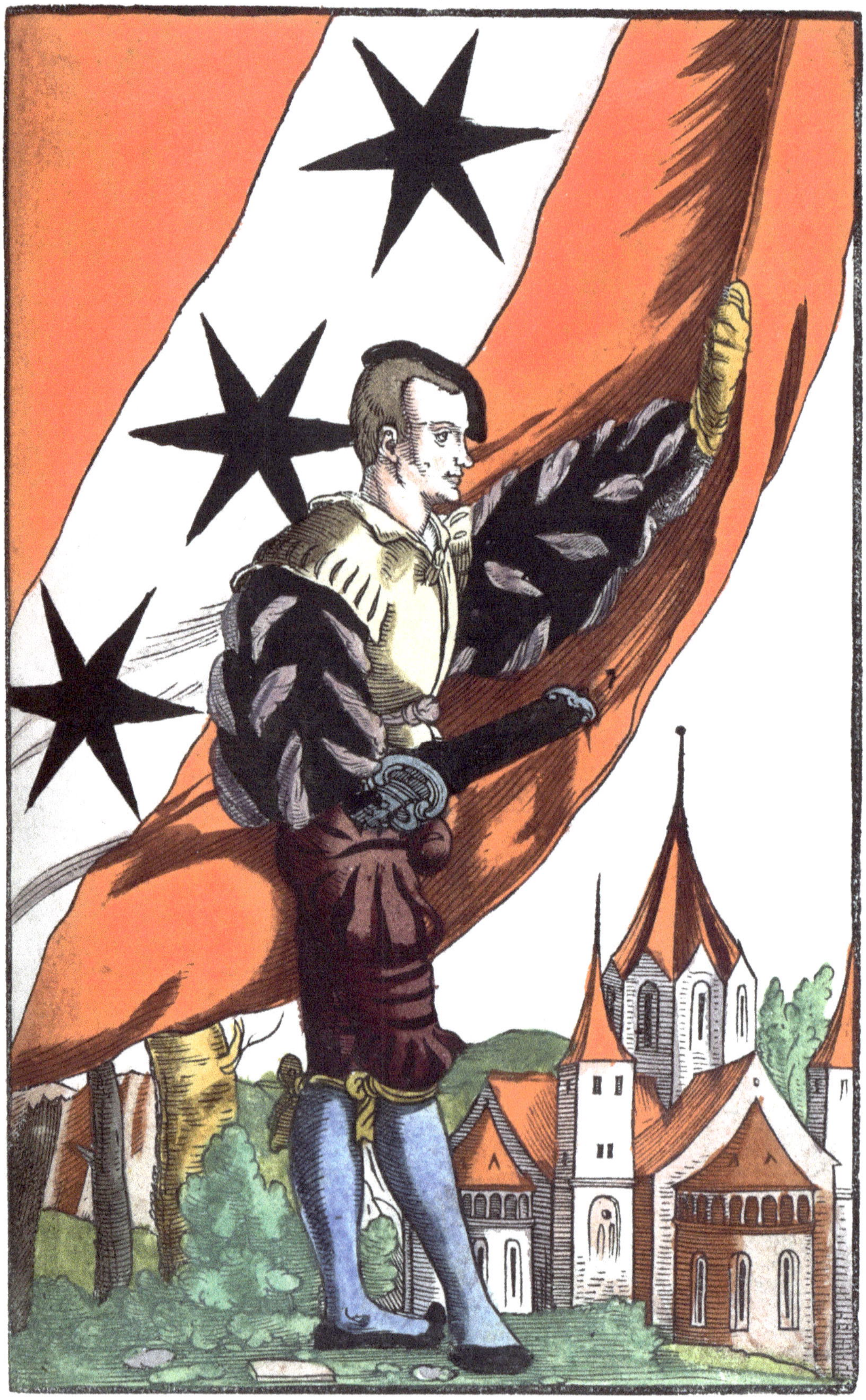

68. Benfelt | Benfeld, France

Mülhaussen.

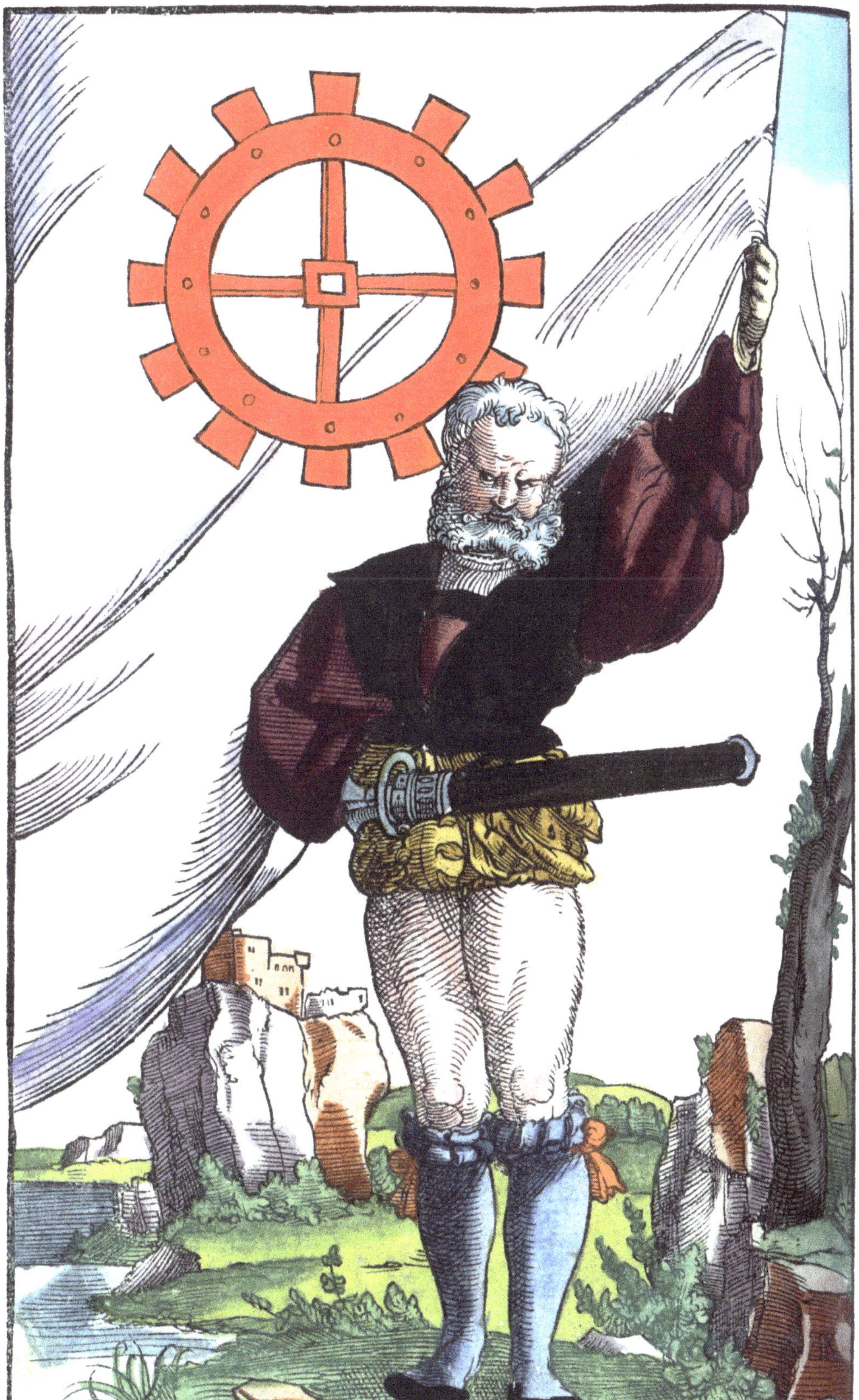

69. Mülhaussen | Muhlhausen

Ole.

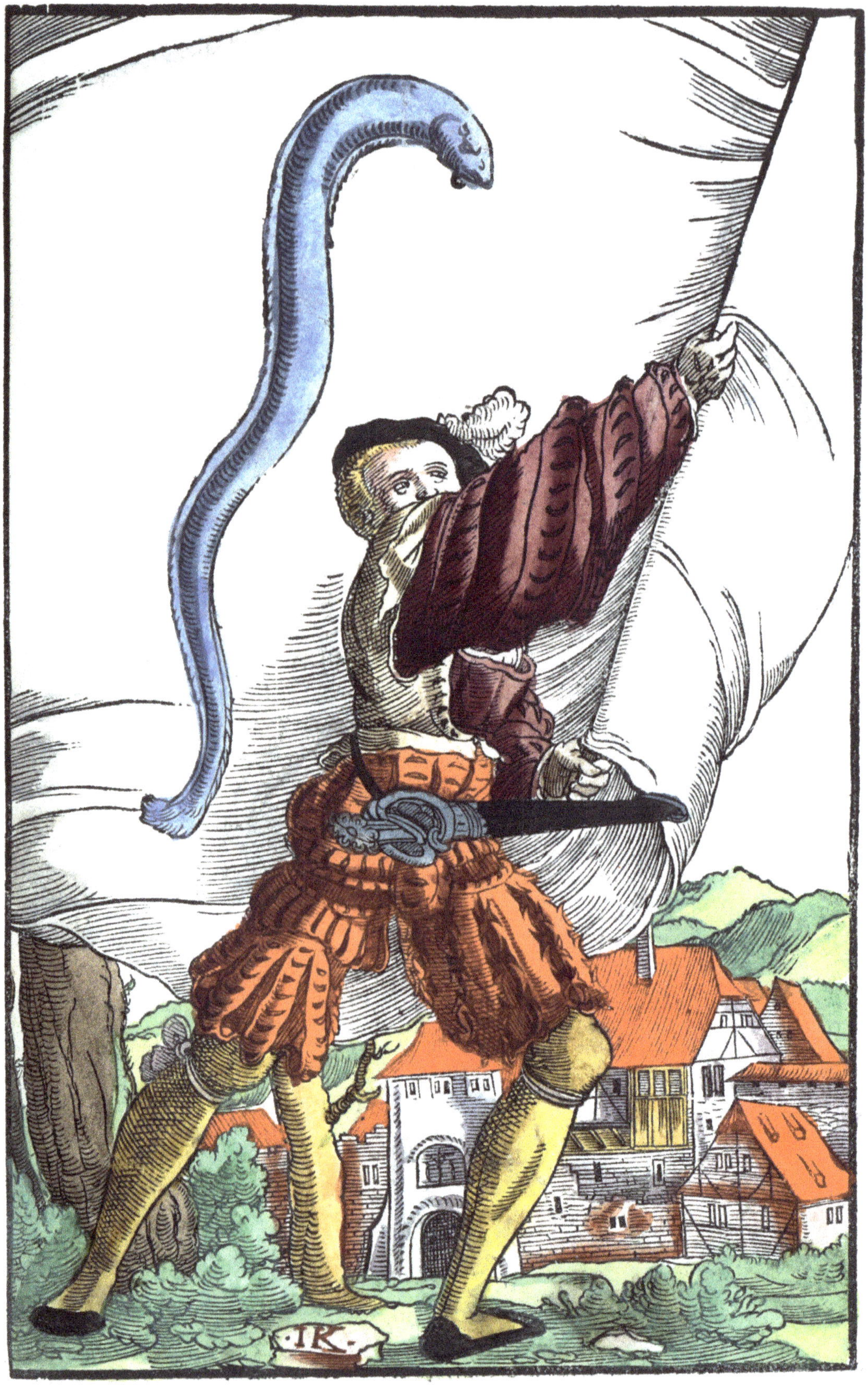

70. Ole | Ala, Halla, Ahl am Etsch, in Trento Italy

Kolmar.

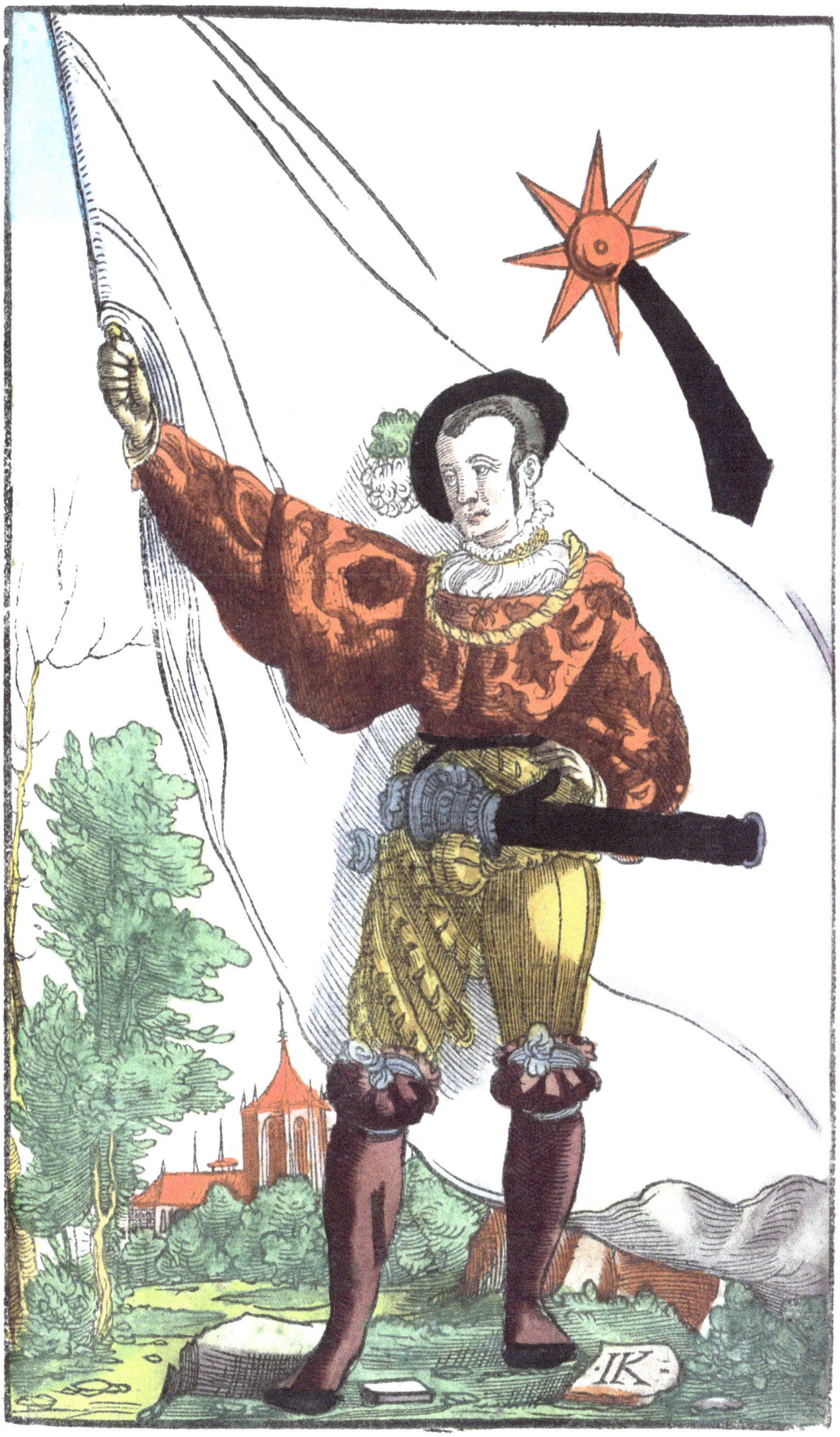

71. Kolmar | Colmar

Biberach.

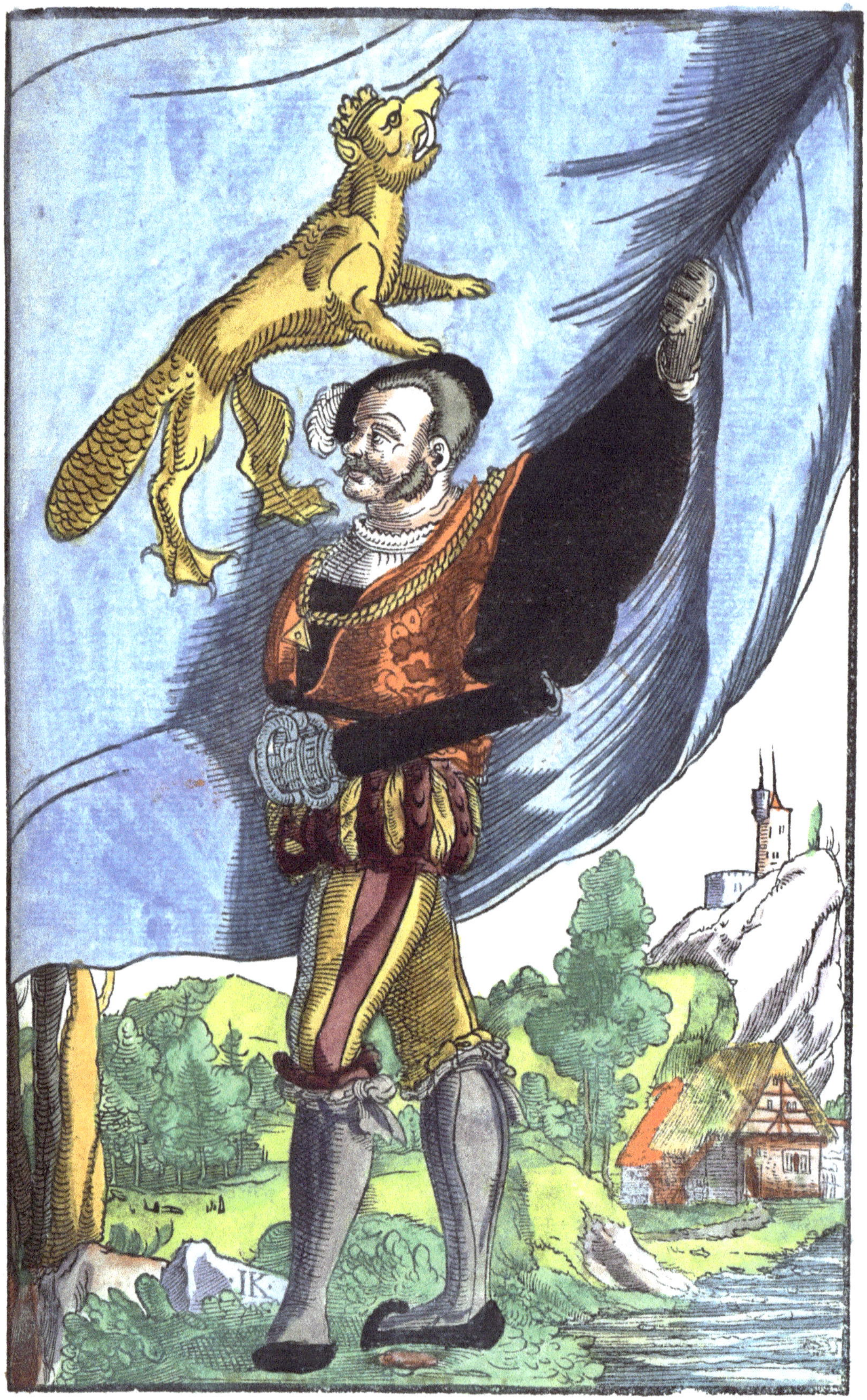

72. Biberach | Biberach an der Riss

Oppenheim.

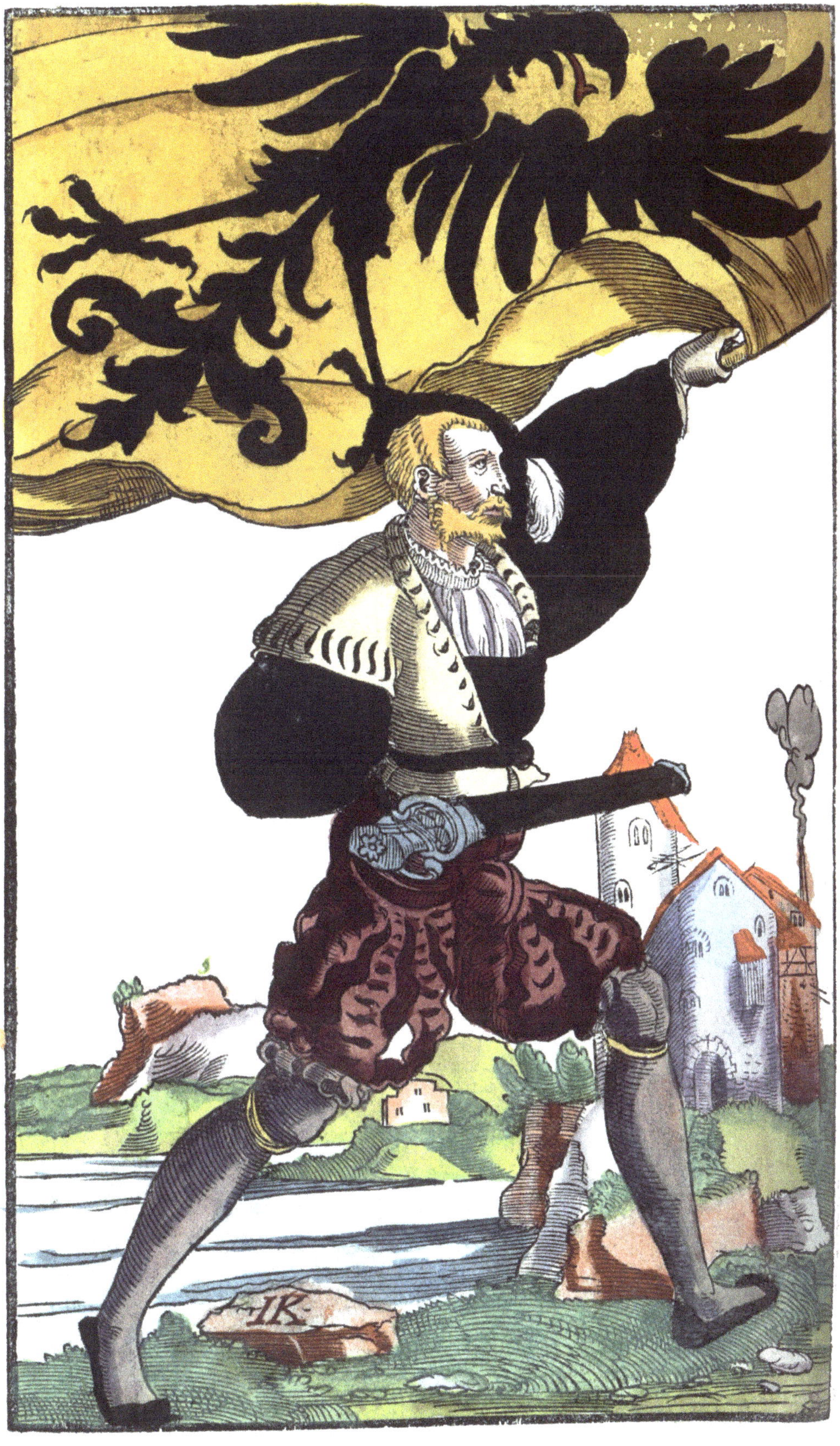

73. Oppenheim | Oppenheim

Breisach.

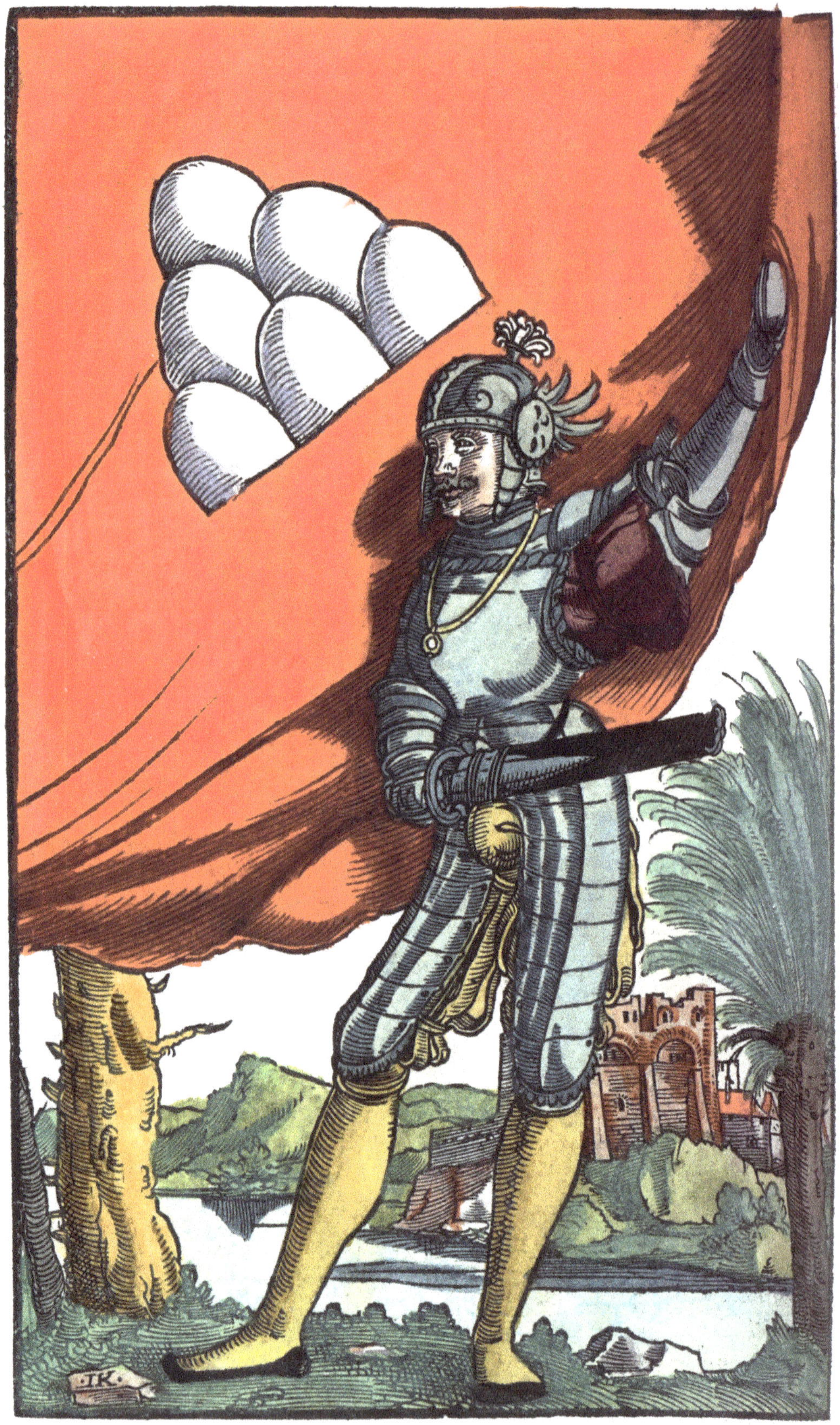

74. Breisach | Breisach

Schwebisch Gemind.

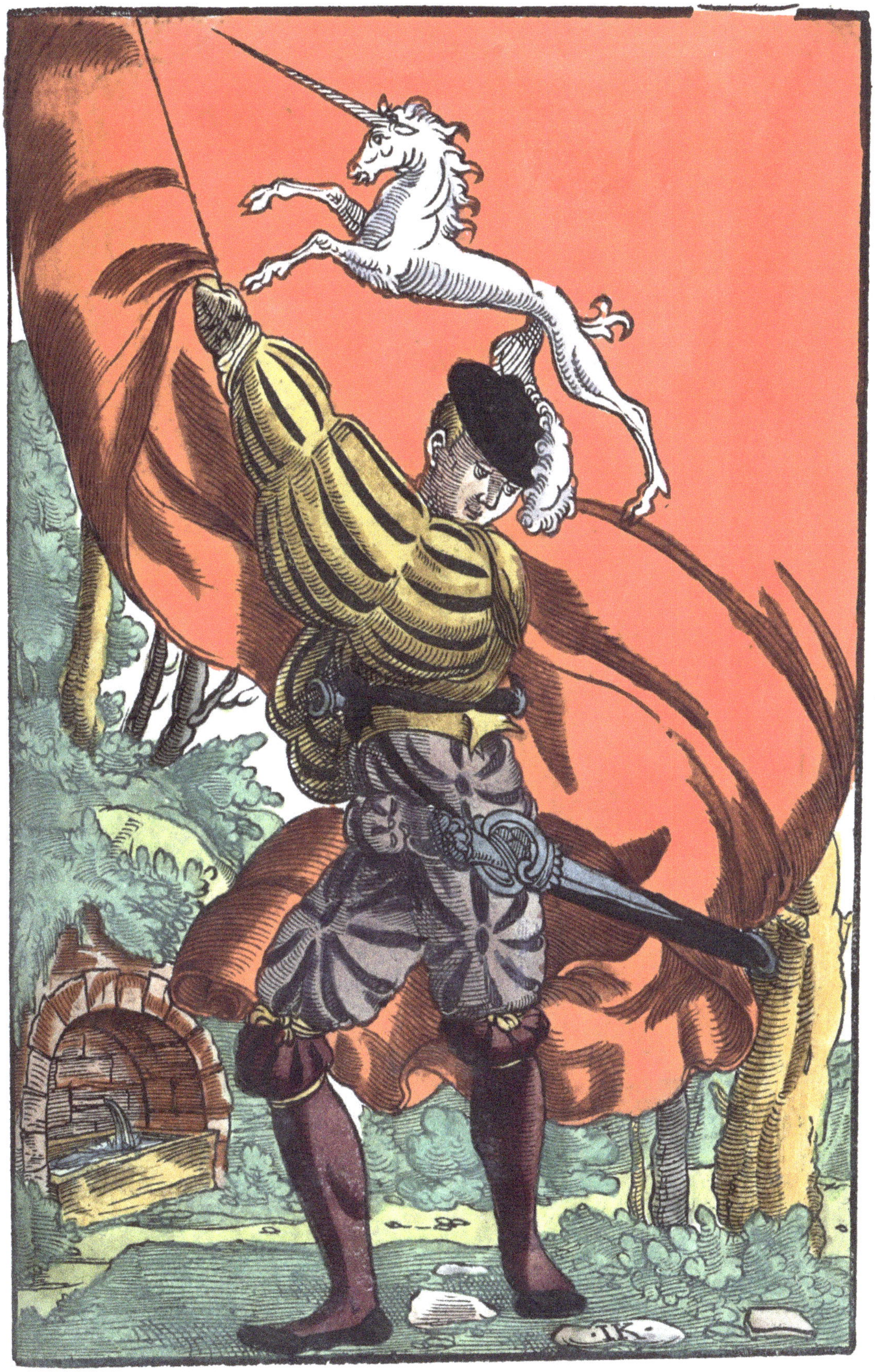

75. Schwebisch Gemind | Schwäbisch Gmünd

Keisers Lautern.

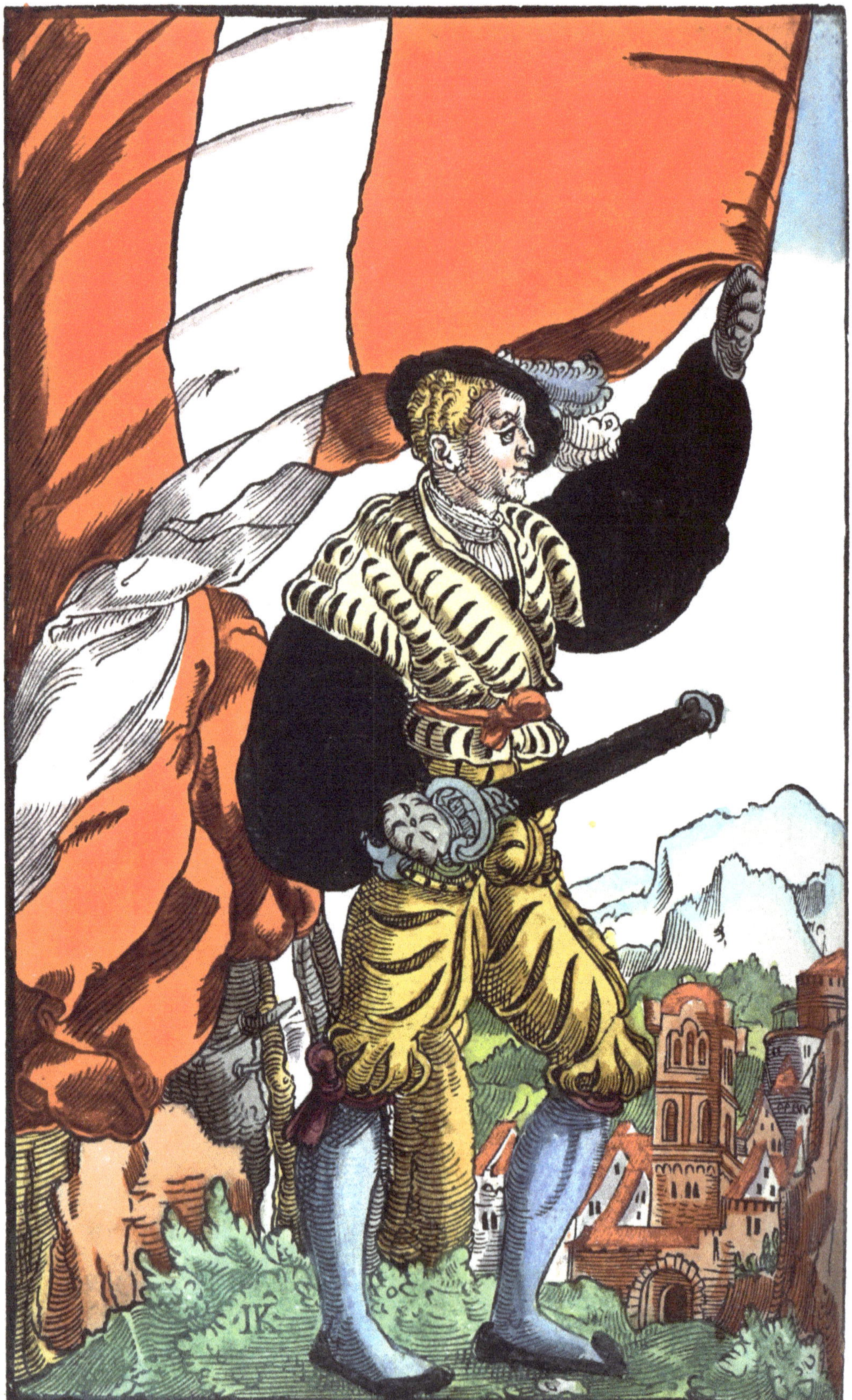

76. Keisers Lautern | Kaiserslautern

Memingen.

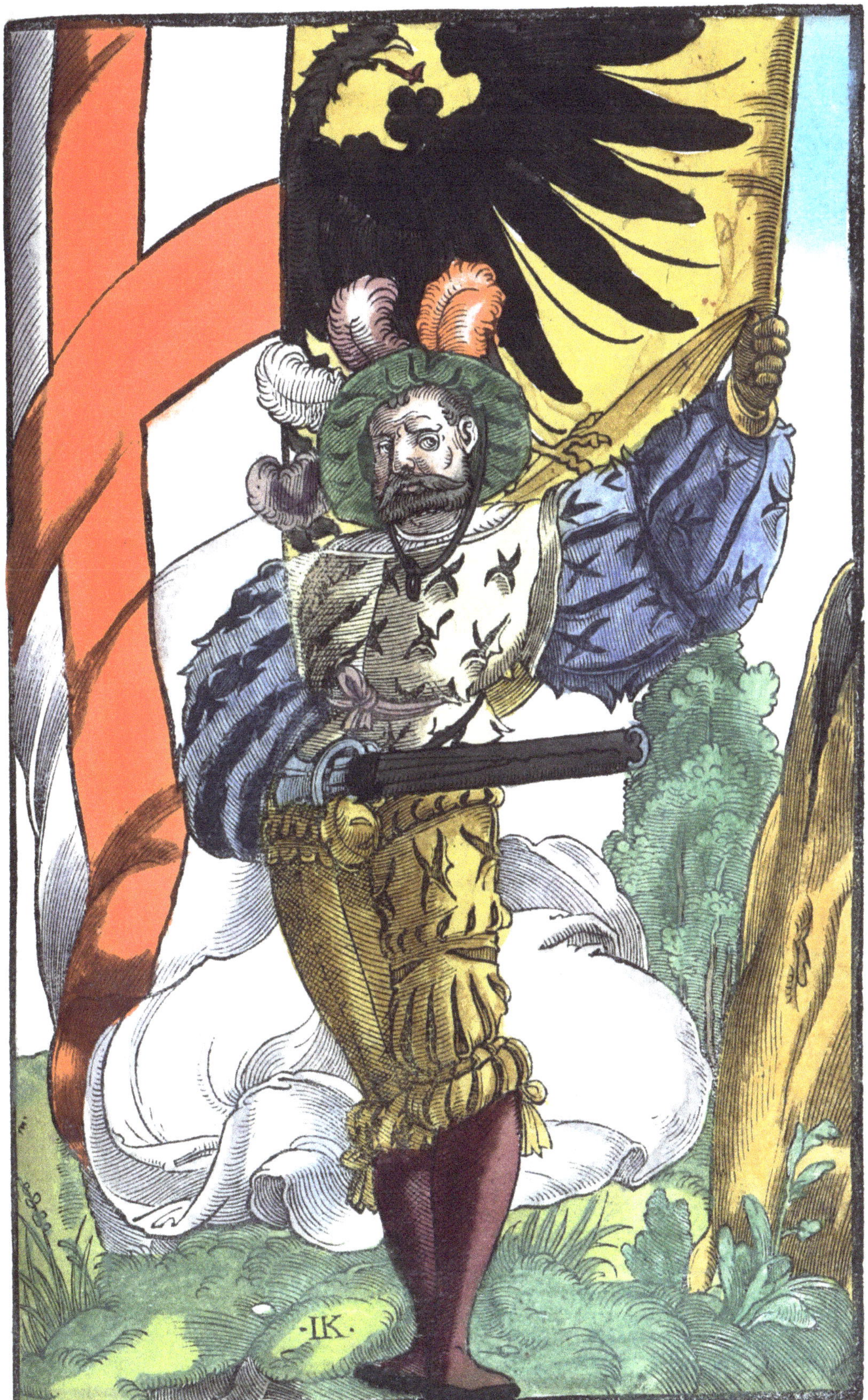

77. Memingen | Memingen

Menz.

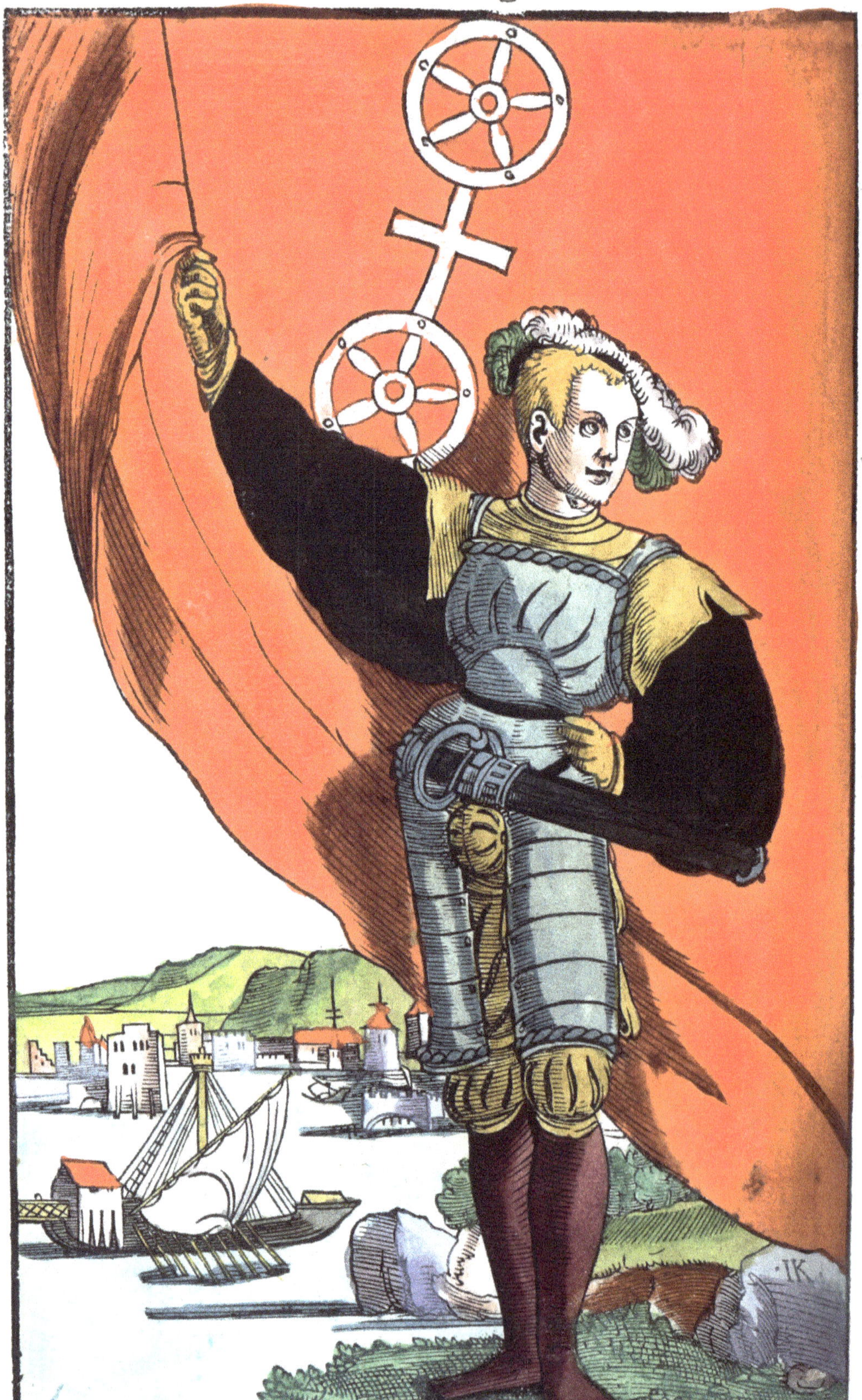

78. Menz | Mainz

Cobelentz.

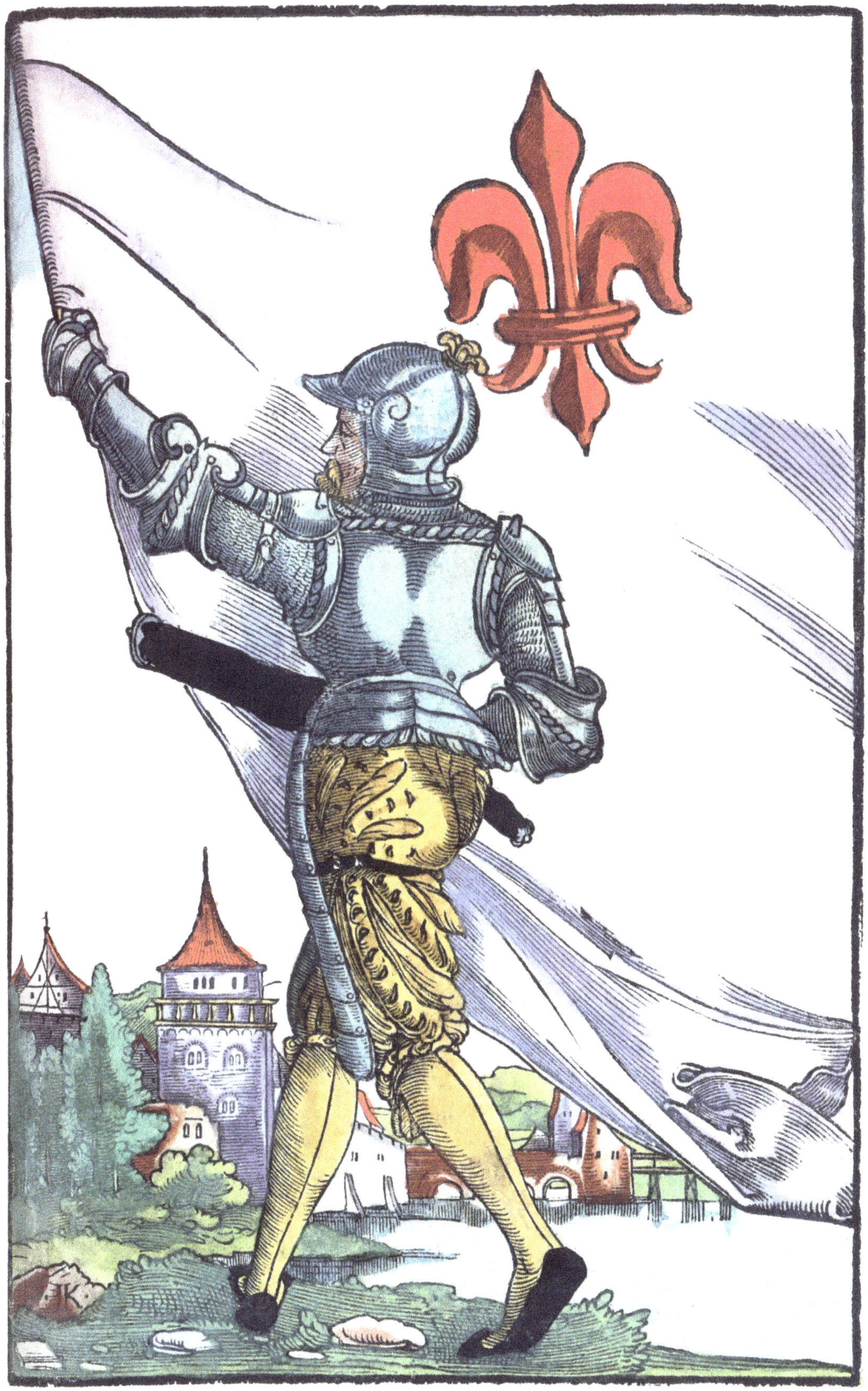

79. Cobelentz | Koblenz

Altdorff.

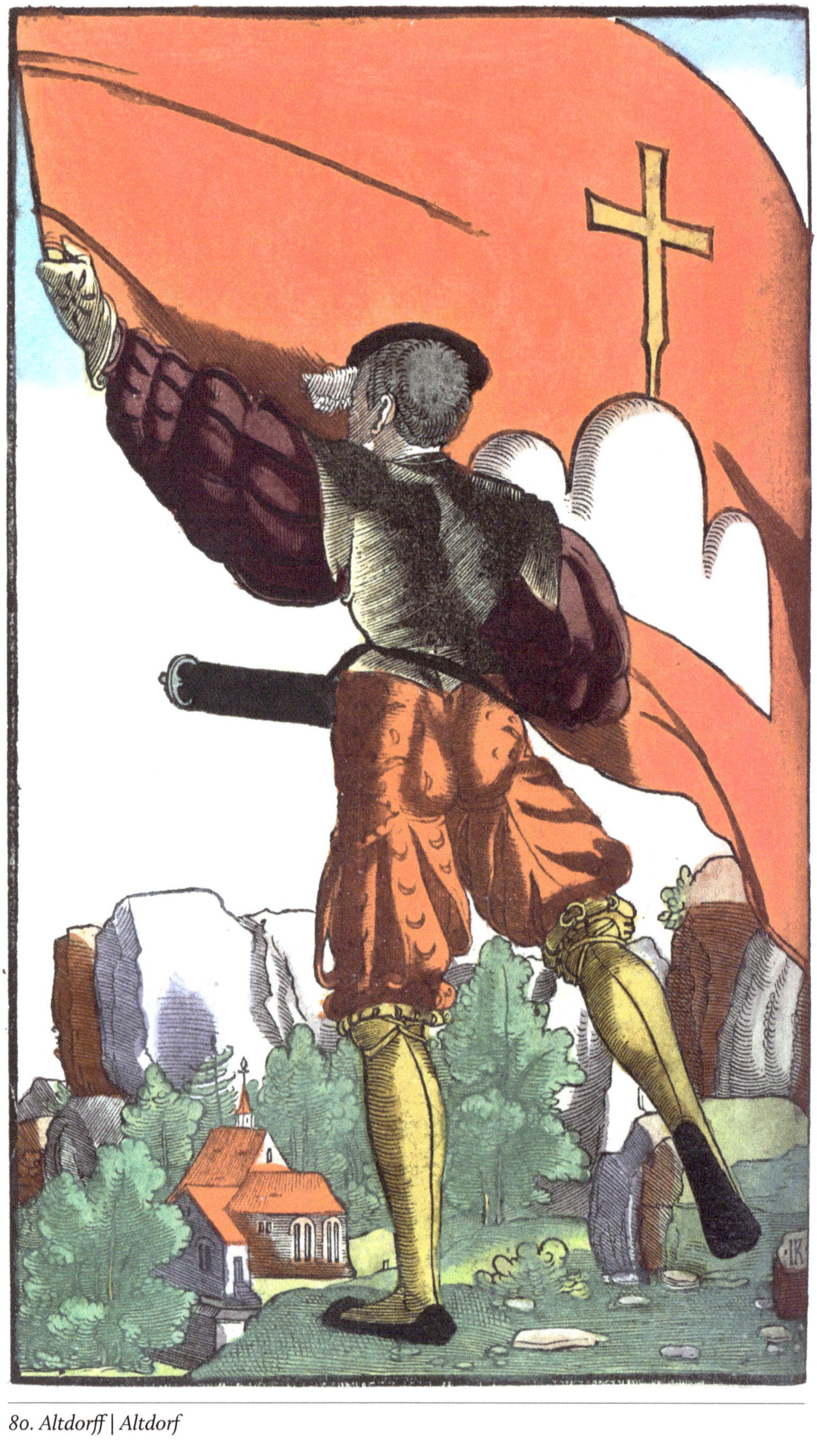

80. Altdorff | Altdorf

Löuen.

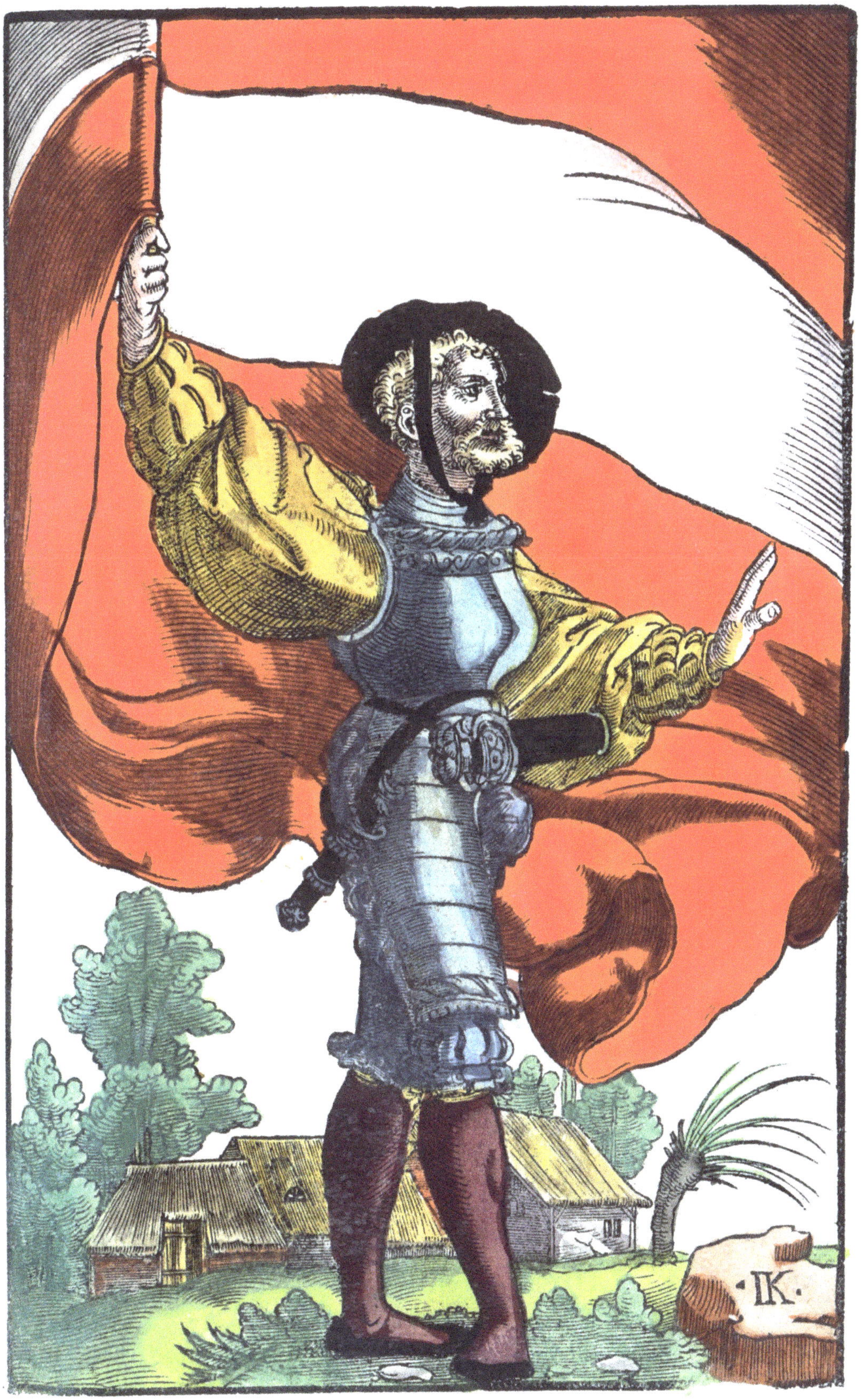

81. Löuen | Leuven, Belgium

Erdfurt.

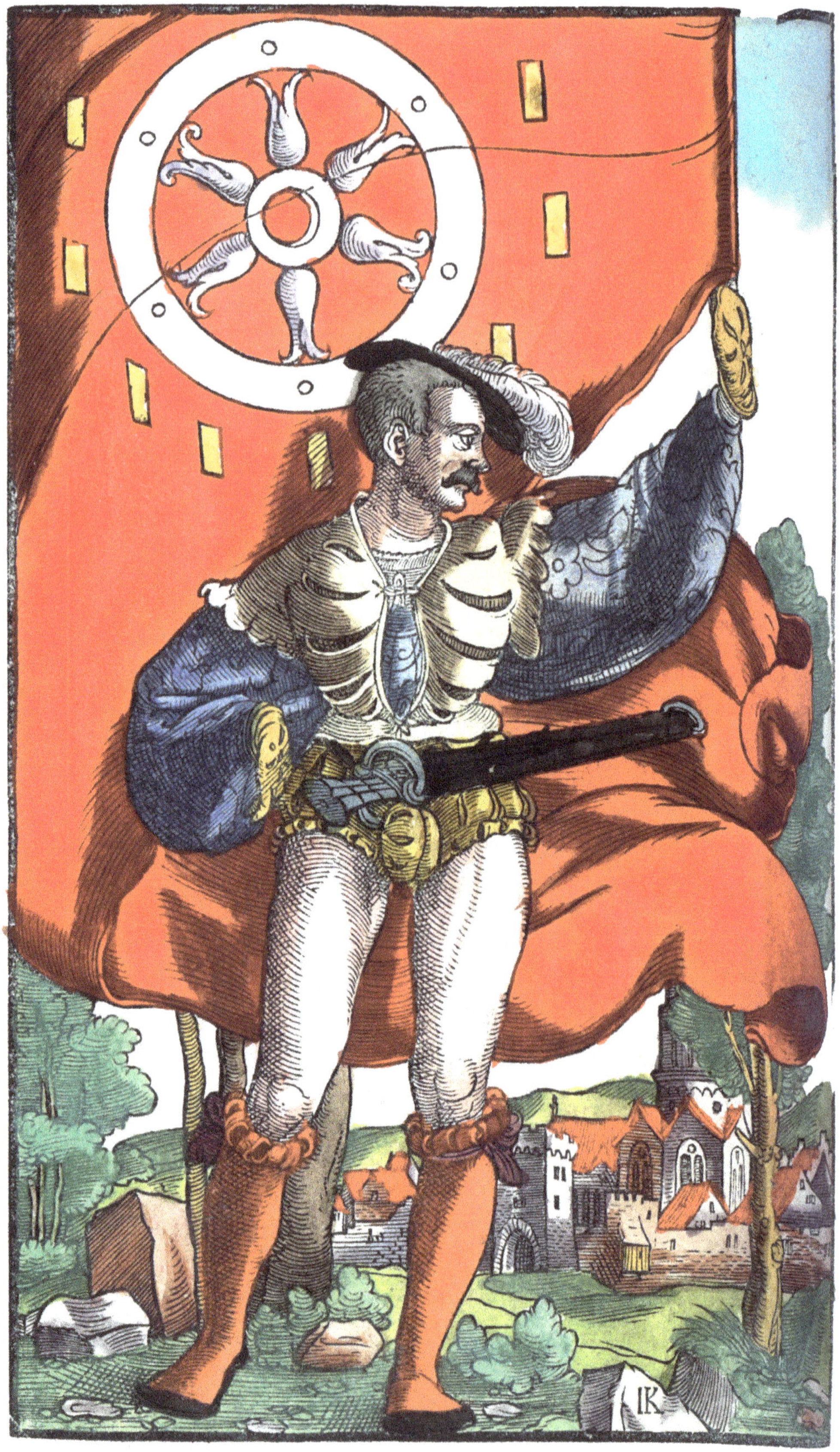

82. *Erdfurt* | *Erfurt*

Lindaw.

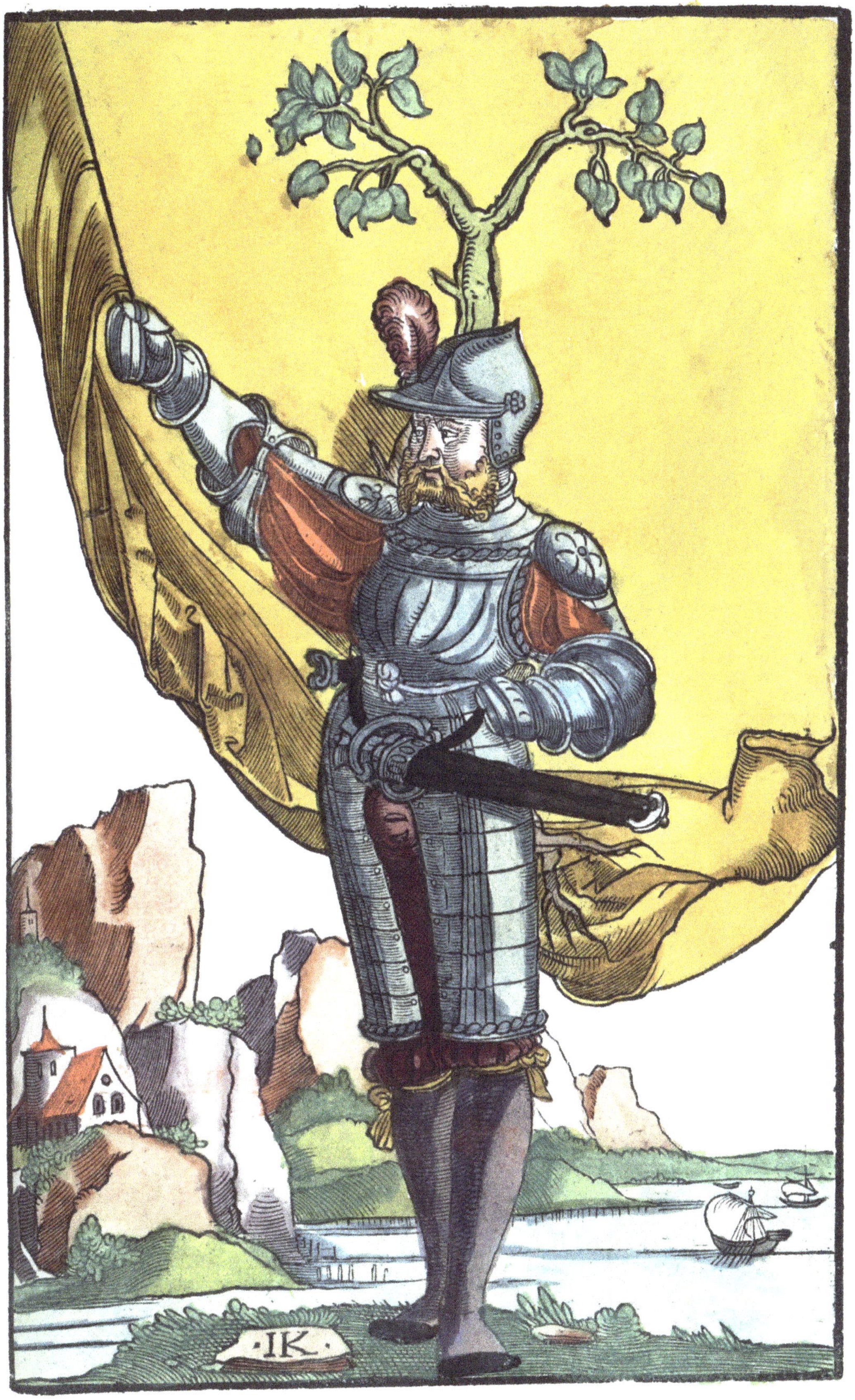

83. Lindaw | Lindau am Bodensee, Lindau by Lake Constance

Halle in Sachssen.

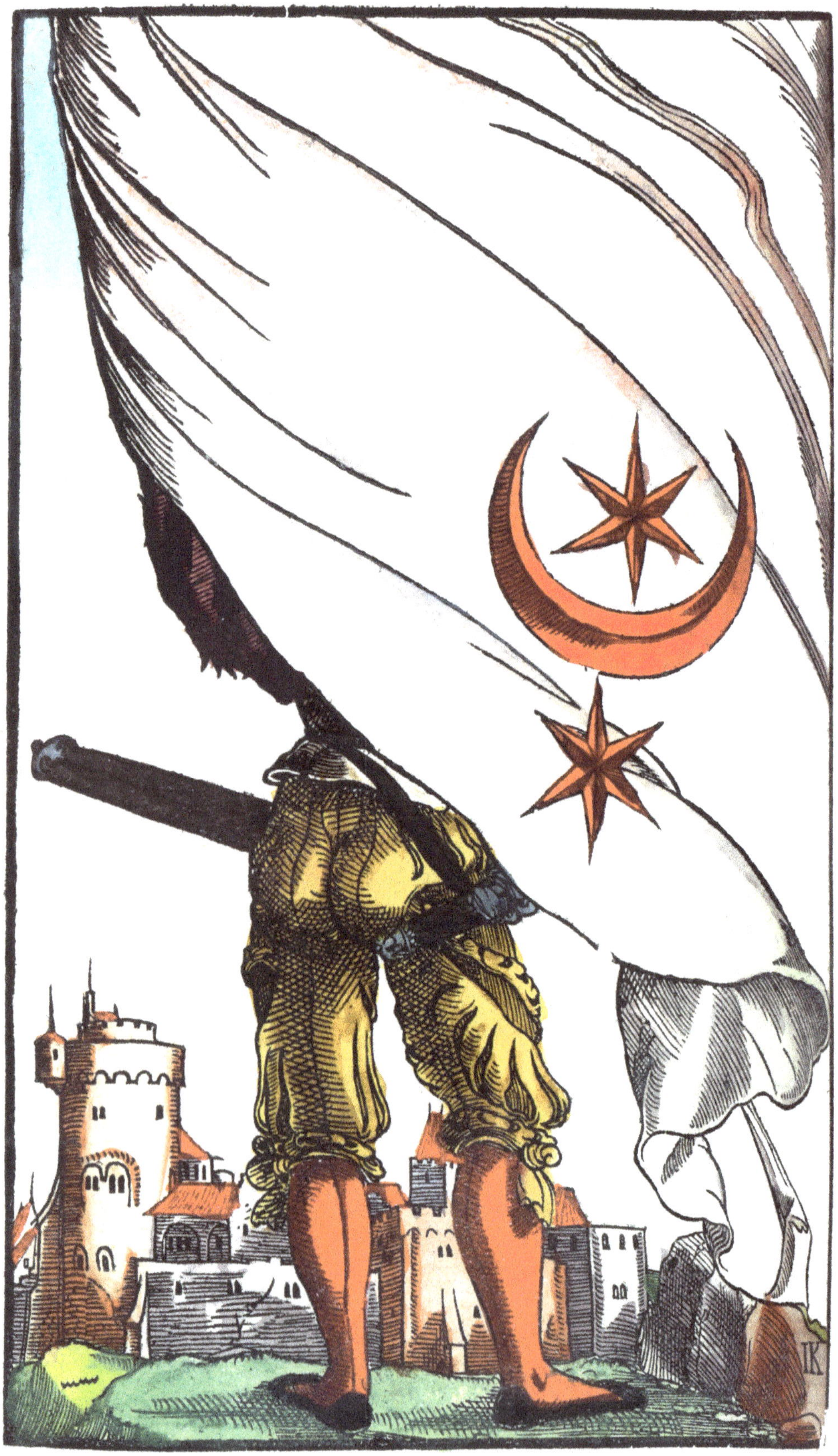

84. Halle in Sachssen | Halle/Saale in Saxony

Dircken.

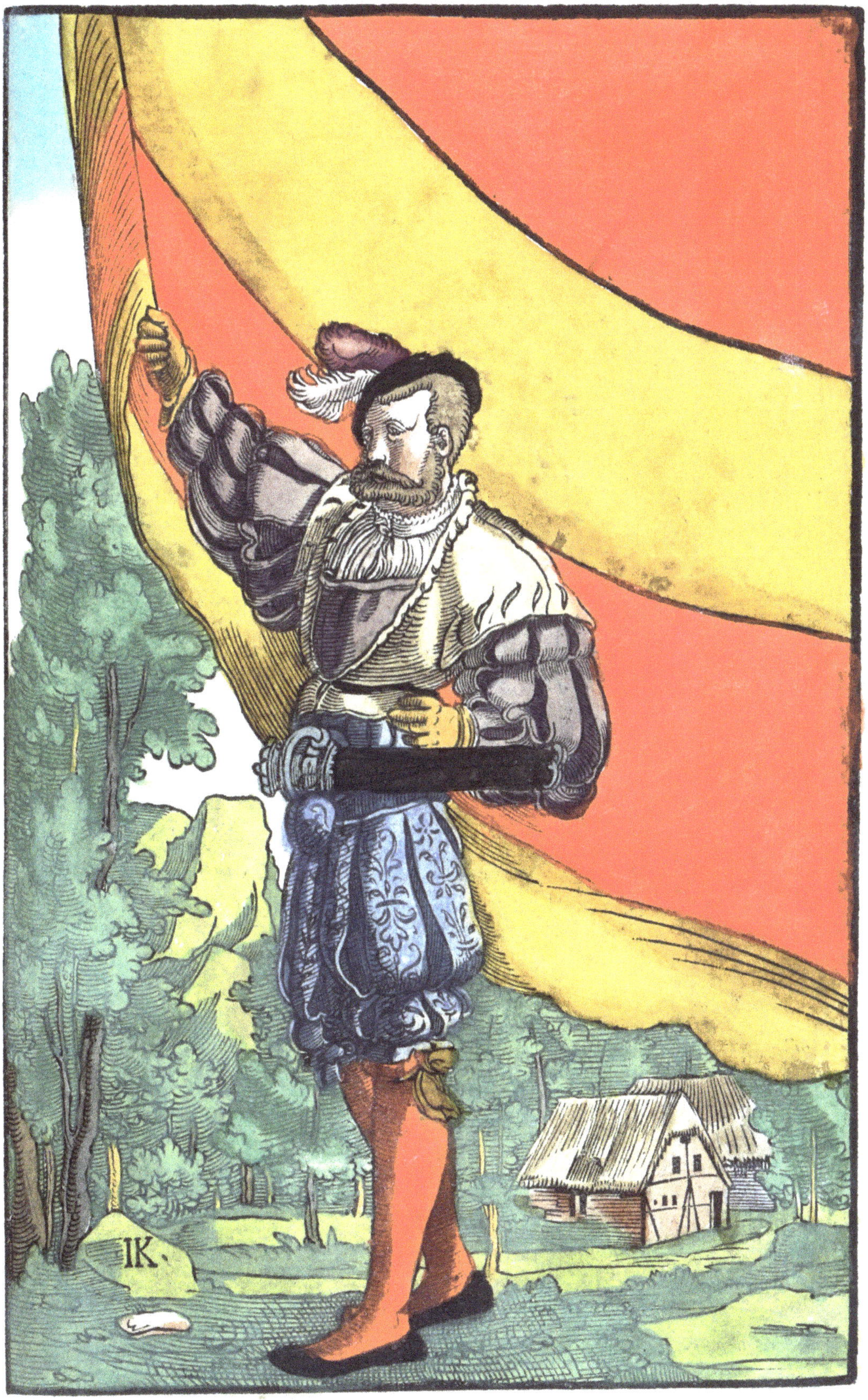

85. Dircken | Turckheim

Newenburg.

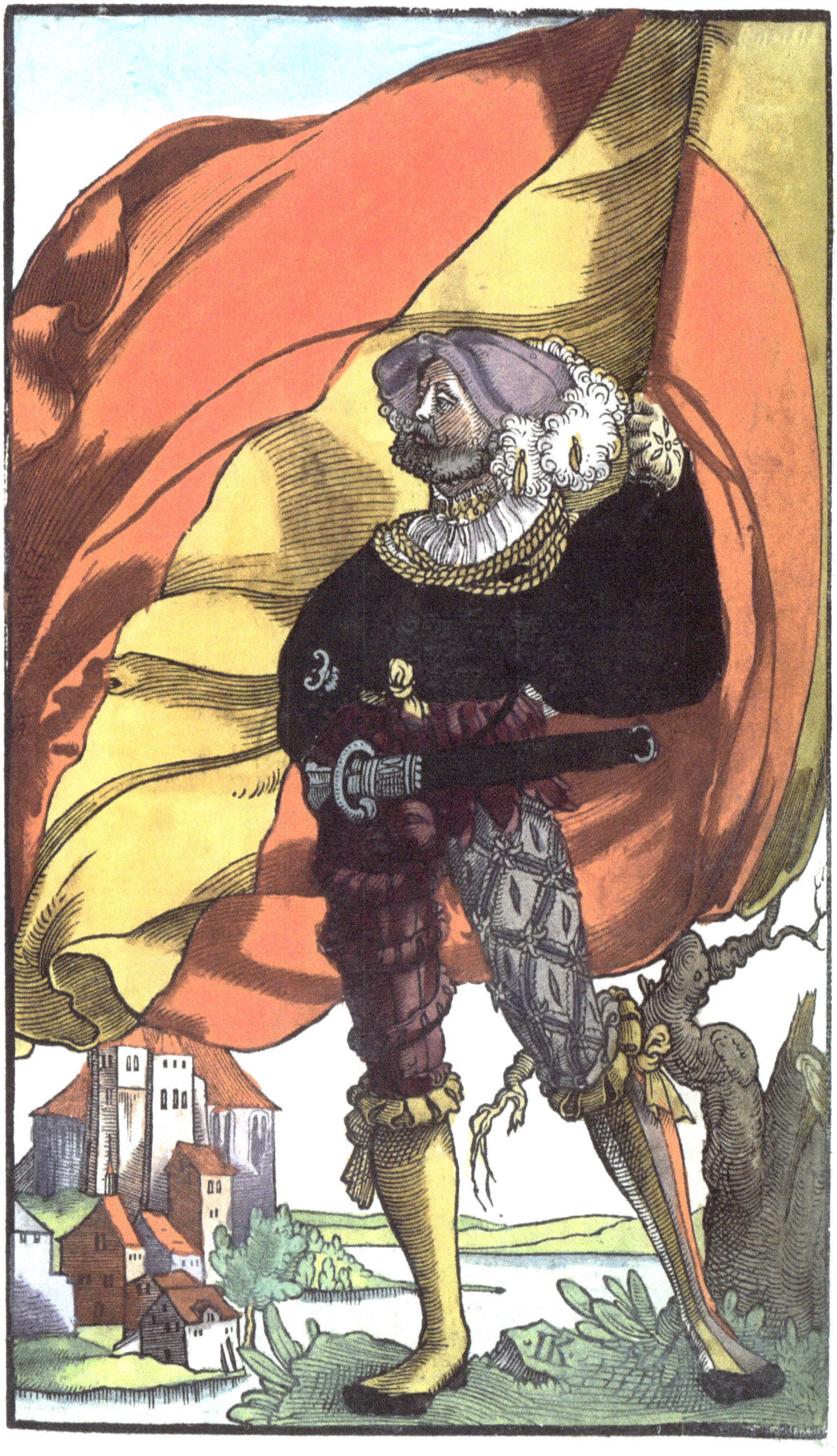

86. Newenburg | Neuenburg am Rhein

Lantzhut.

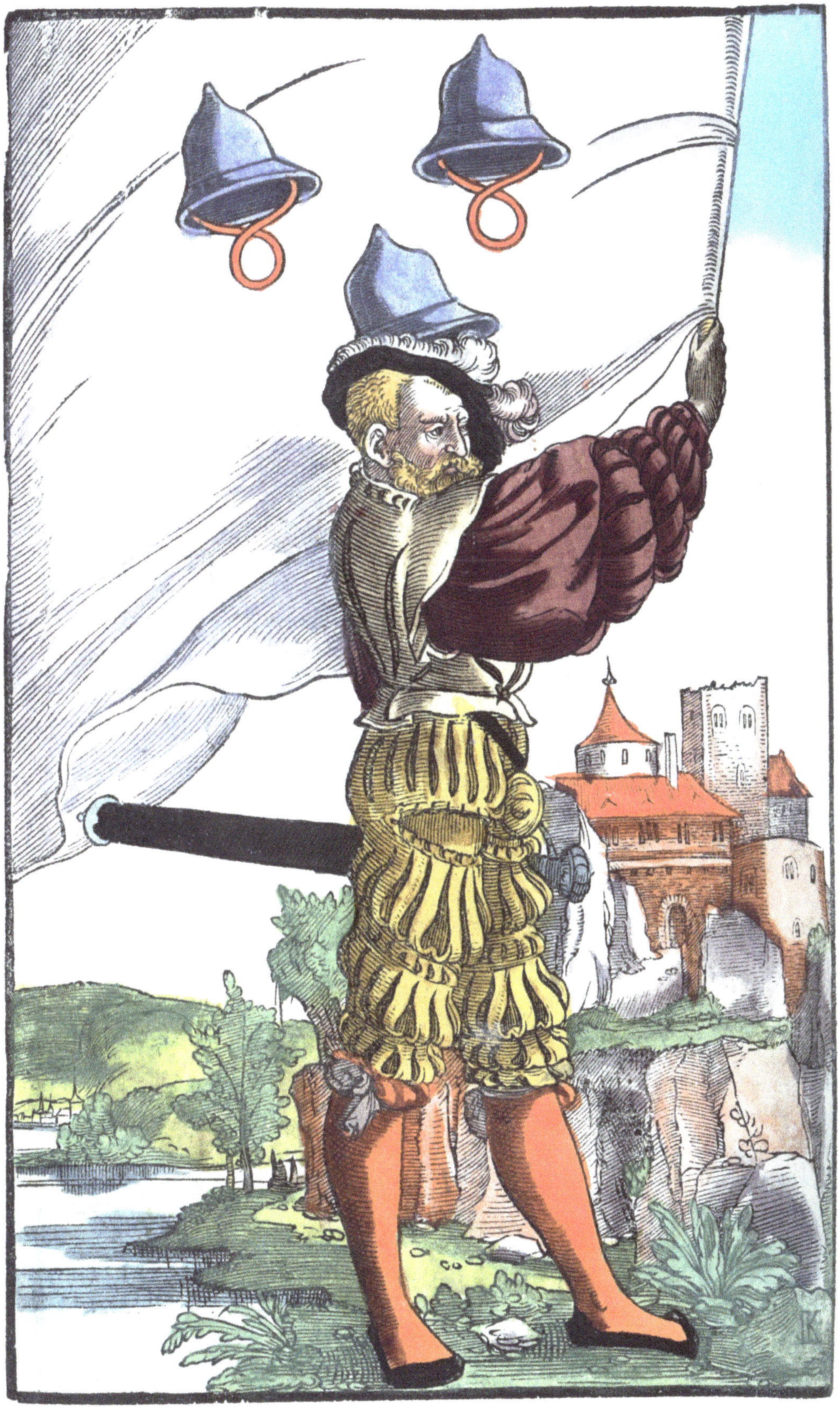

87. *Lantzhut* | *Landshut*

Schweinfurt.

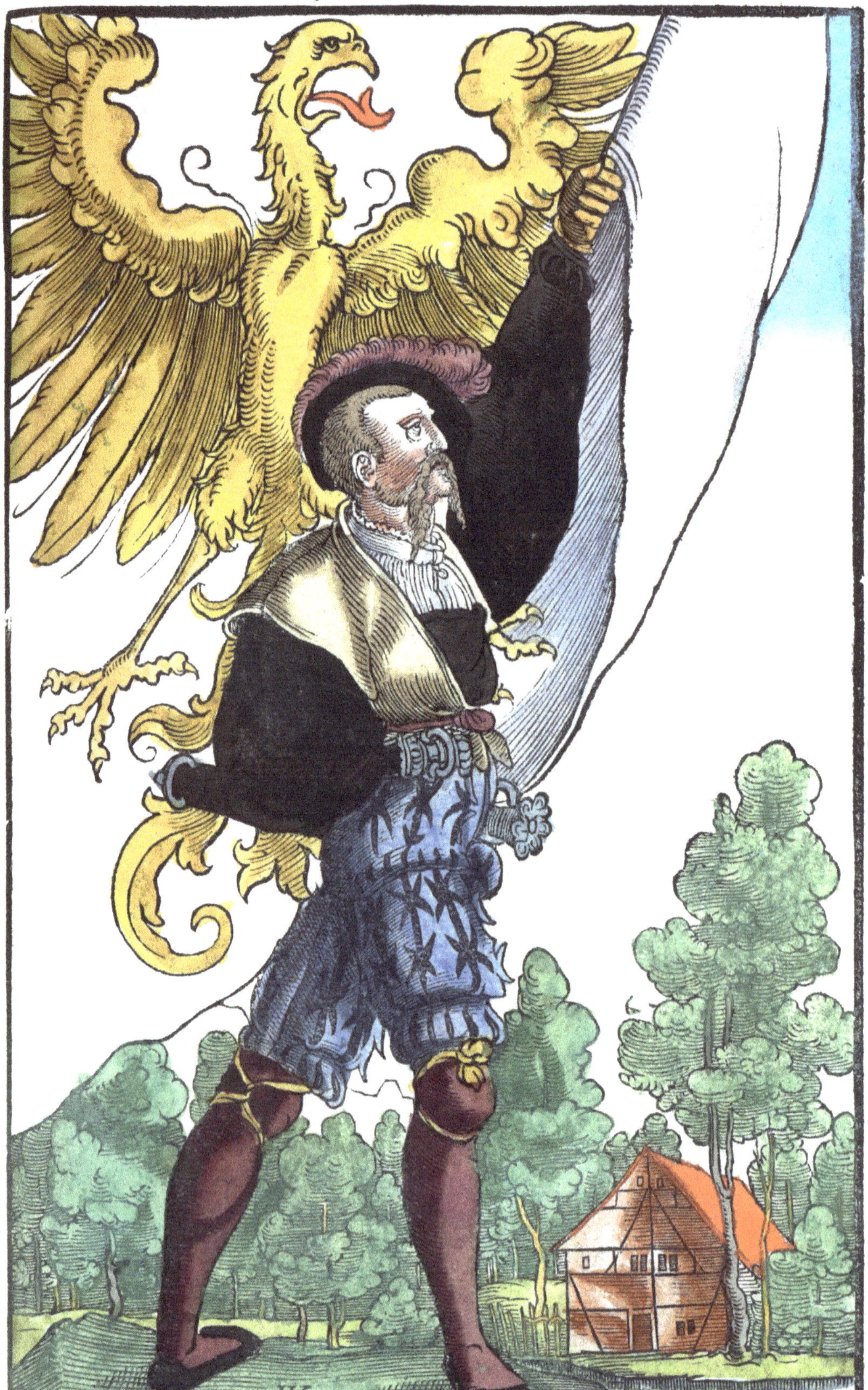

88. *Schweinfurt* | *Schweinfurt*

Liechtennaw.

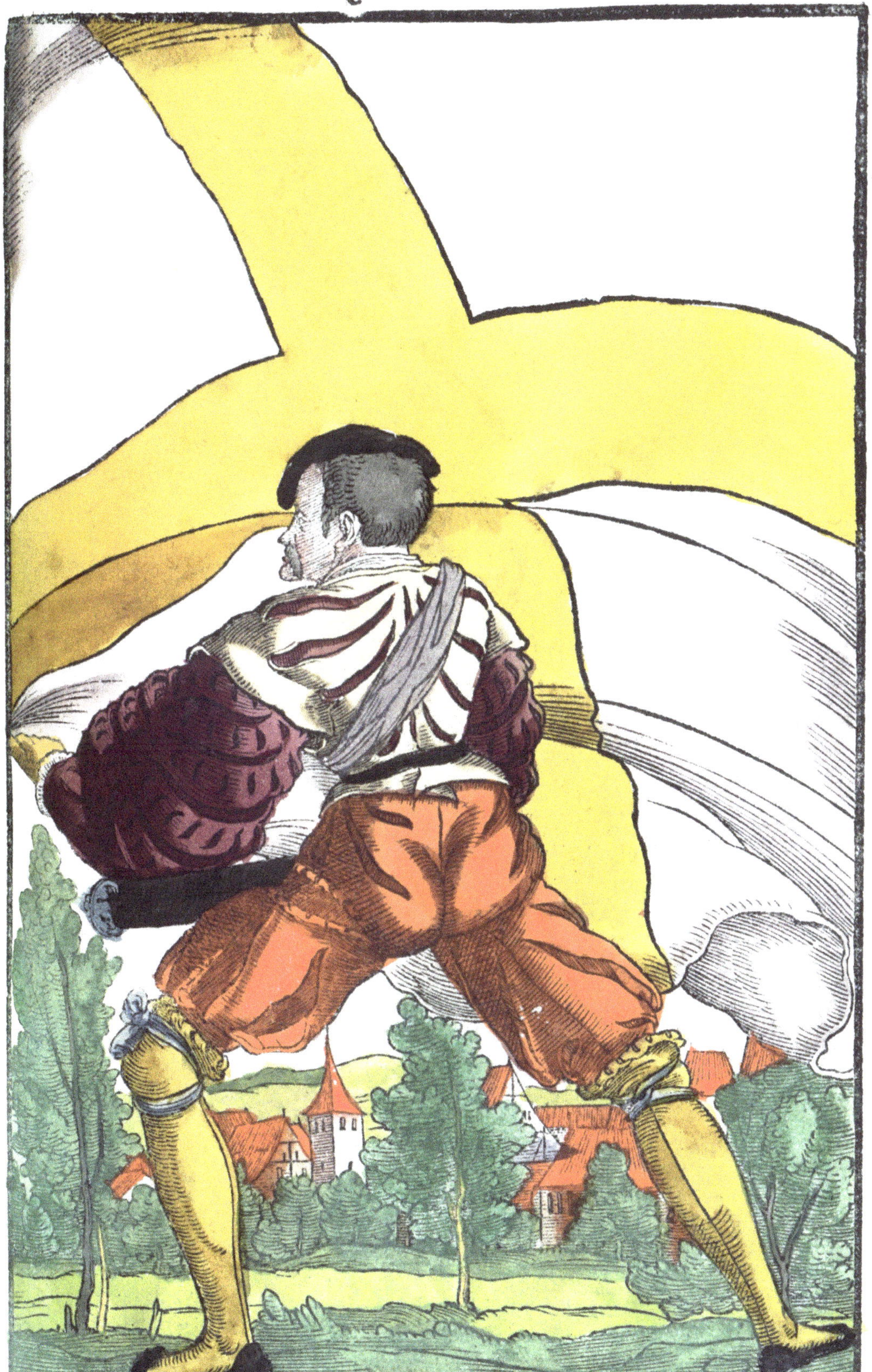

89. Leichtennaw | Lichtenau

Maſtricht.

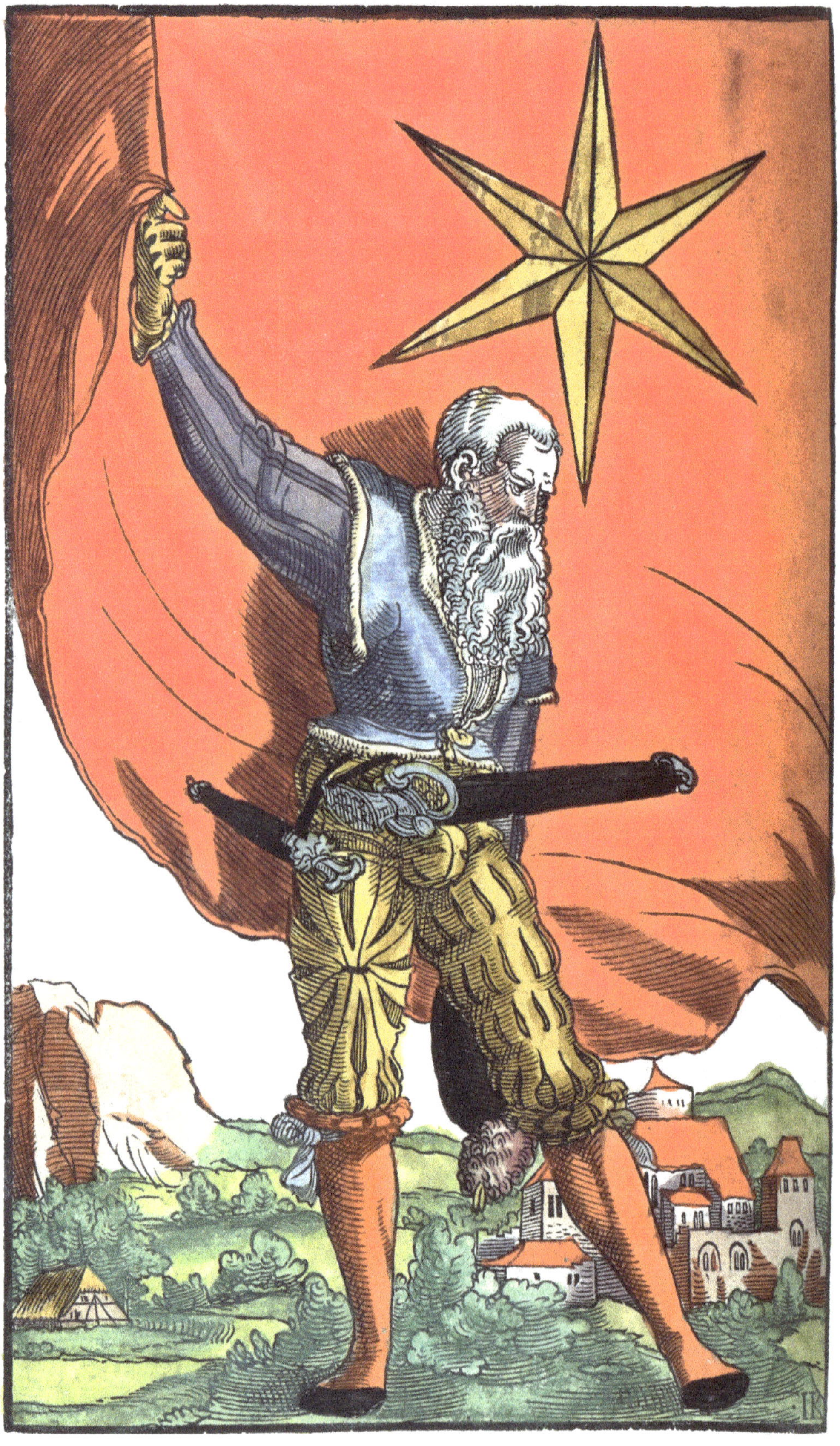

90. Mastricht | Maastricht, Netherlands

Halberstat.

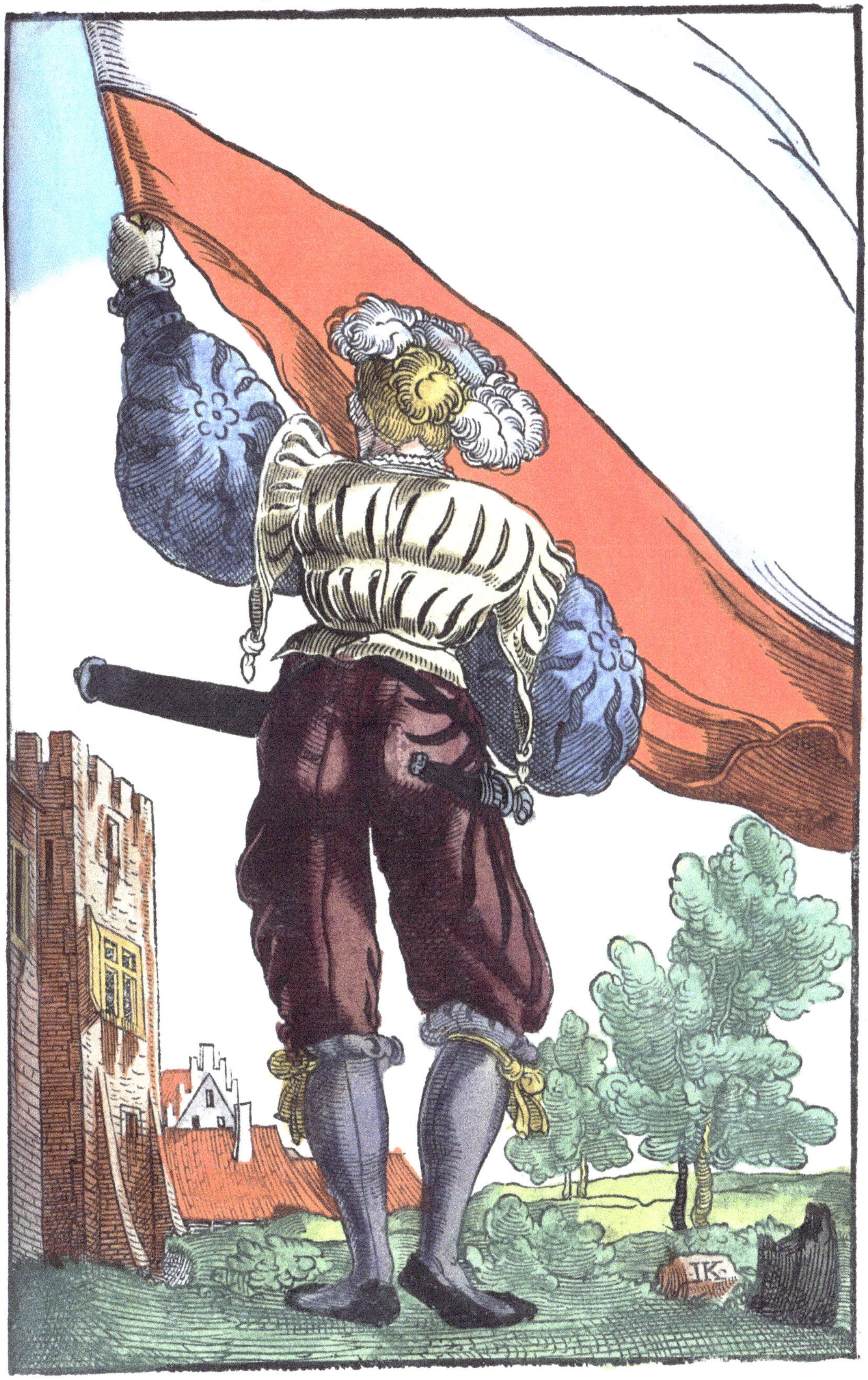

91. Halberstat | *Halberstadt*

Lagingen.

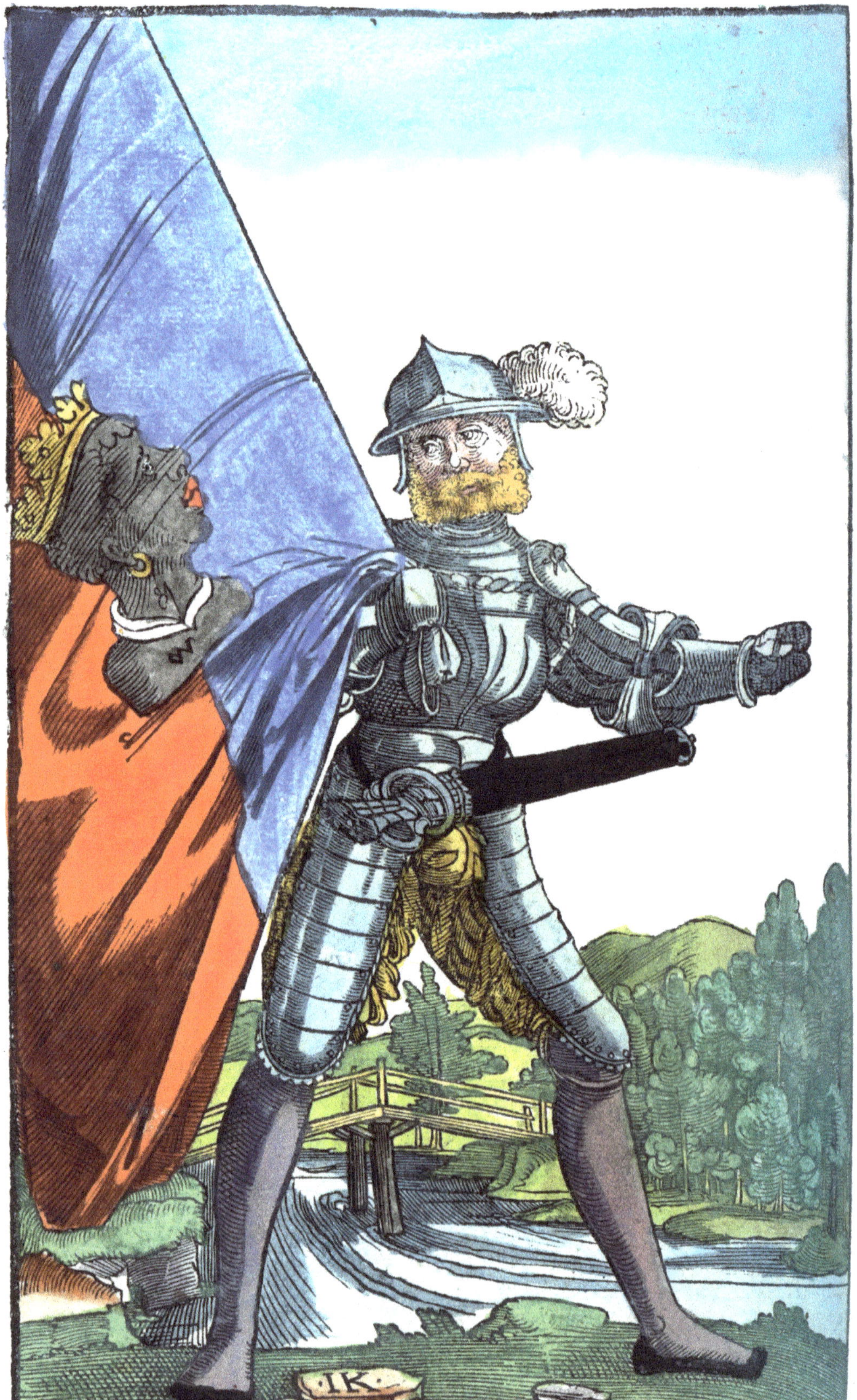

92. Lagingen | Lauingen

Straßburg.

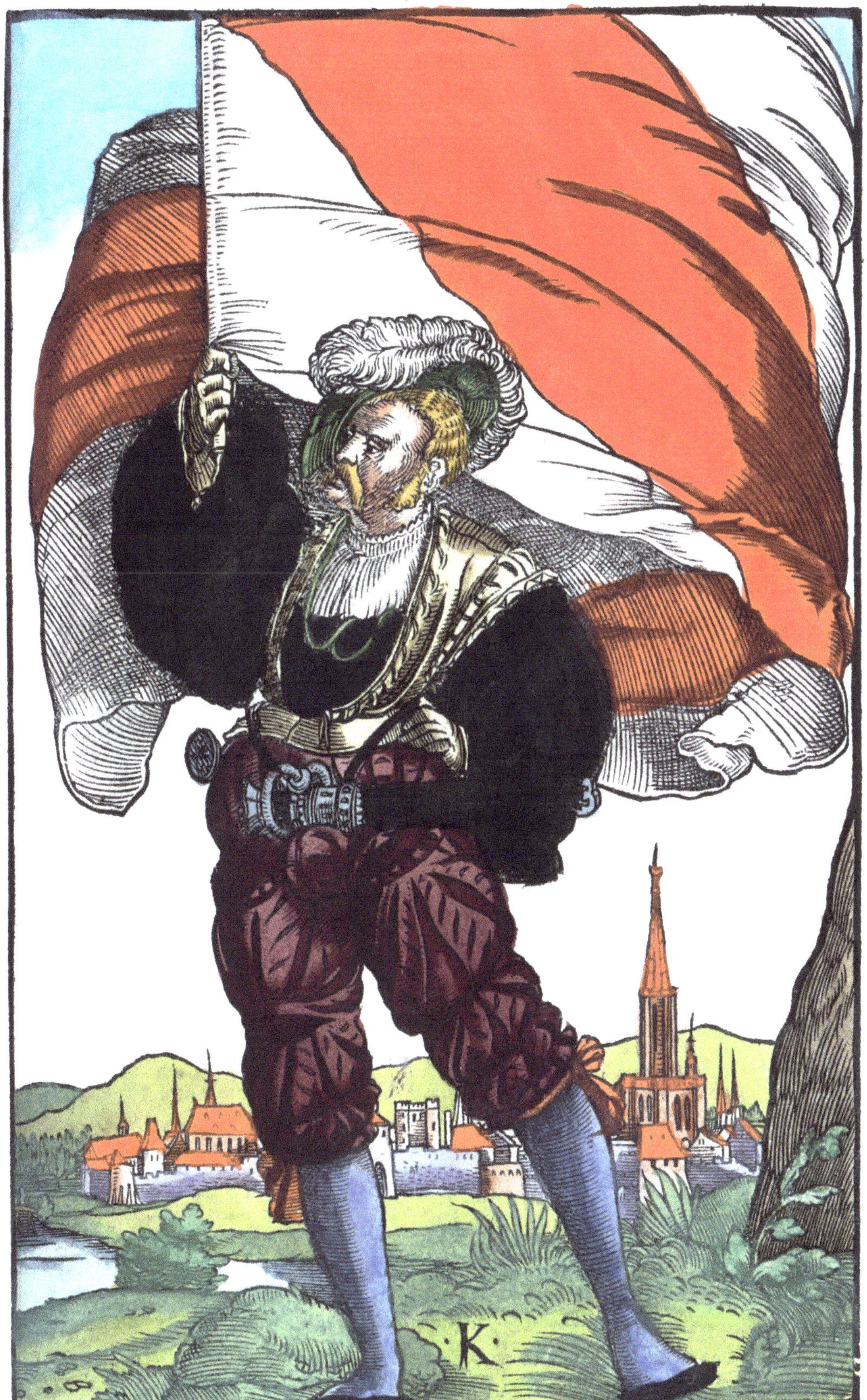

93. Straszburg | Strasbourg, France

Franckfurt.

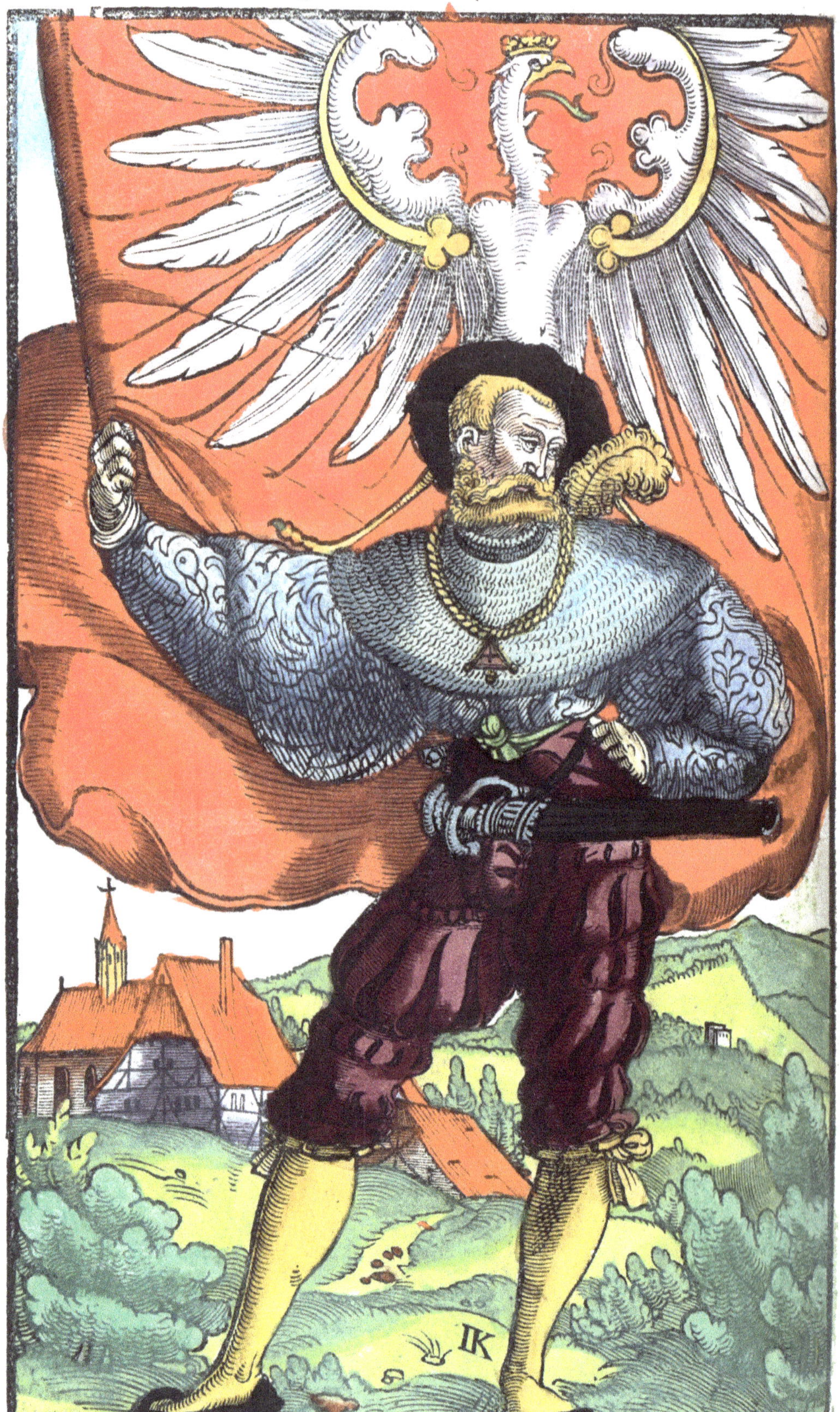

94. Franckfurt | Frankfurt

Buchhorn.

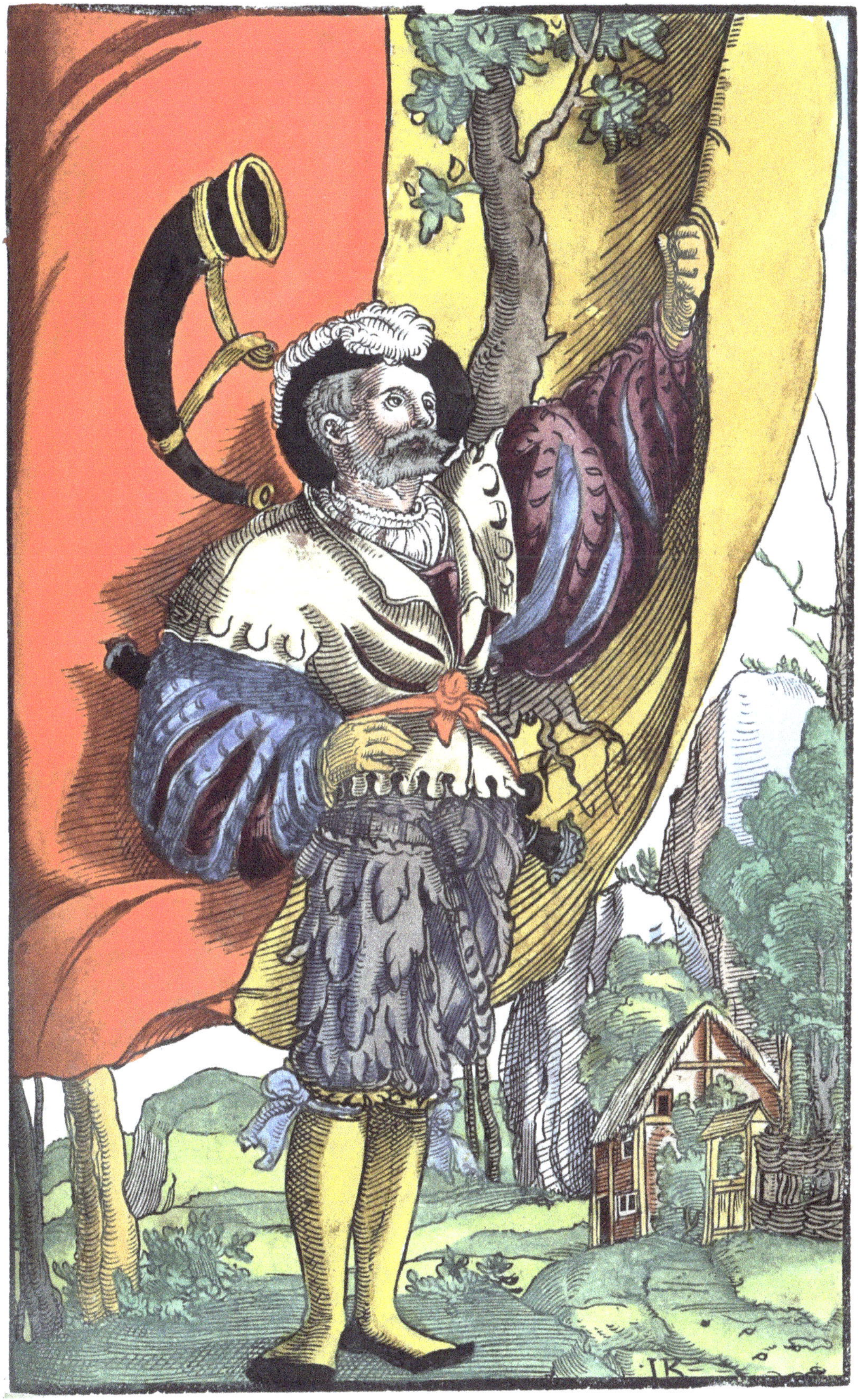

95. Buchhorn | Friedrichshafen

Gengen.

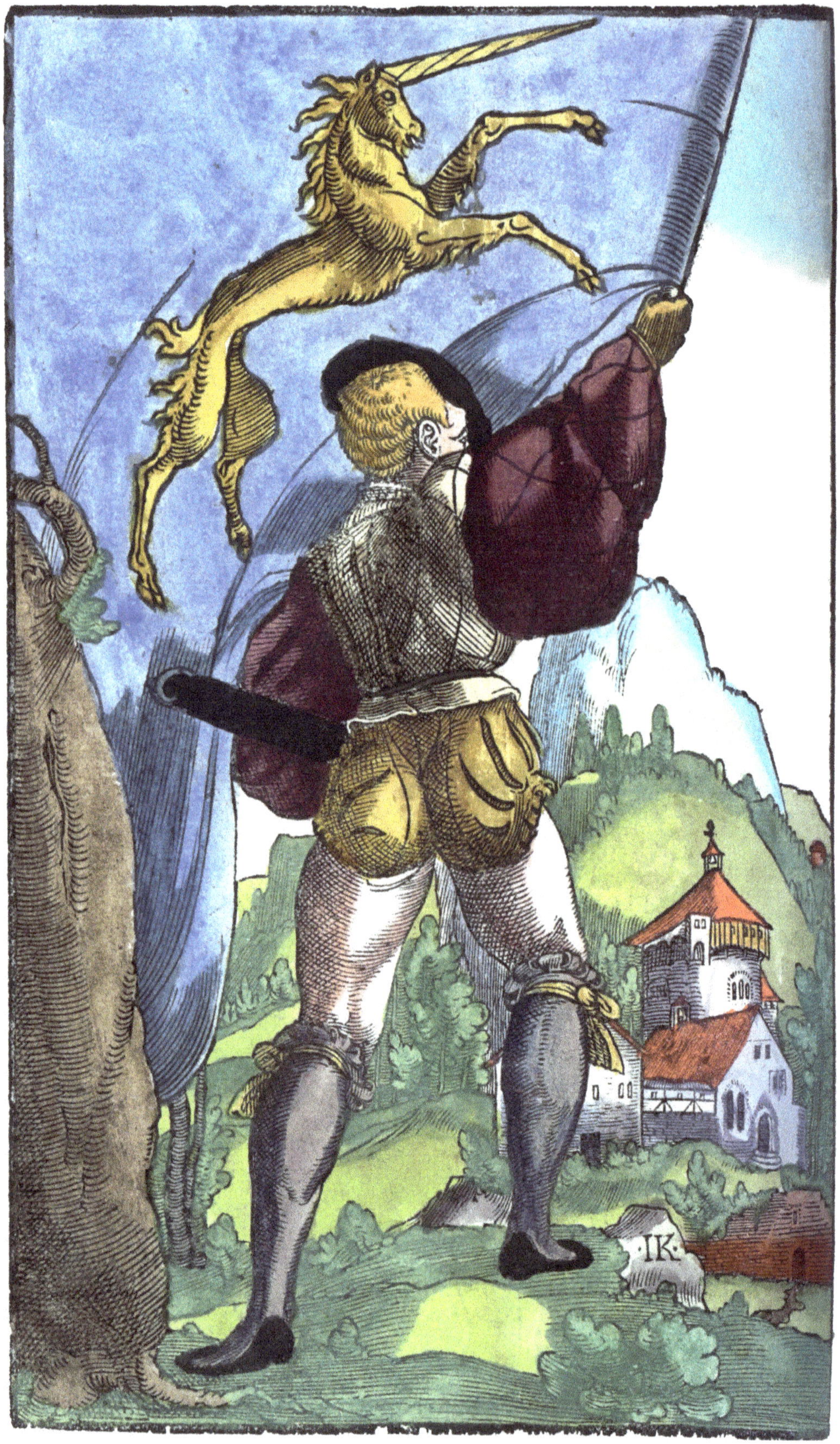

96. Gengen | Giengen an der Brenz

Halle in Schwaben.

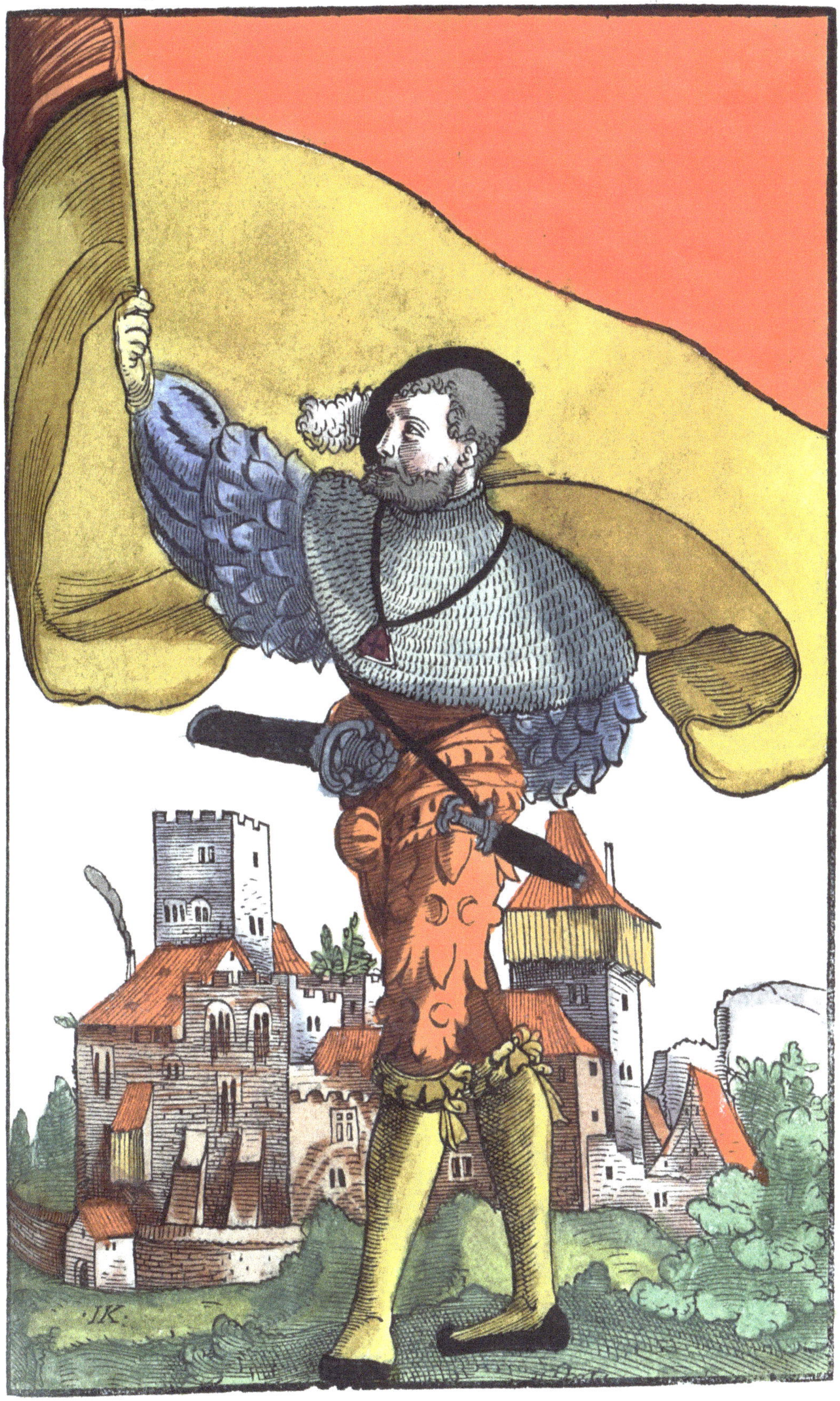

97. Halle in Schwaben | Schwäbisch Hall

Gengenbach.

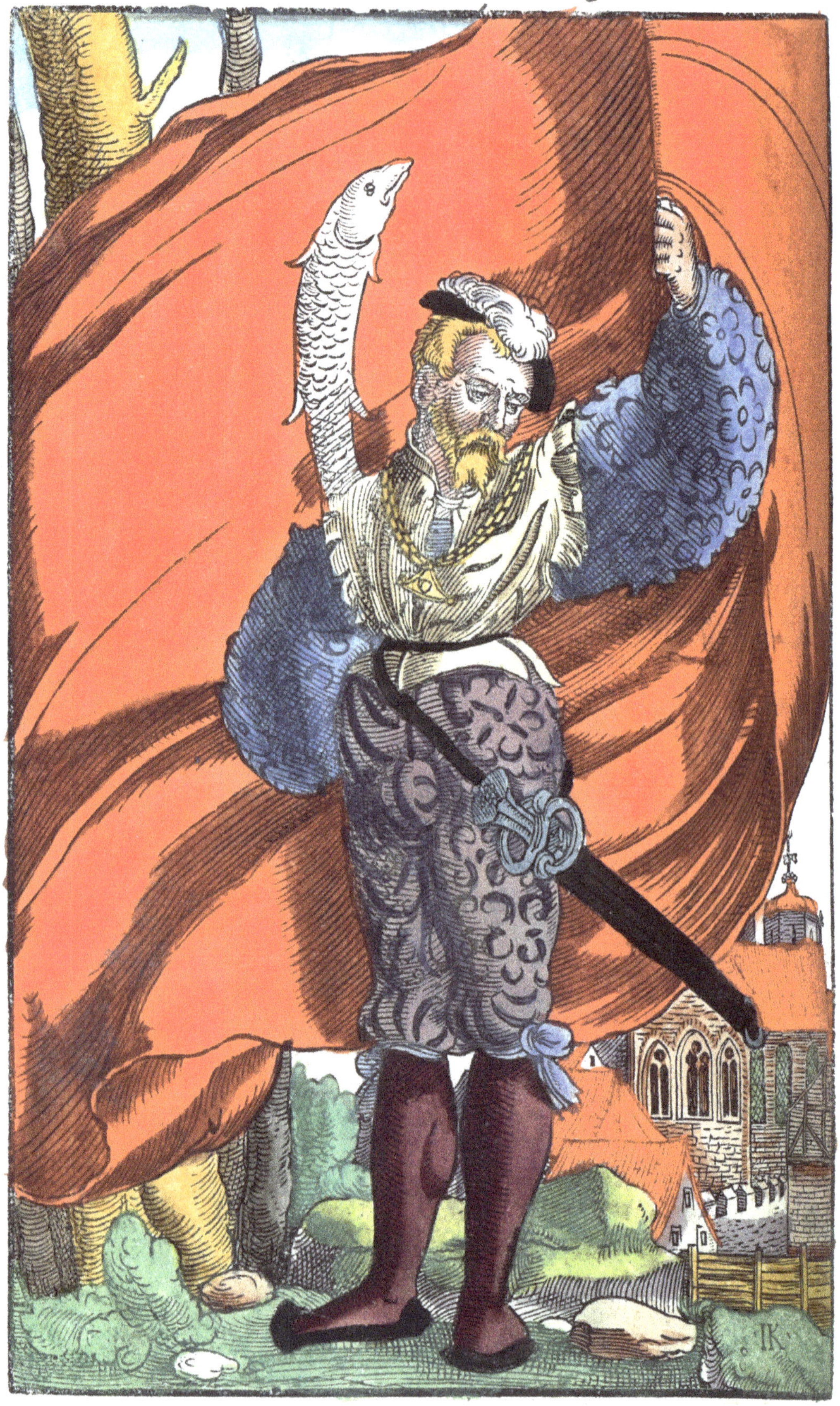

98. Gengenbach | Gengenbach

Keiſerberg.

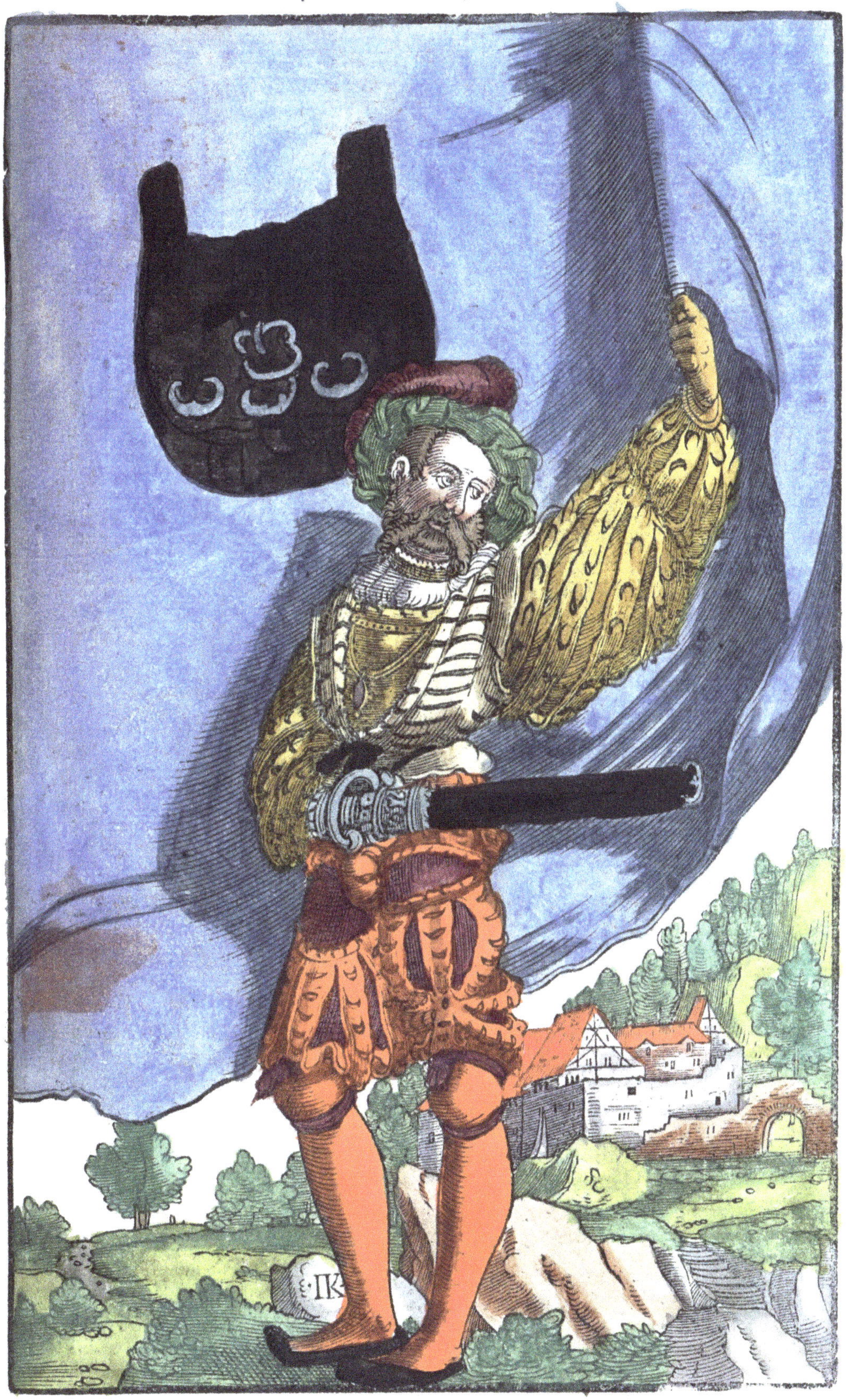

99. Keiserberg | Kaysersberg

Raffenspurg.

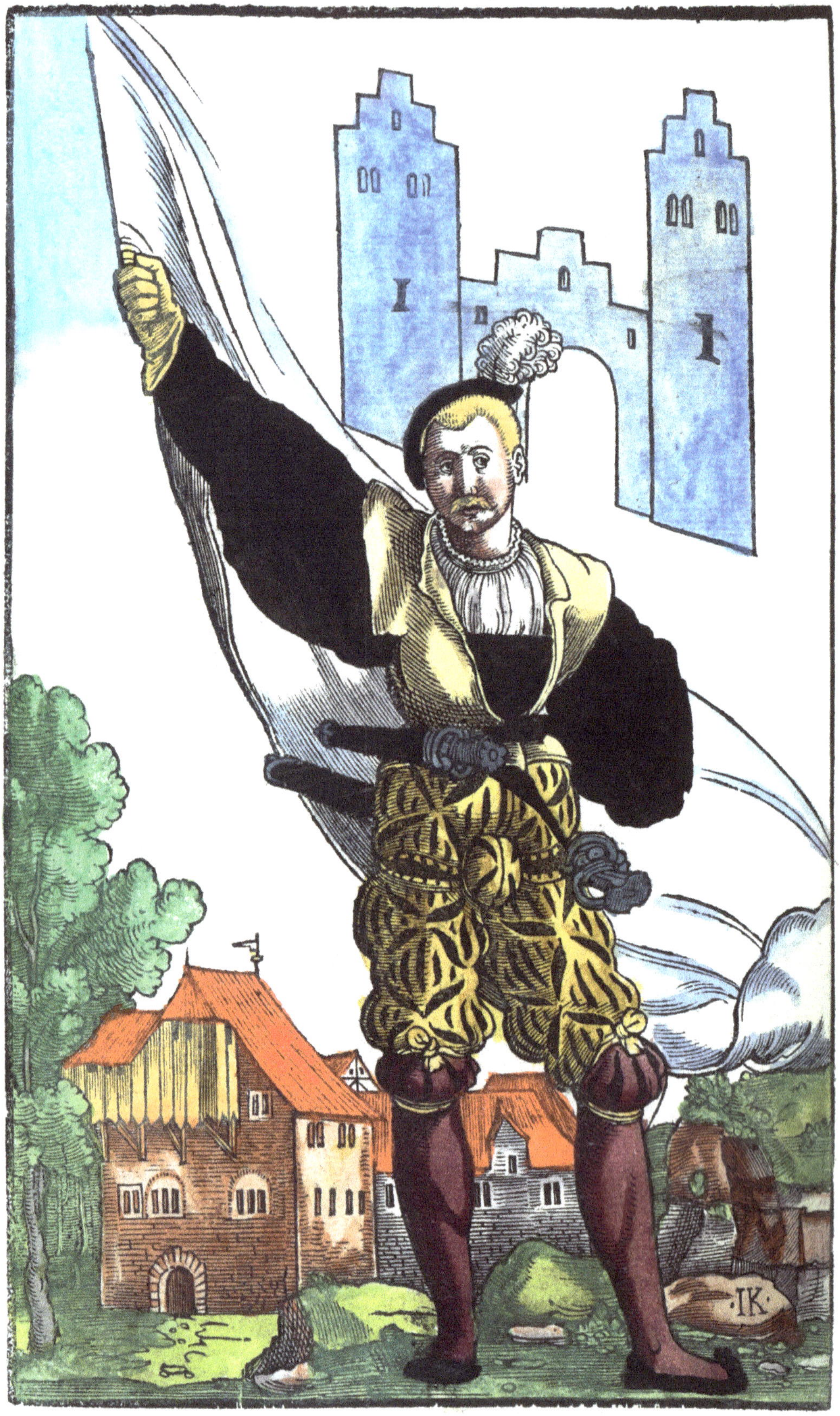

100. Raffenspurg | Ravensburg

Pfullendorff.

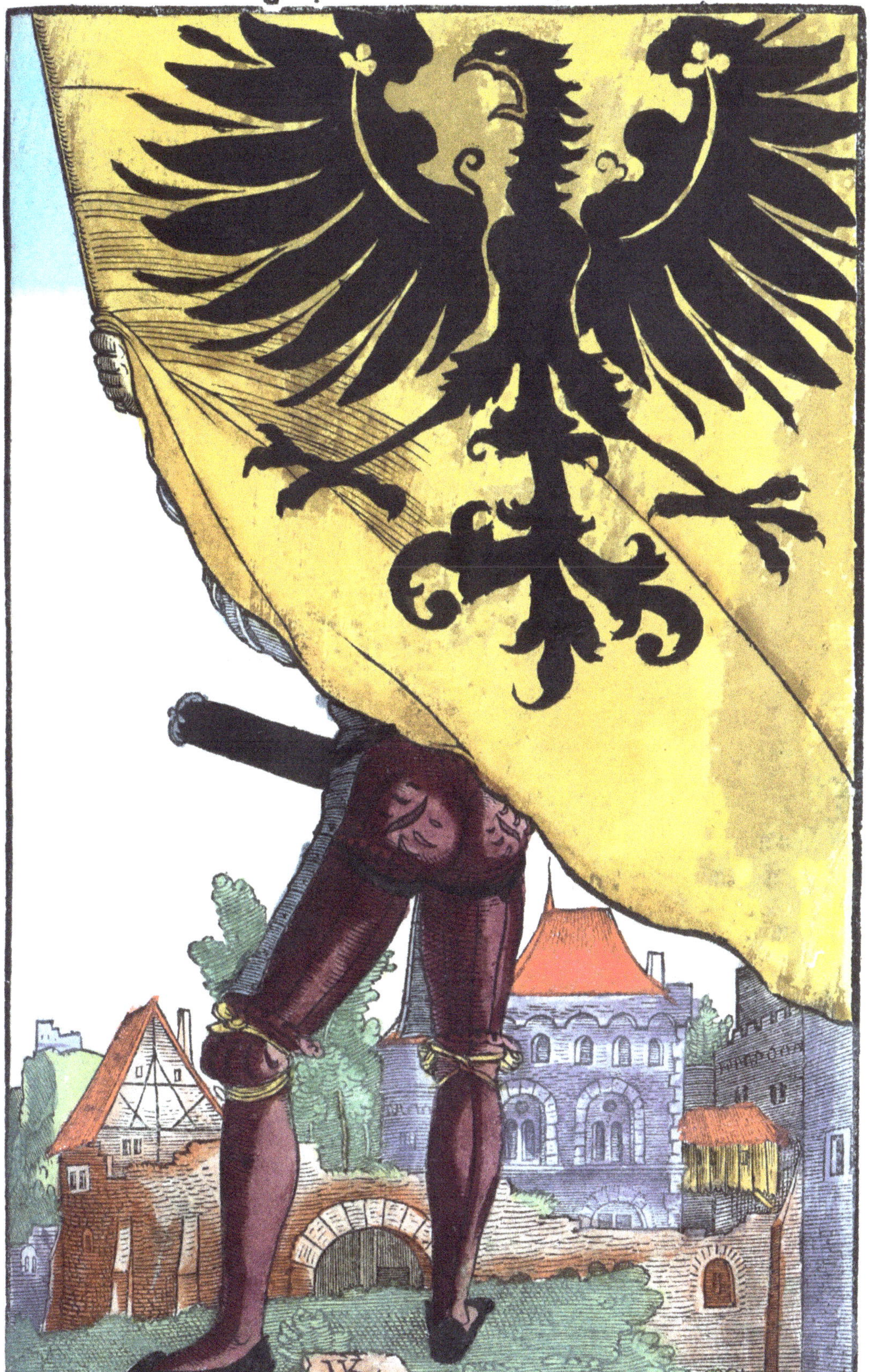

101. Pfullendorff | Pfullendorf

Iſznach.

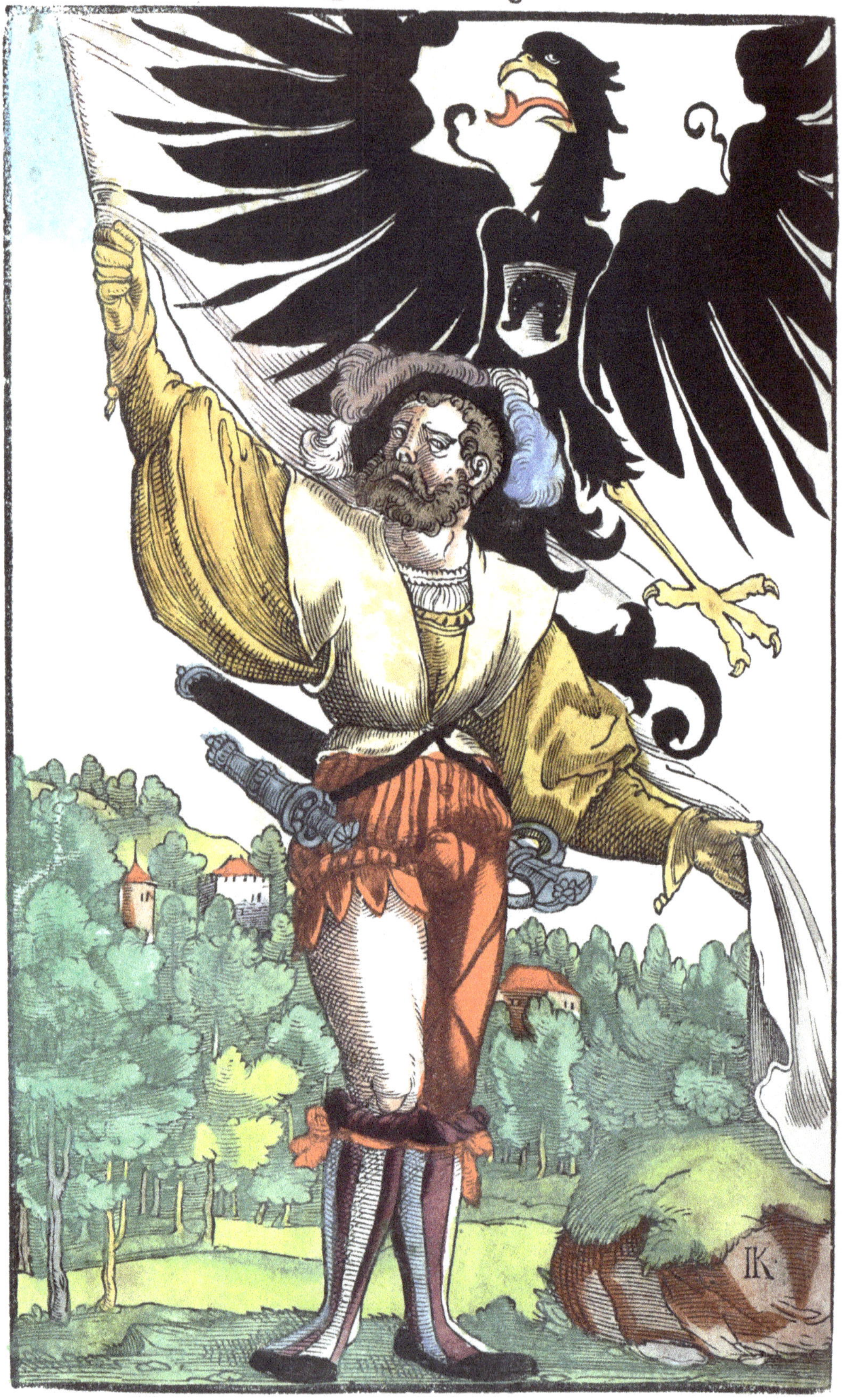

102. Issnach | Isny im Allgäu

Nürenberg.

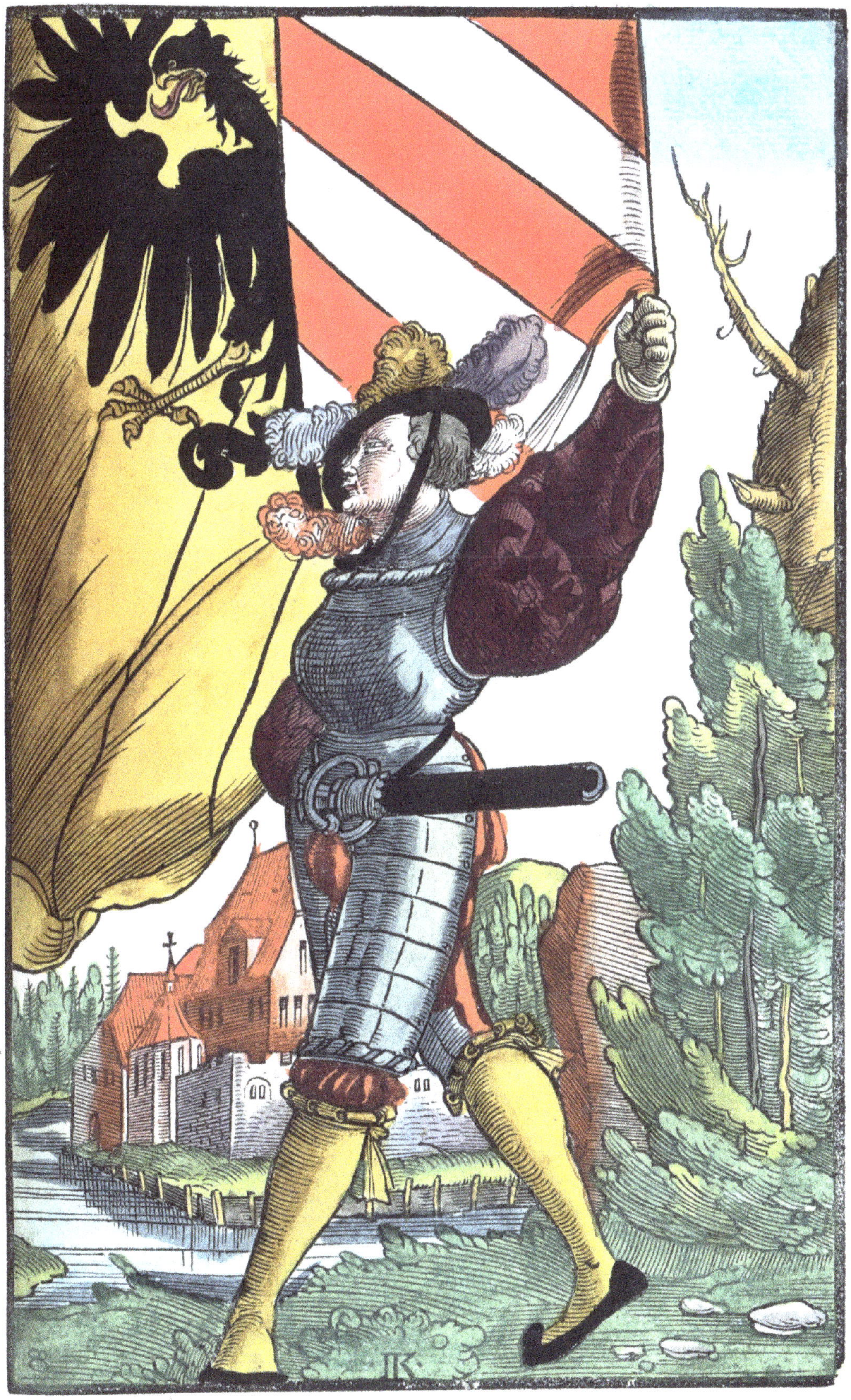

103. Nürenberg | Nürnberg, Nuremberg Flag of the Imperial City

Eßlingen.

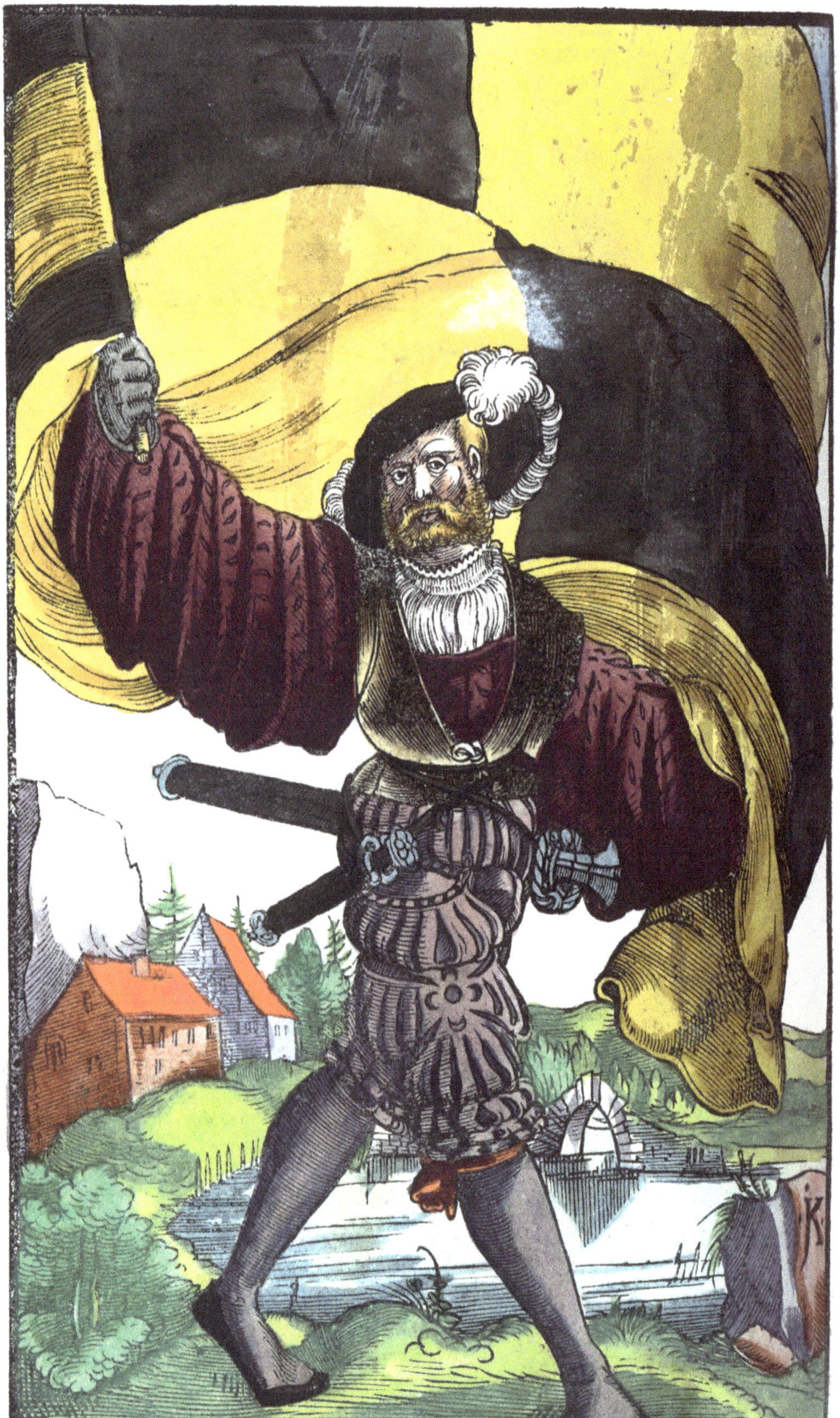

104. Esslingen | Esslingen

Wirzburg.

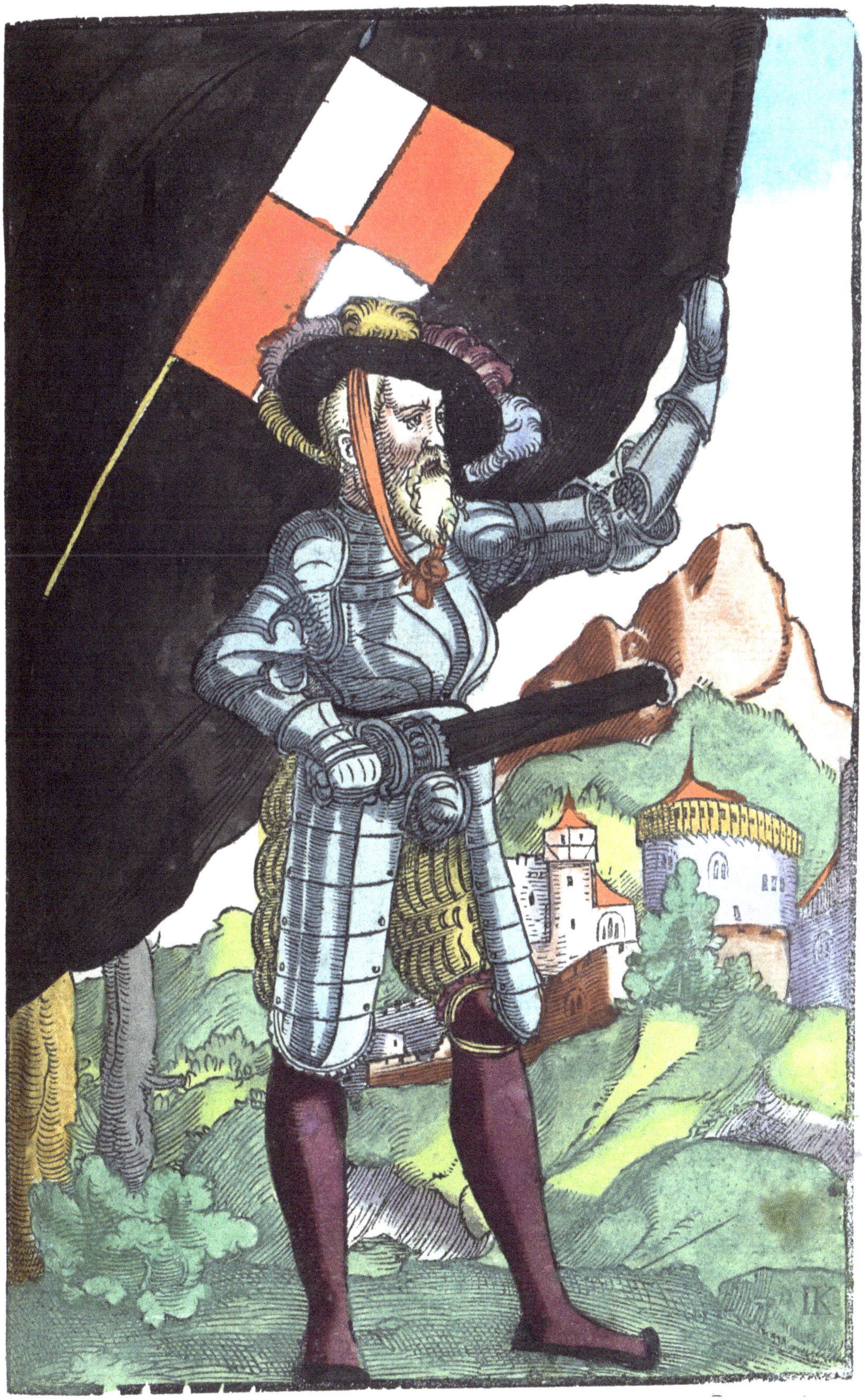

105. Wirzburg | Wurzburg

Kauffbeüren.

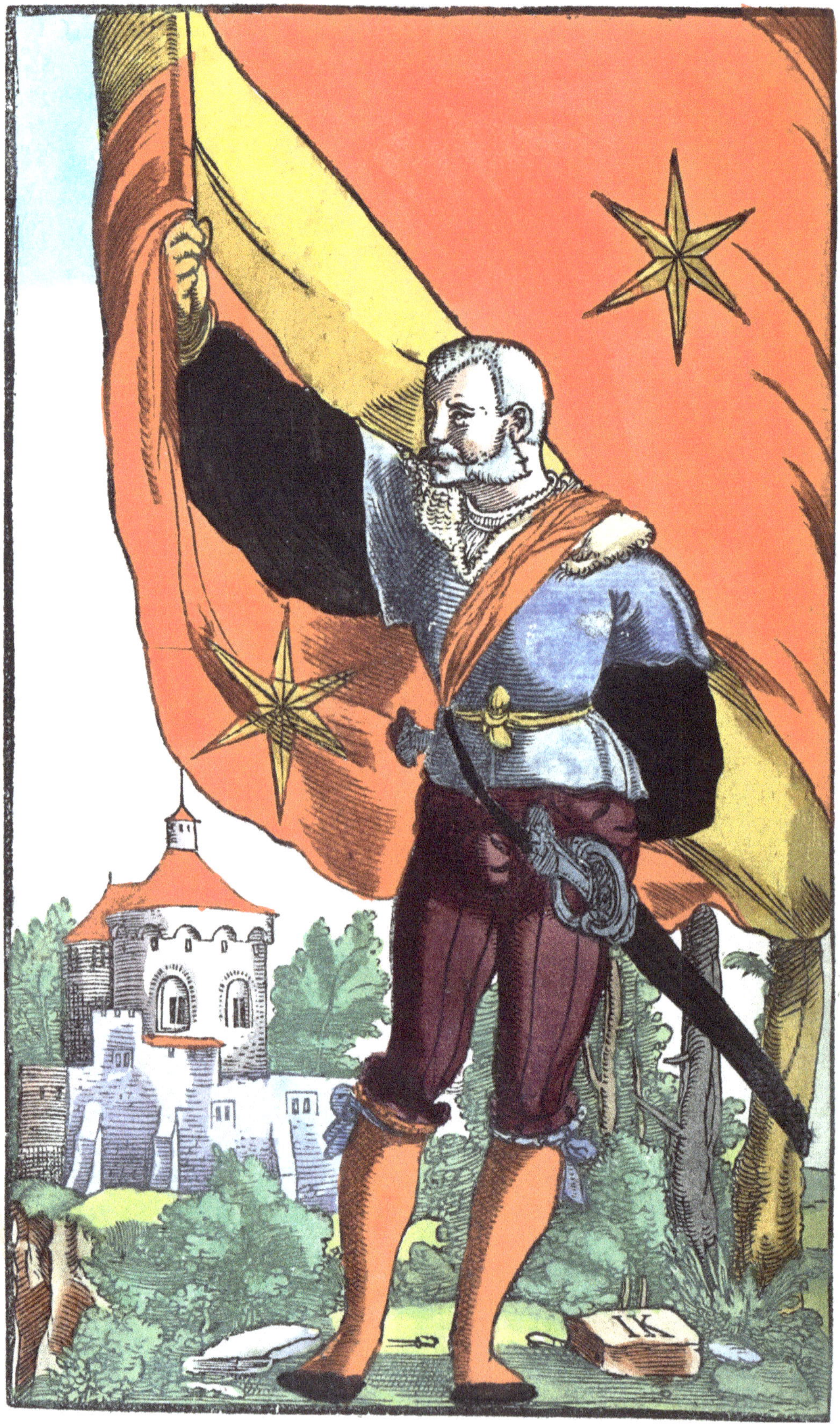

106. Kauffbeüren | Kaufbeuren

Rottenburg an der Thauber.

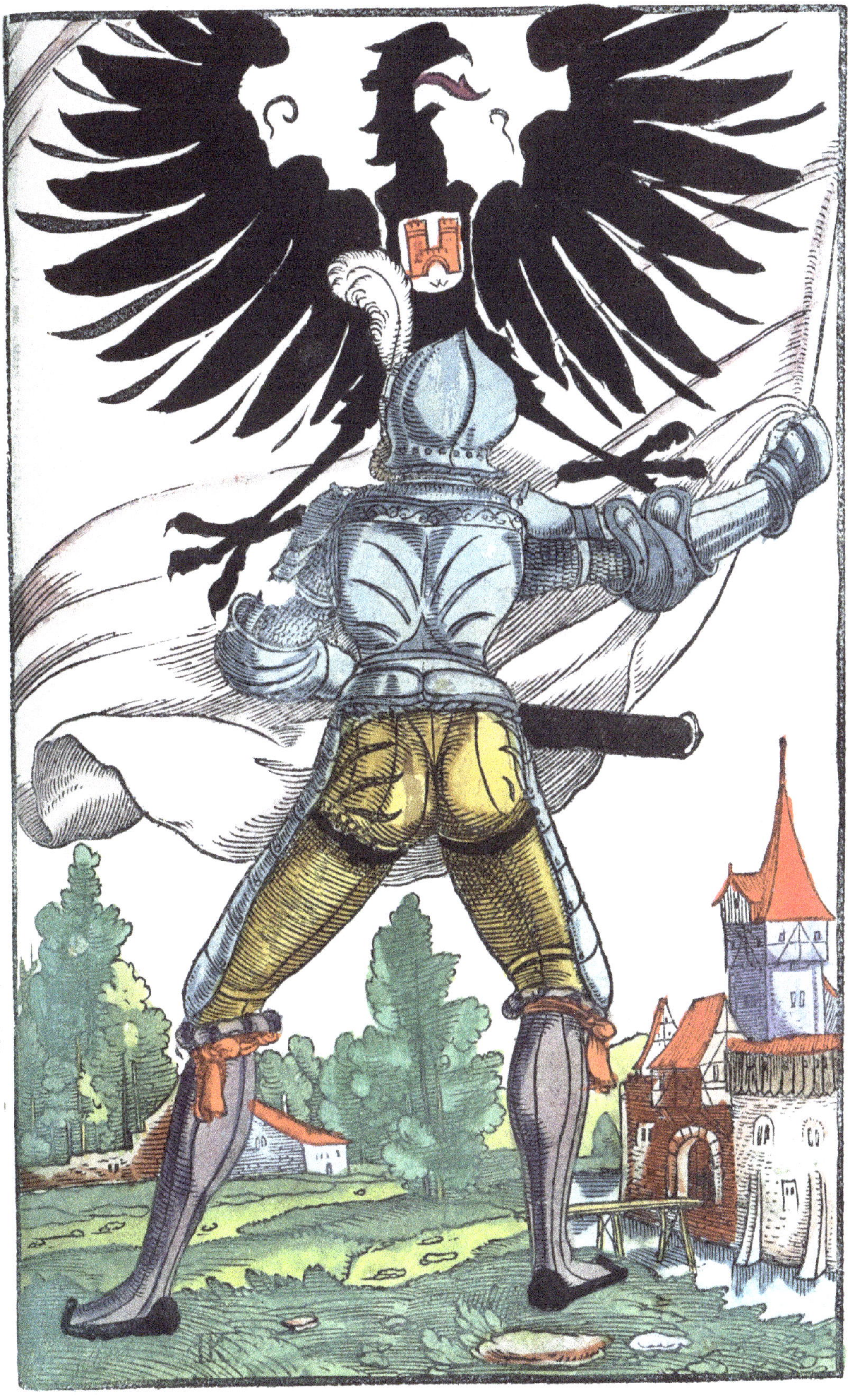

107. Rottenburg an der Thauber | Rothenburg ob der Tauber

Speir.

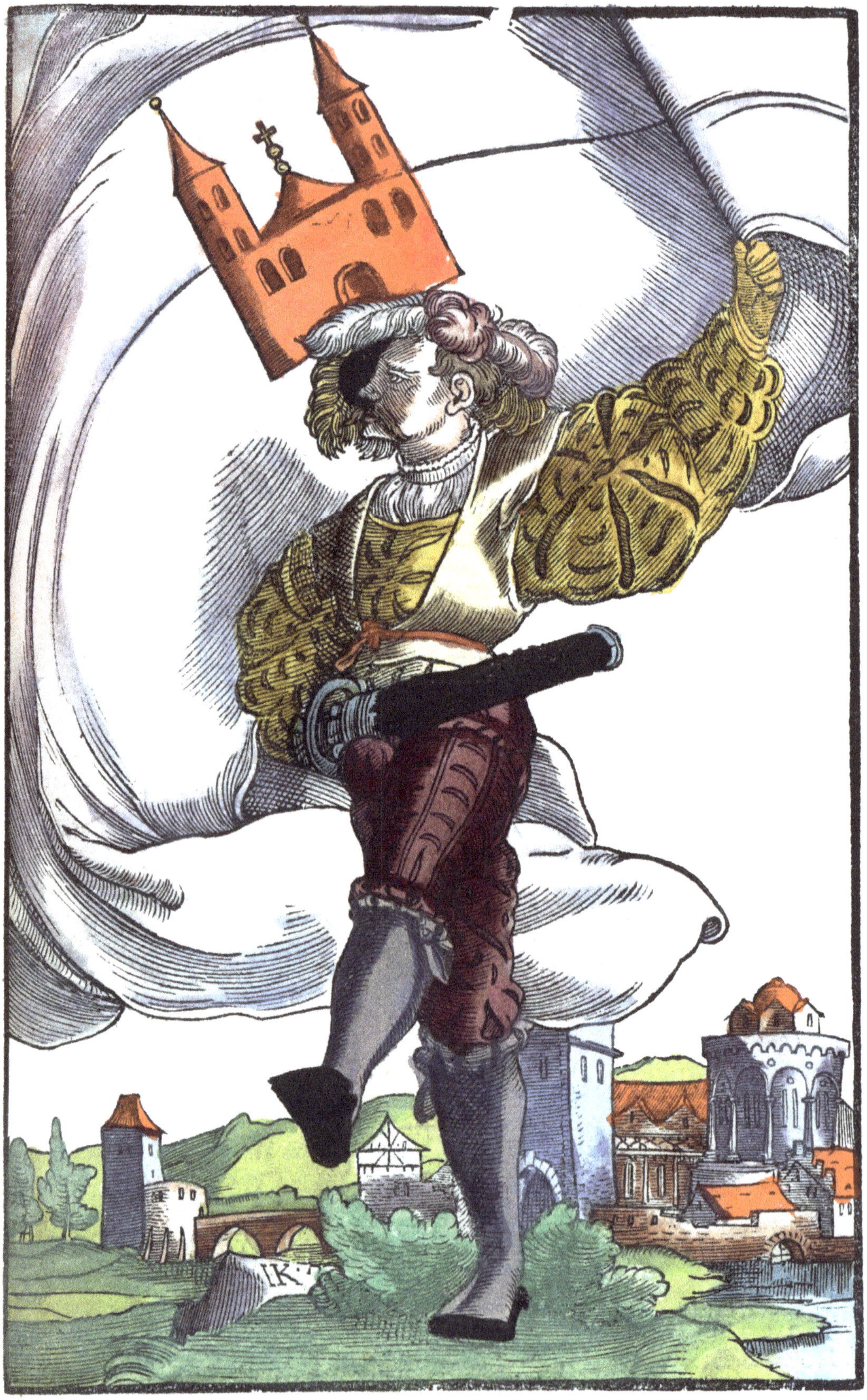

108. Speir | Speyer

Wurmbs.

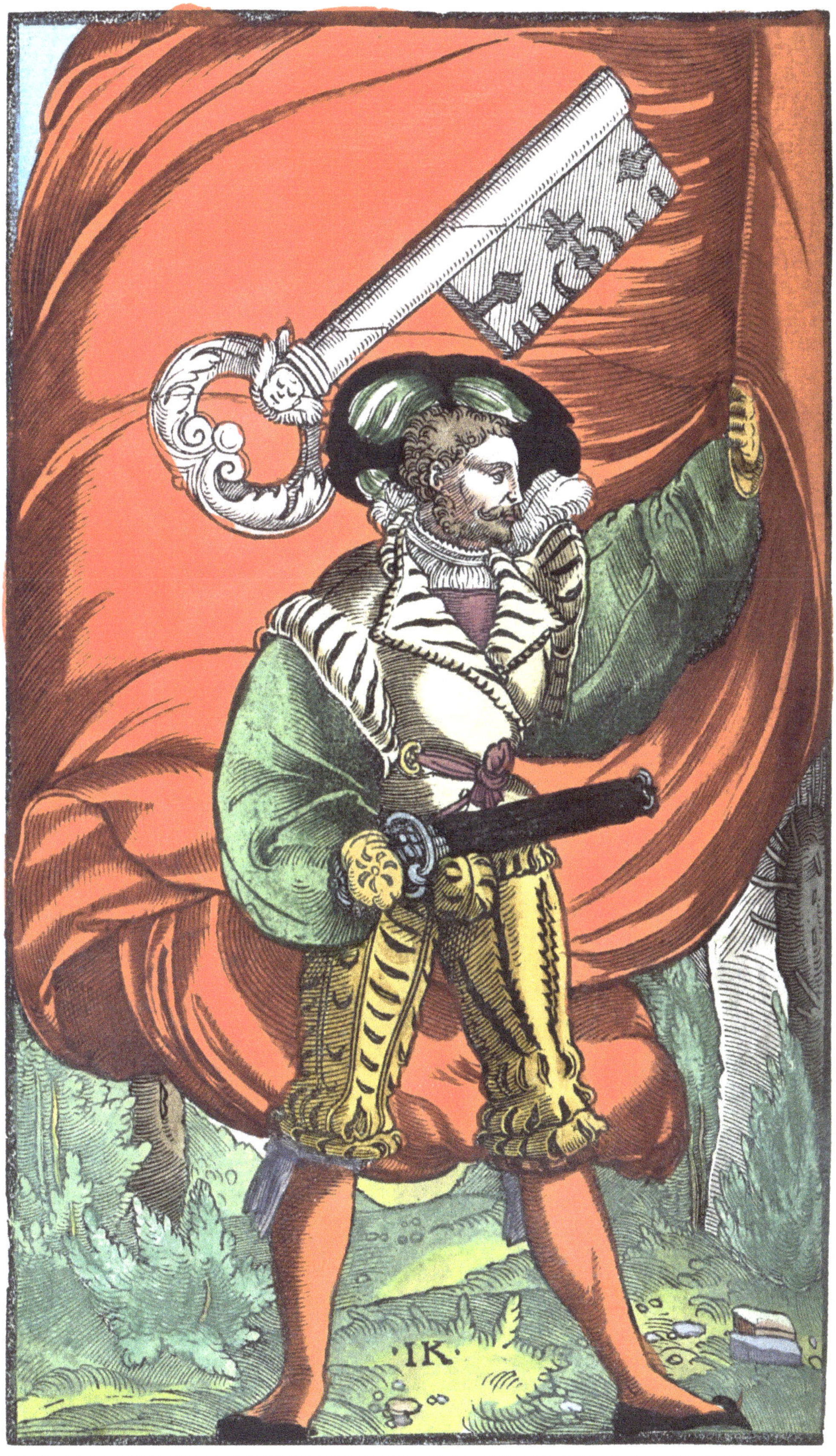

109. Wurmbs | Worms

Sarburg.

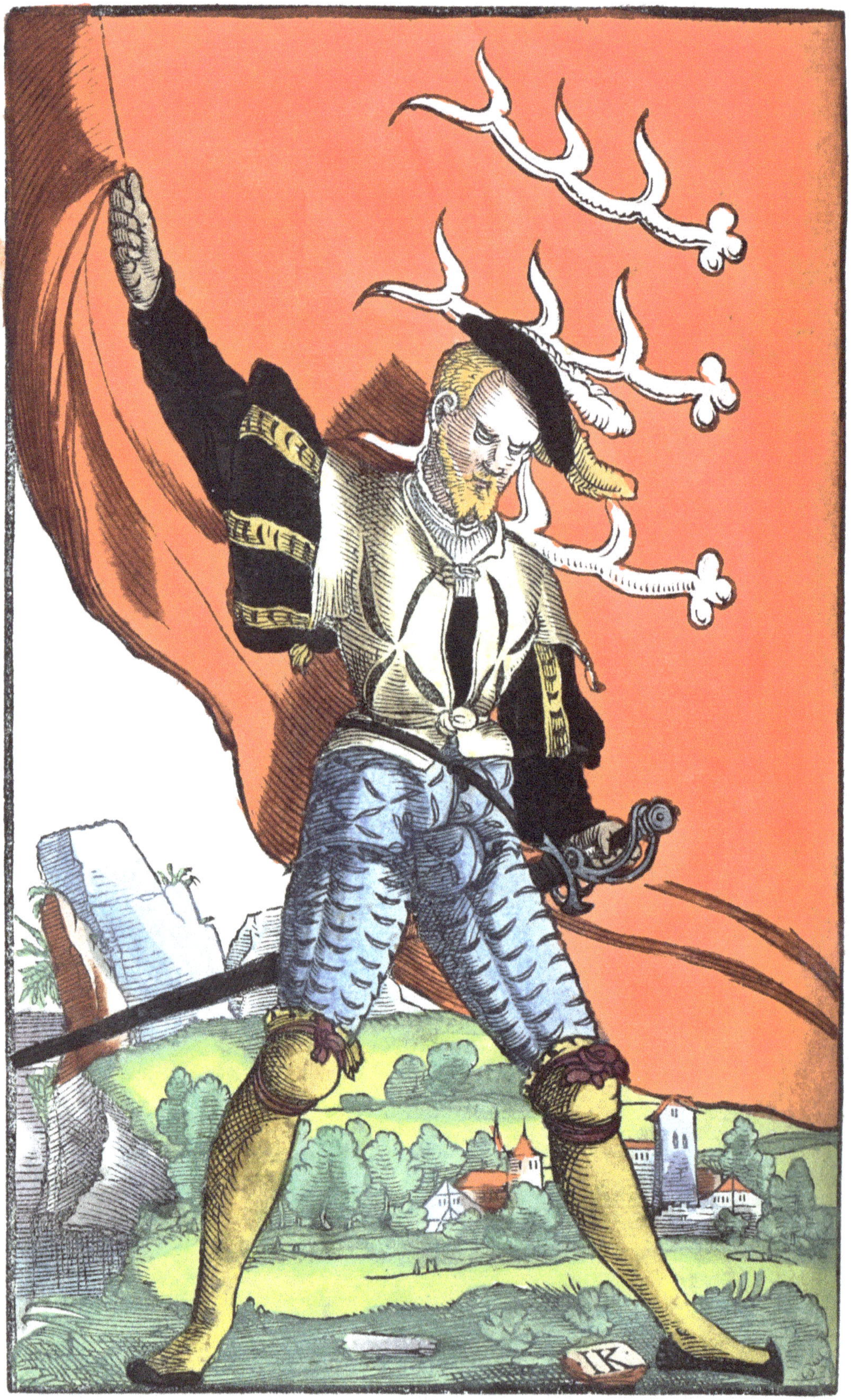

110. Sarburg | Sarrebourg, Lorraine France

Ritlingen.

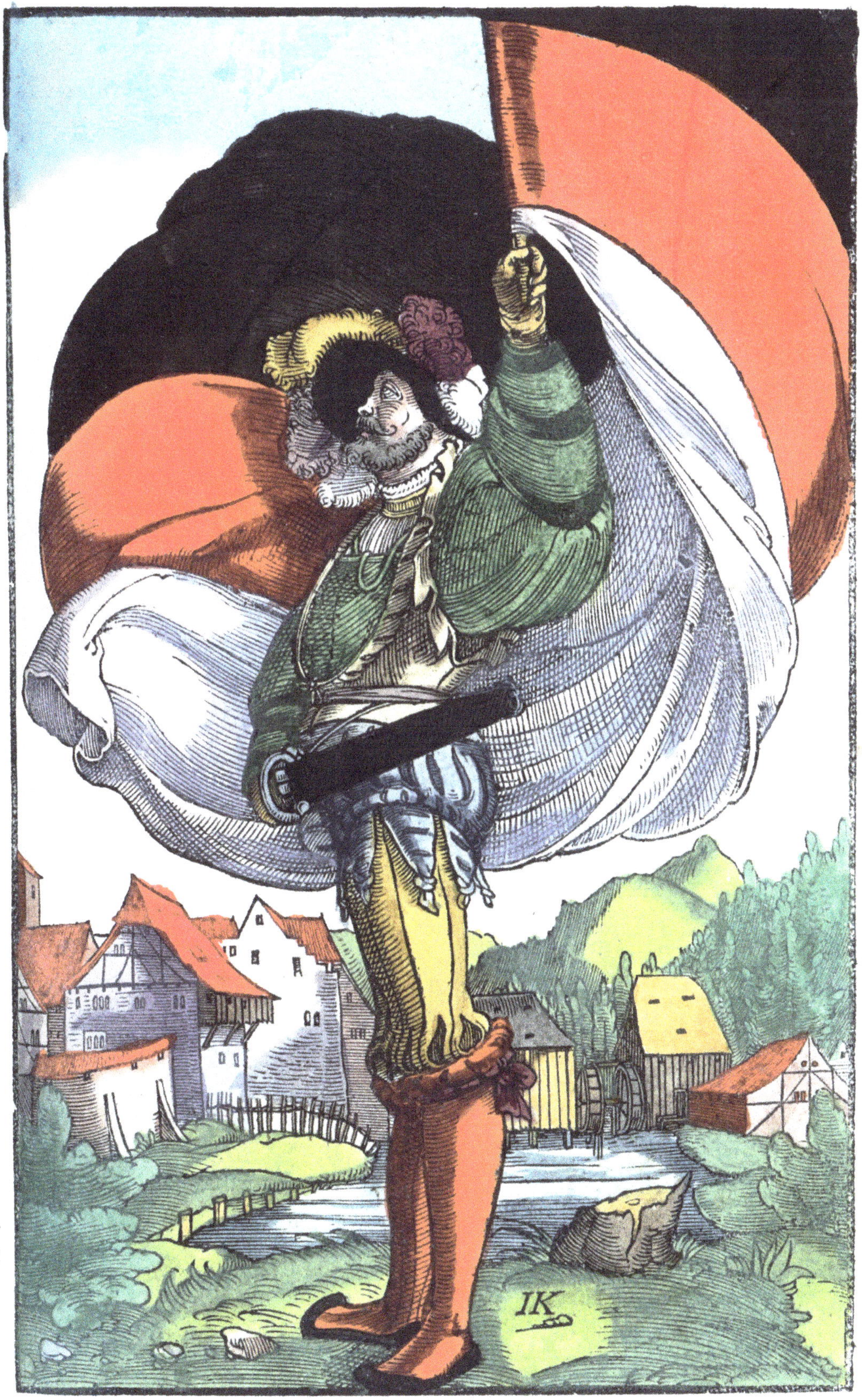

III. Ritlingen | Reutlingen

Weissenburg.

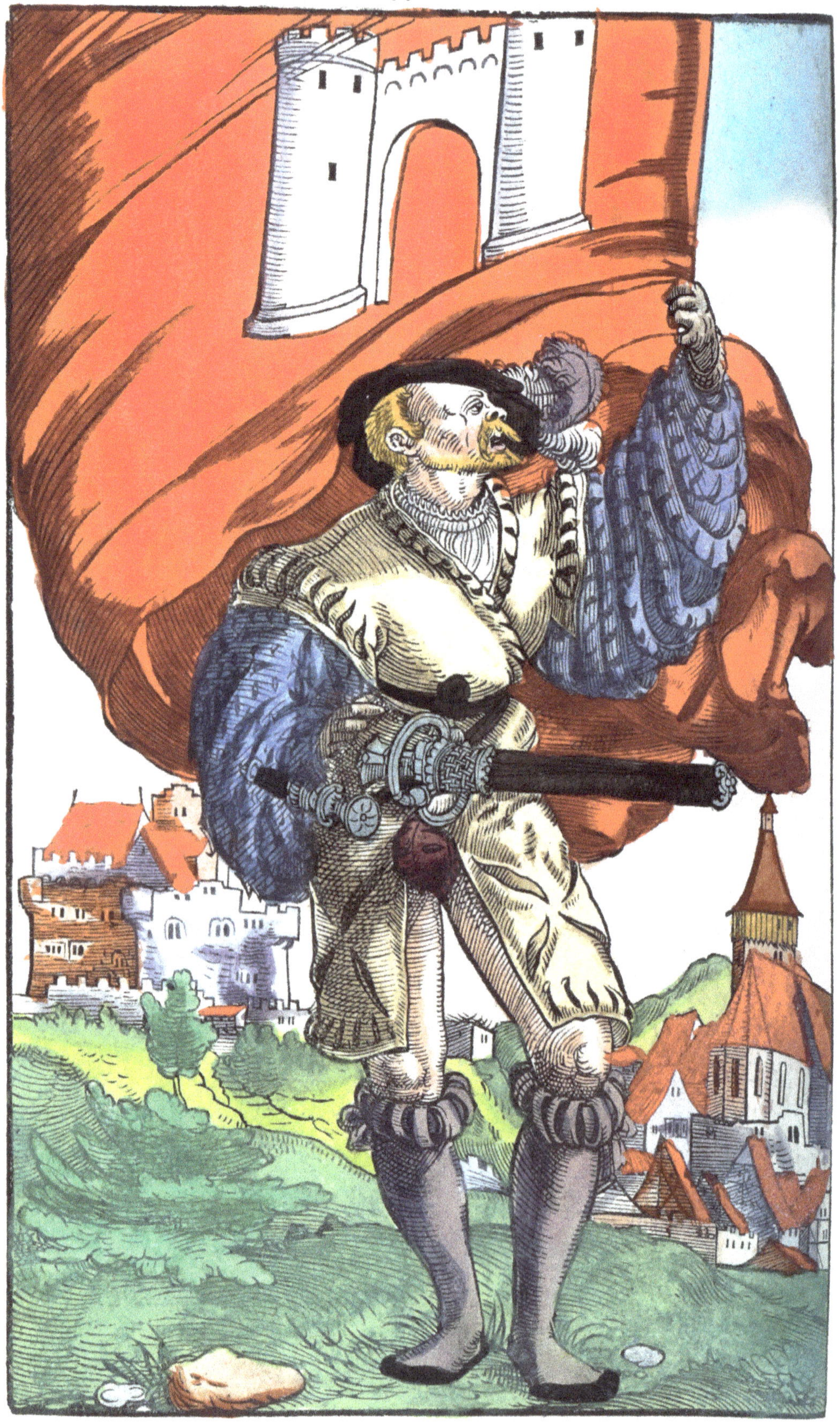

112. *Weissenburg das erst | Weißenburg in Bavaria*

Weiſenburg.

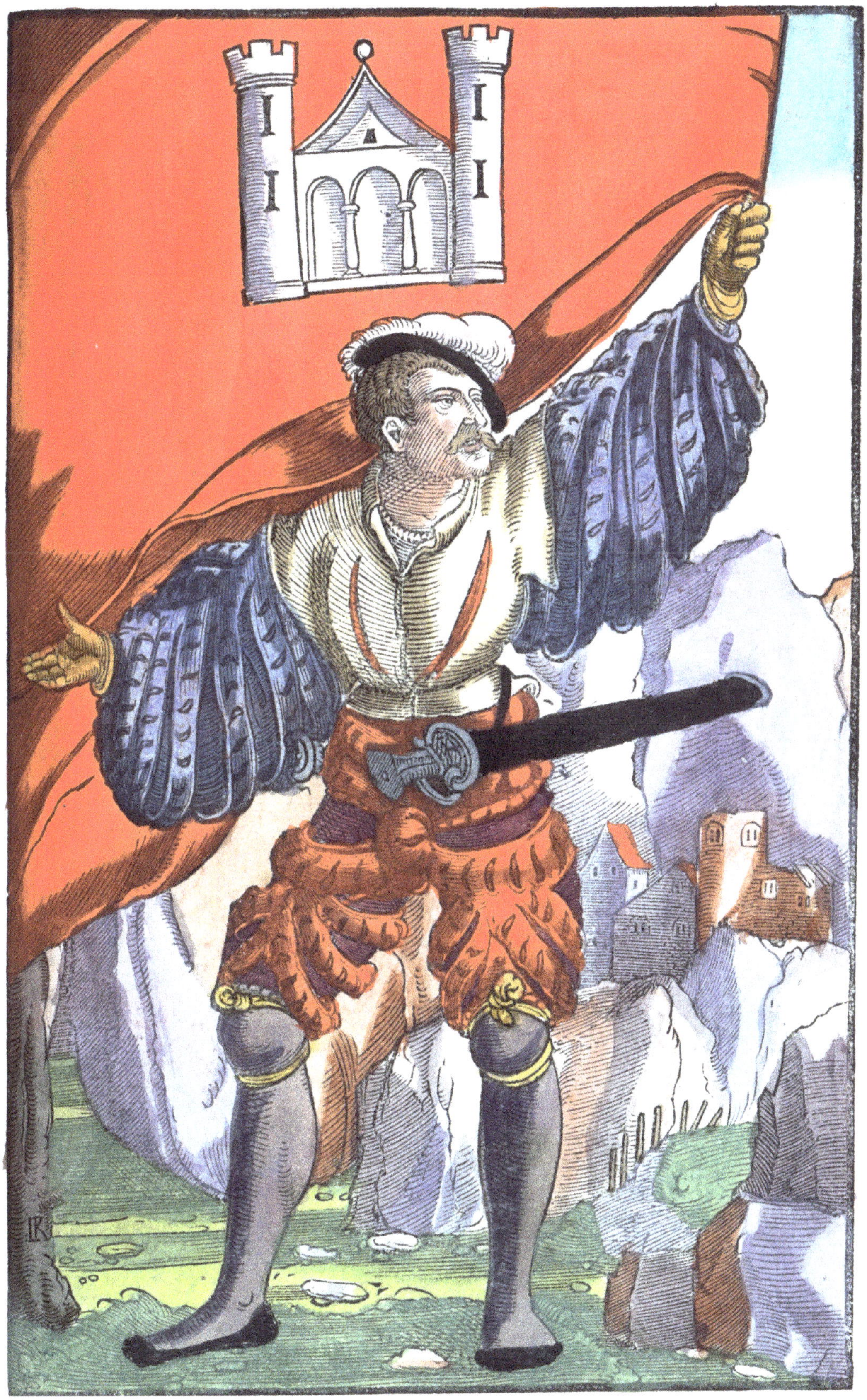

113. Weisenburg das ander | Weißenburg in Alsace, Wissembourg, France

Ingelnheim.

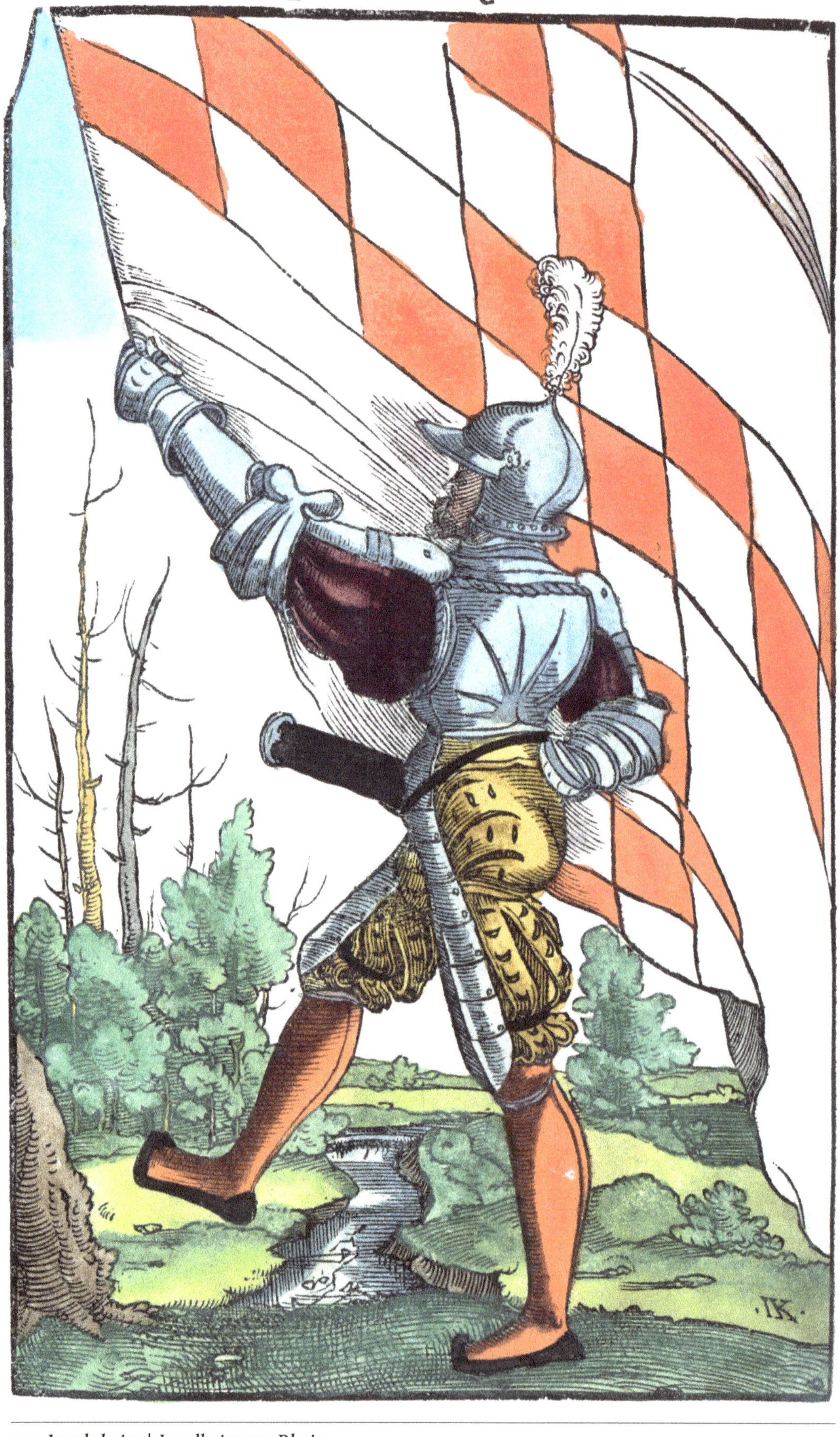

114. Ingelnheim | Ingelheim am Rhein

Mechel.

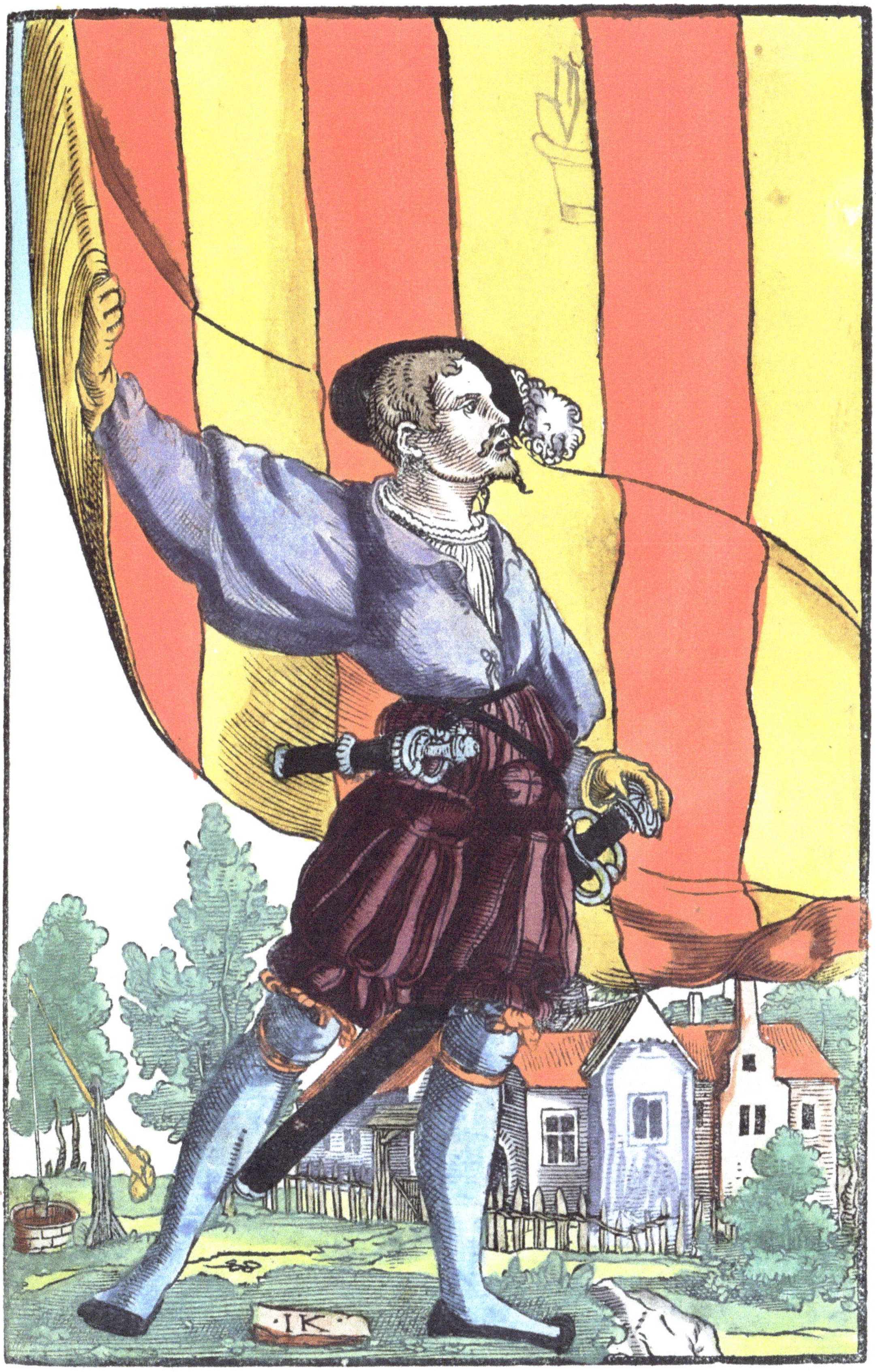

115. Mechel | Mechelen, Belgium

Lůttich.

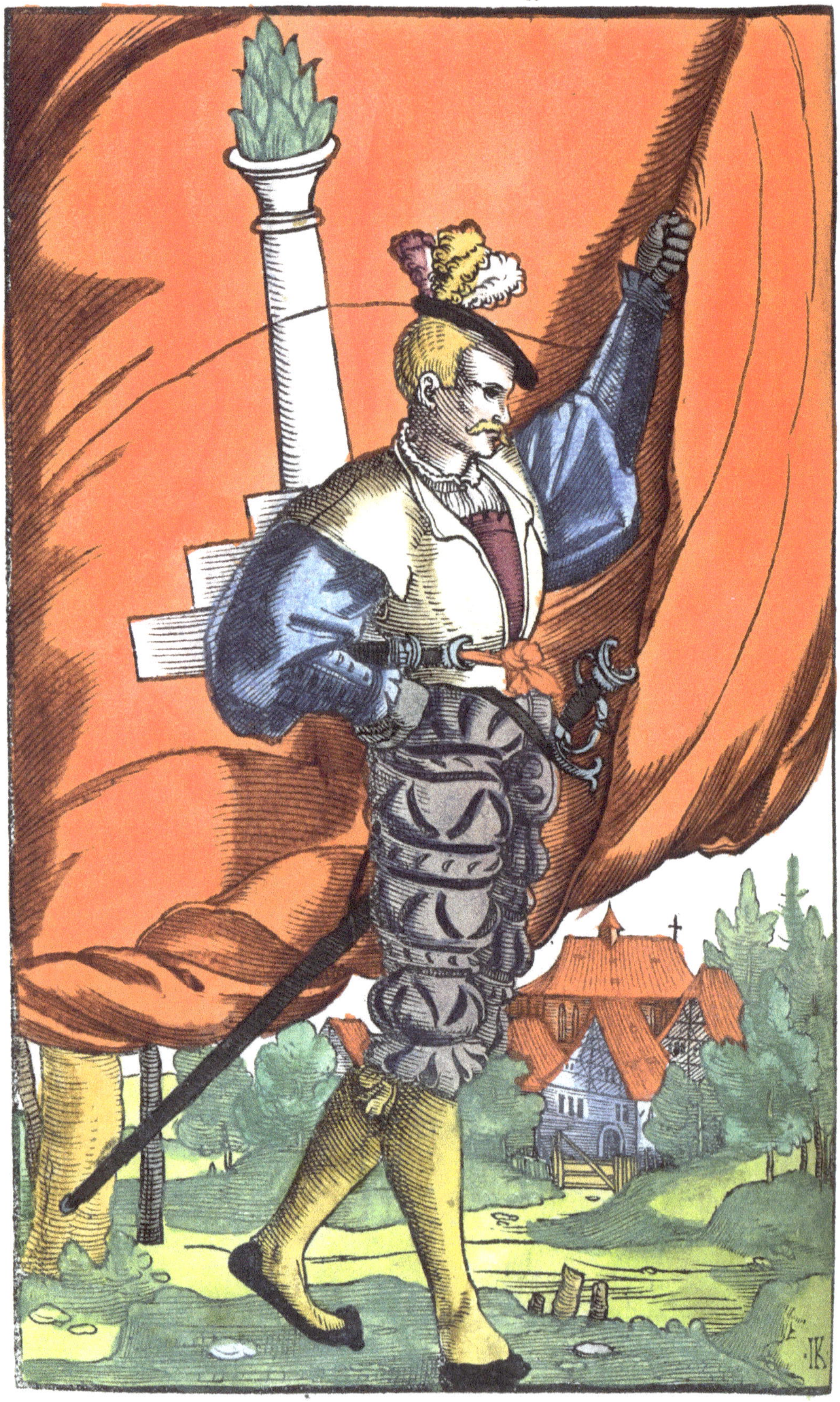

116. Lüttich | Liège, Belgium

Lübeck.

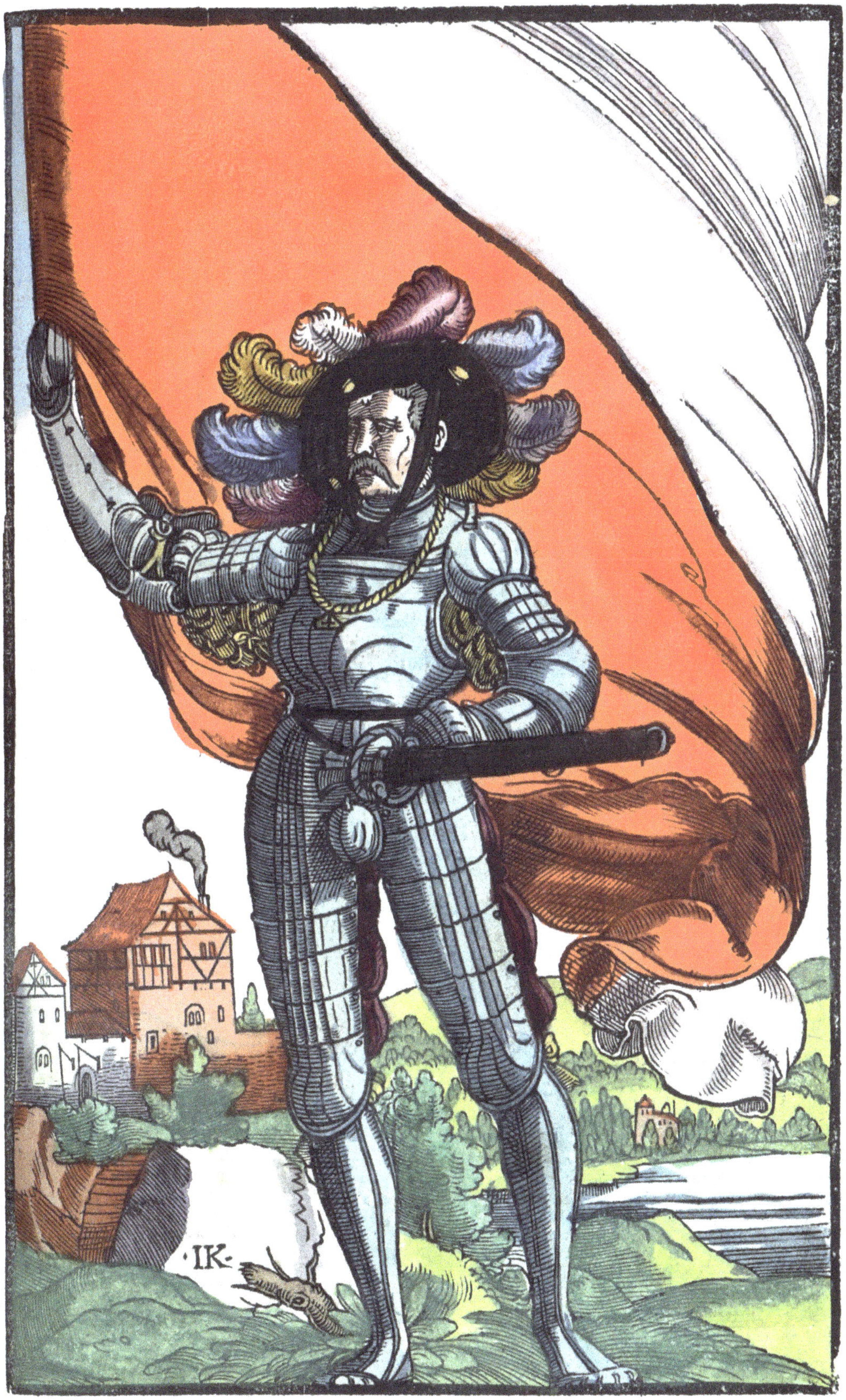

117. Lübeck | Lübeck

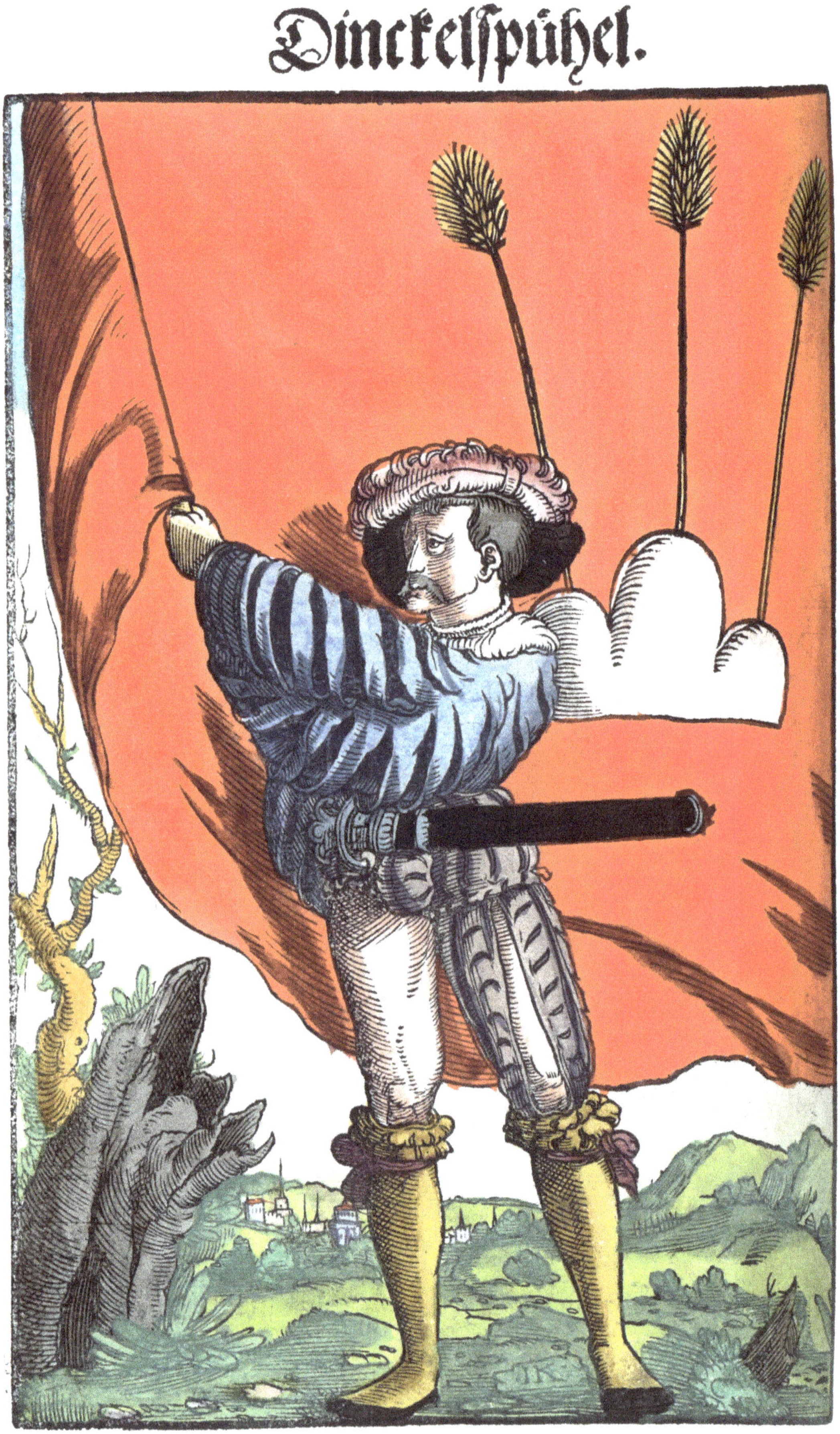

118. *Dinckelspühel | Dinkelsbühl*

Trier

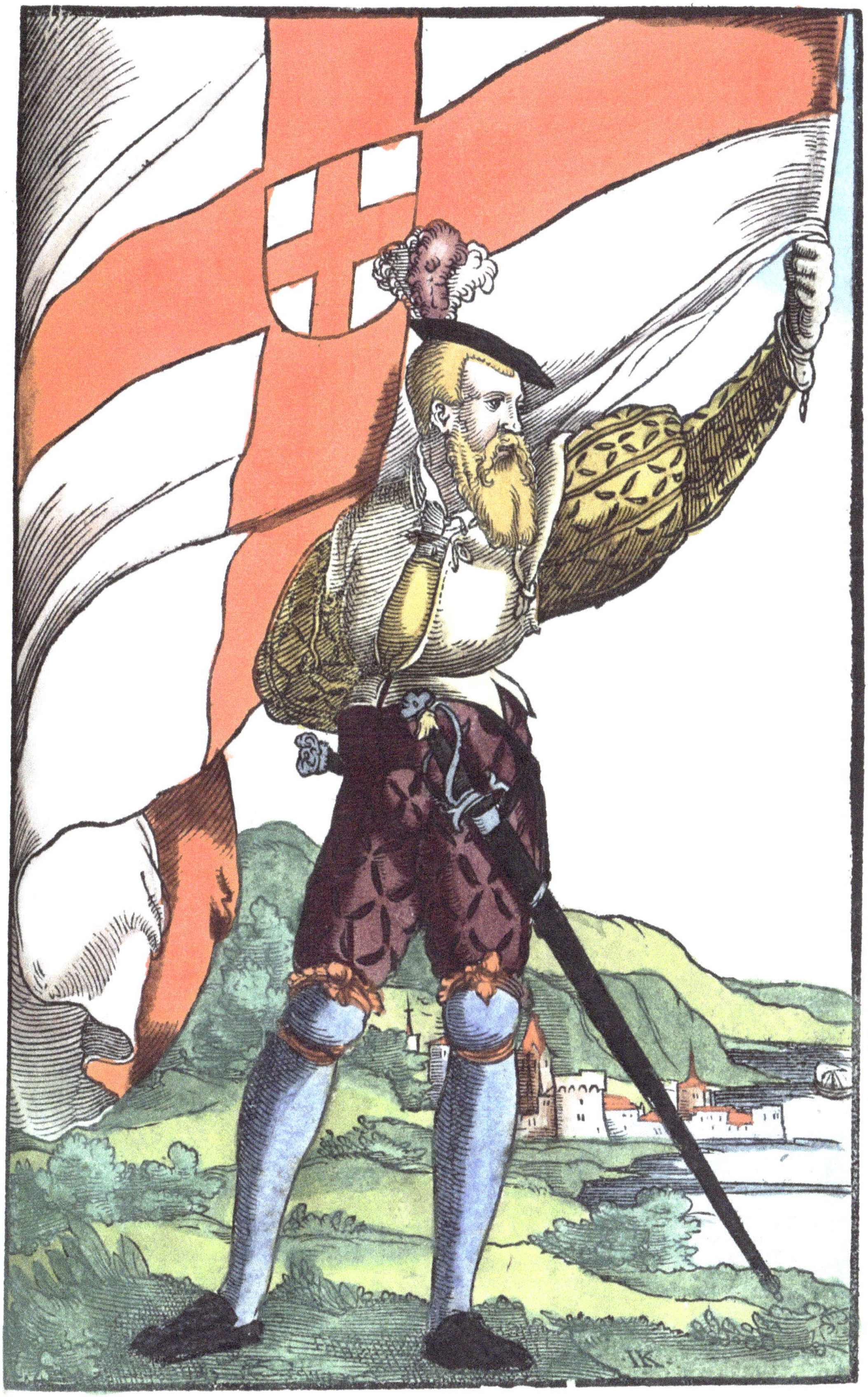

119. Trier | City of Trier

Kitzingen.

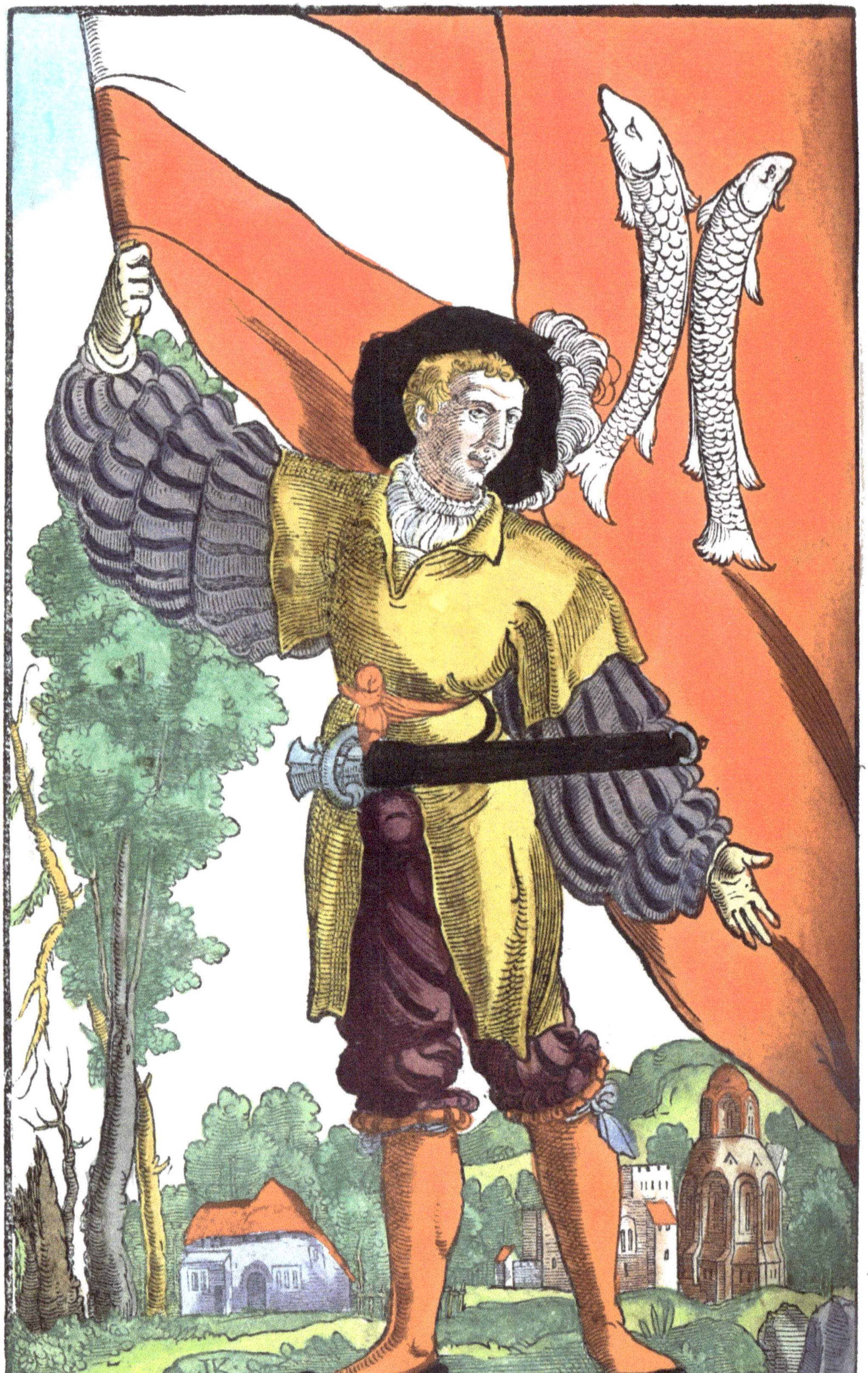

120. Kitzingen | Kenzingen

Wiensßheym.

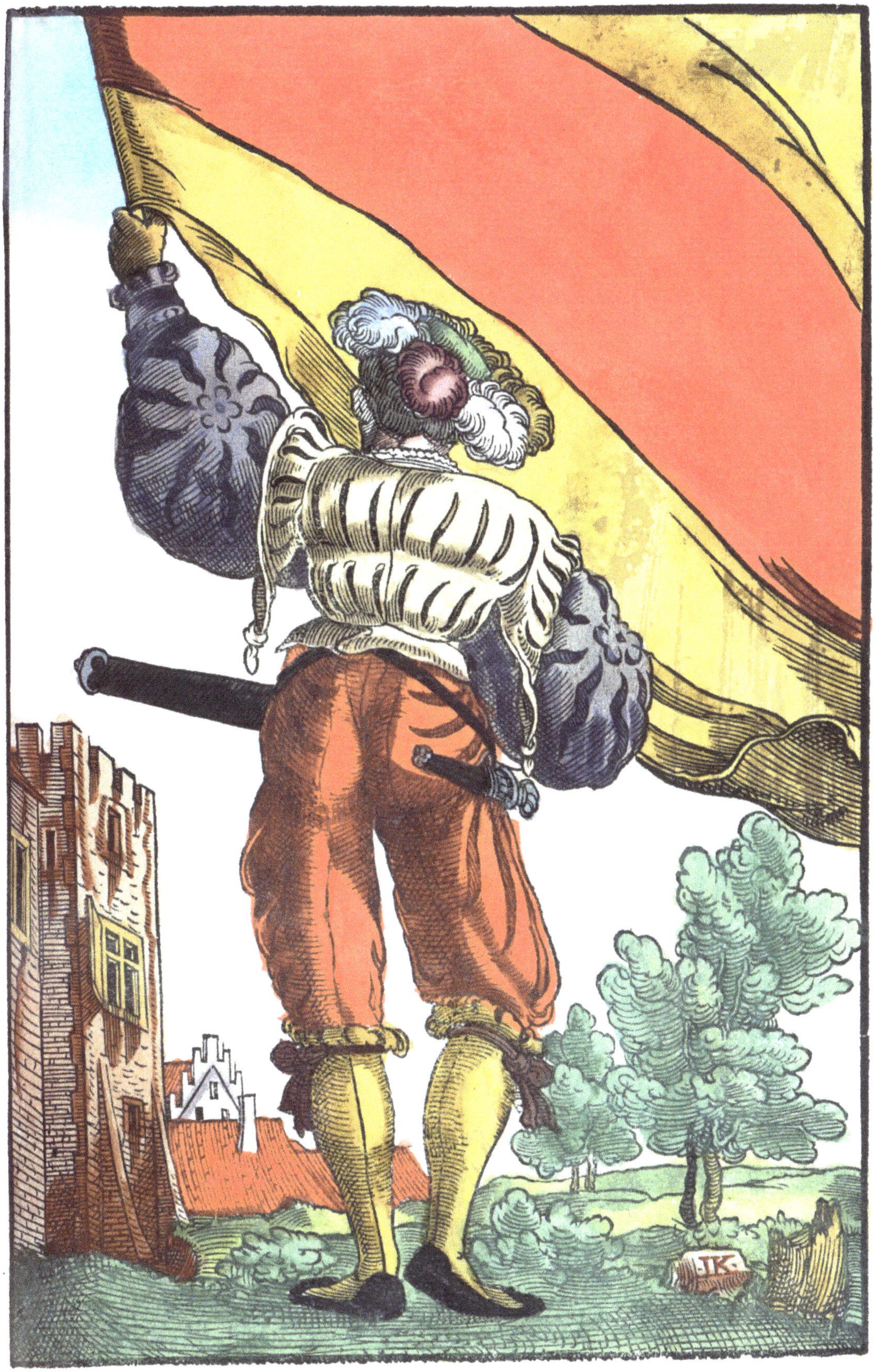

121. Wiensshheym | Windsheim

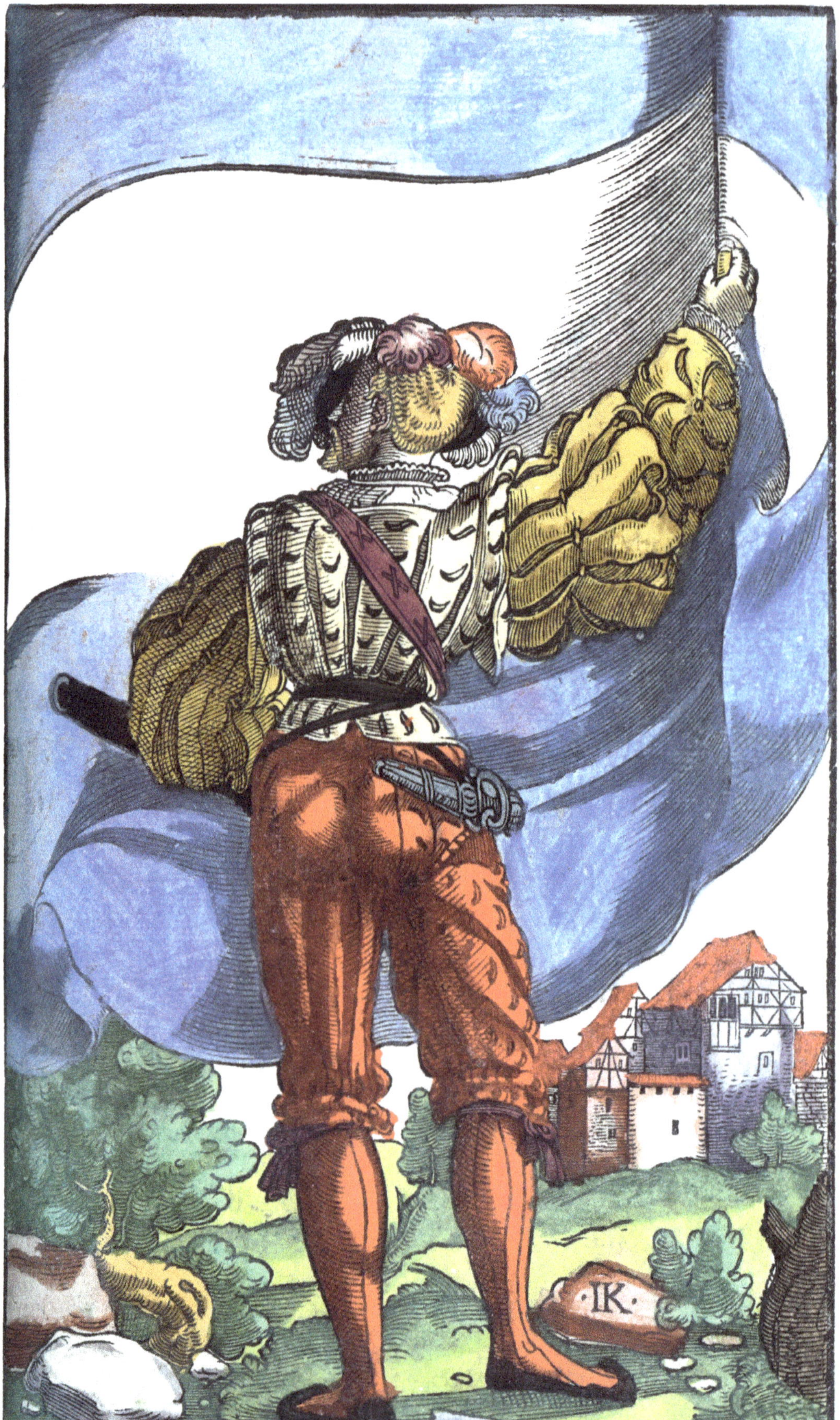

122. Unassigned Flag

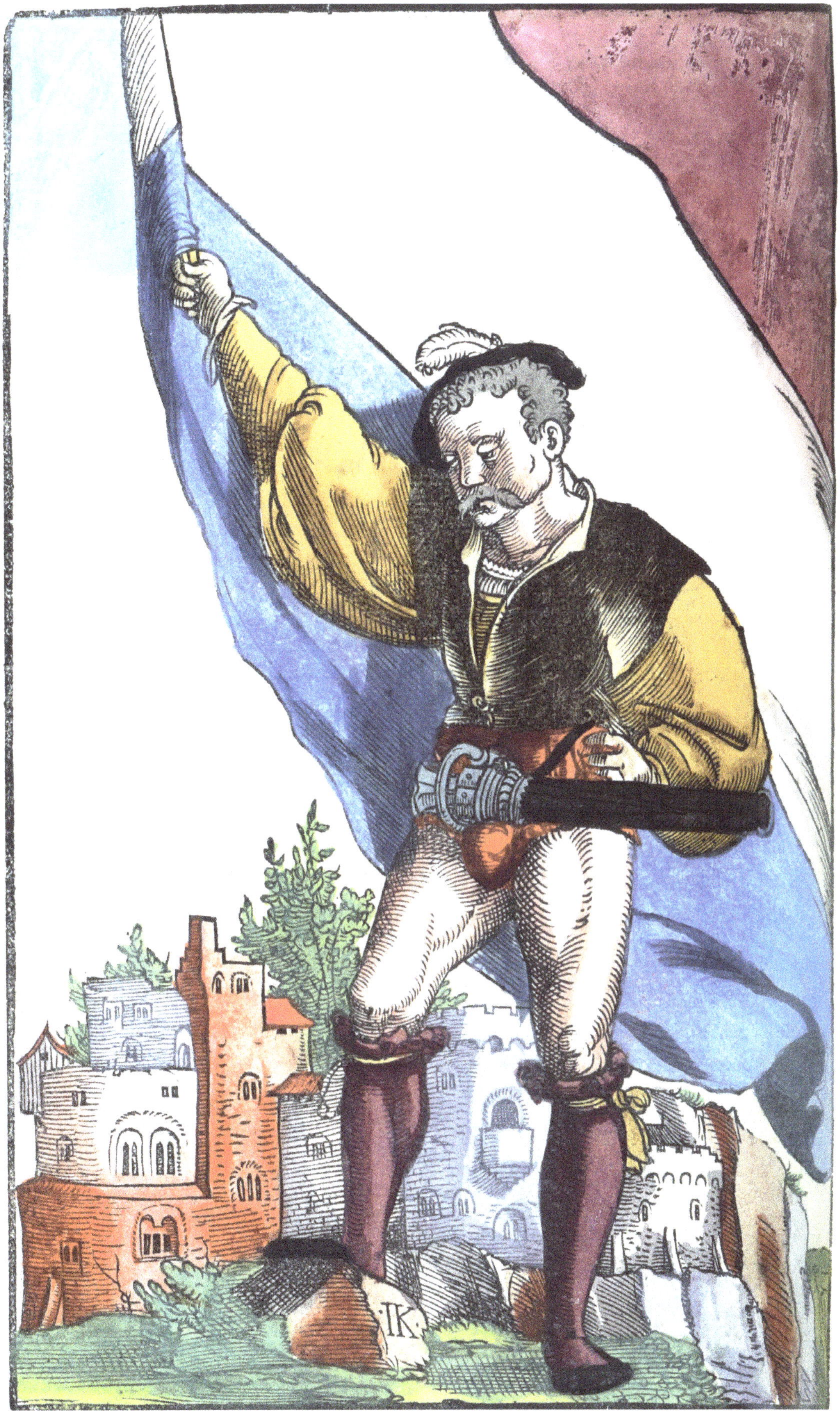

123. Unassigned Flag

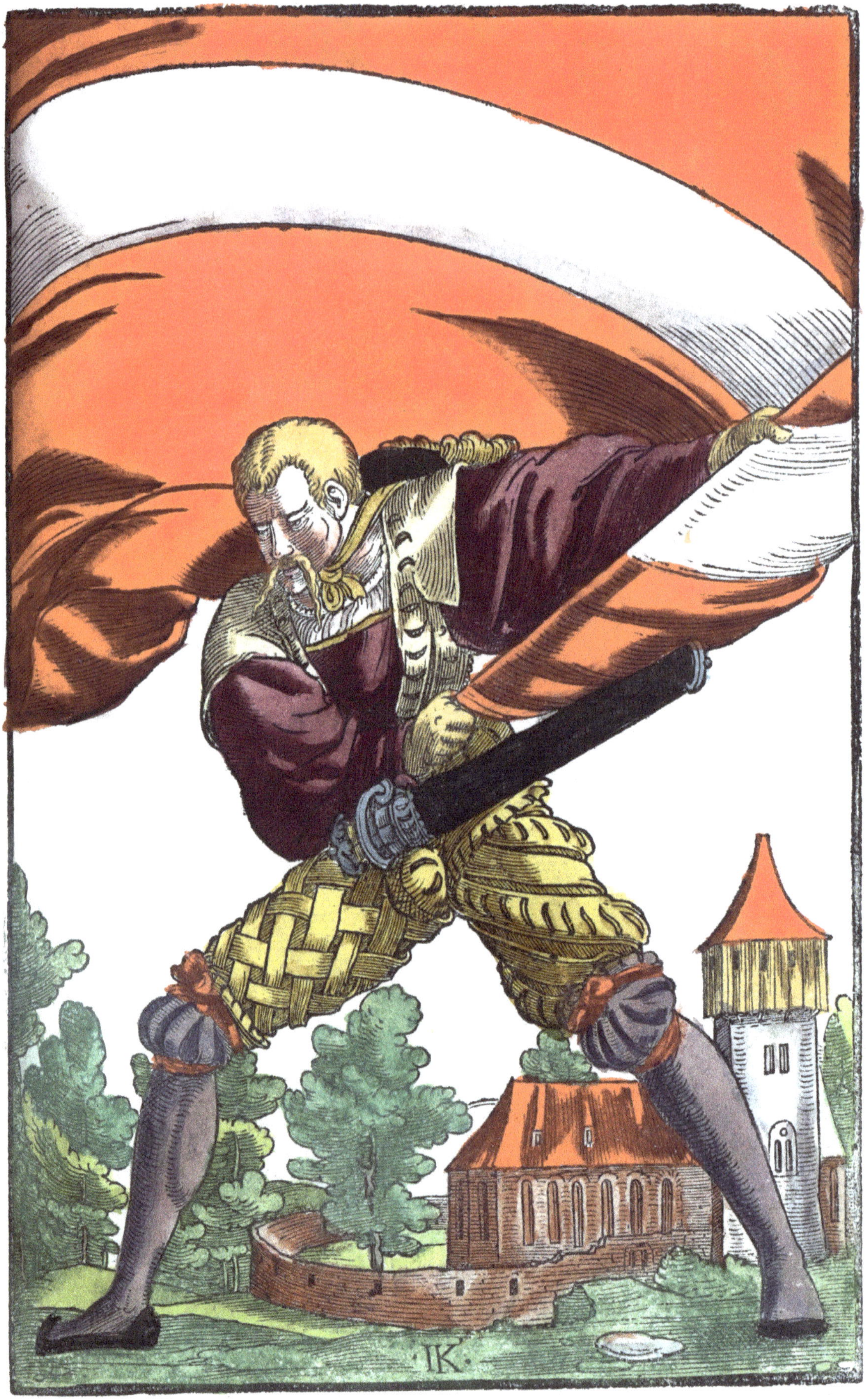

124. Unassigned Flag

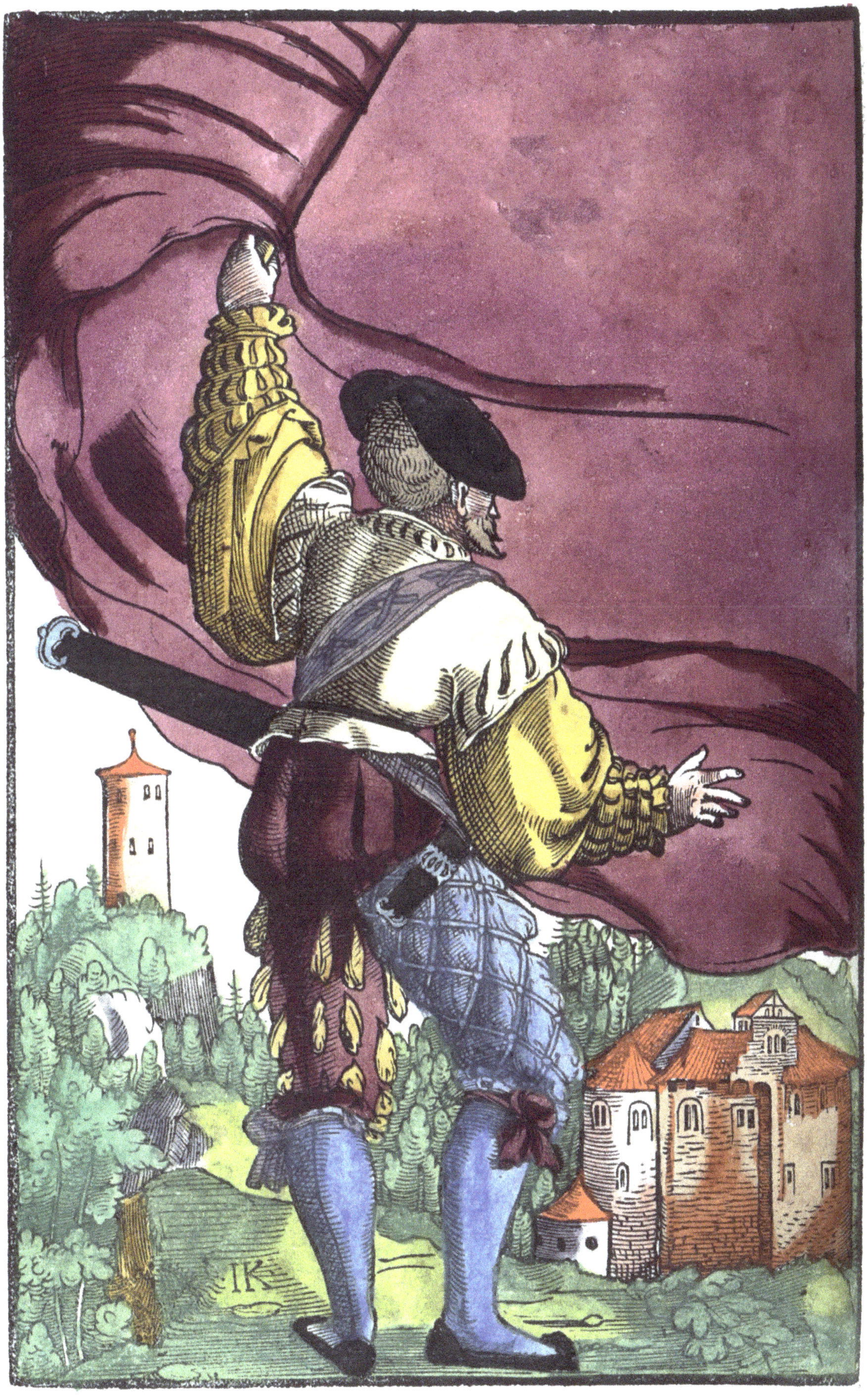

125. Unassigned Flag

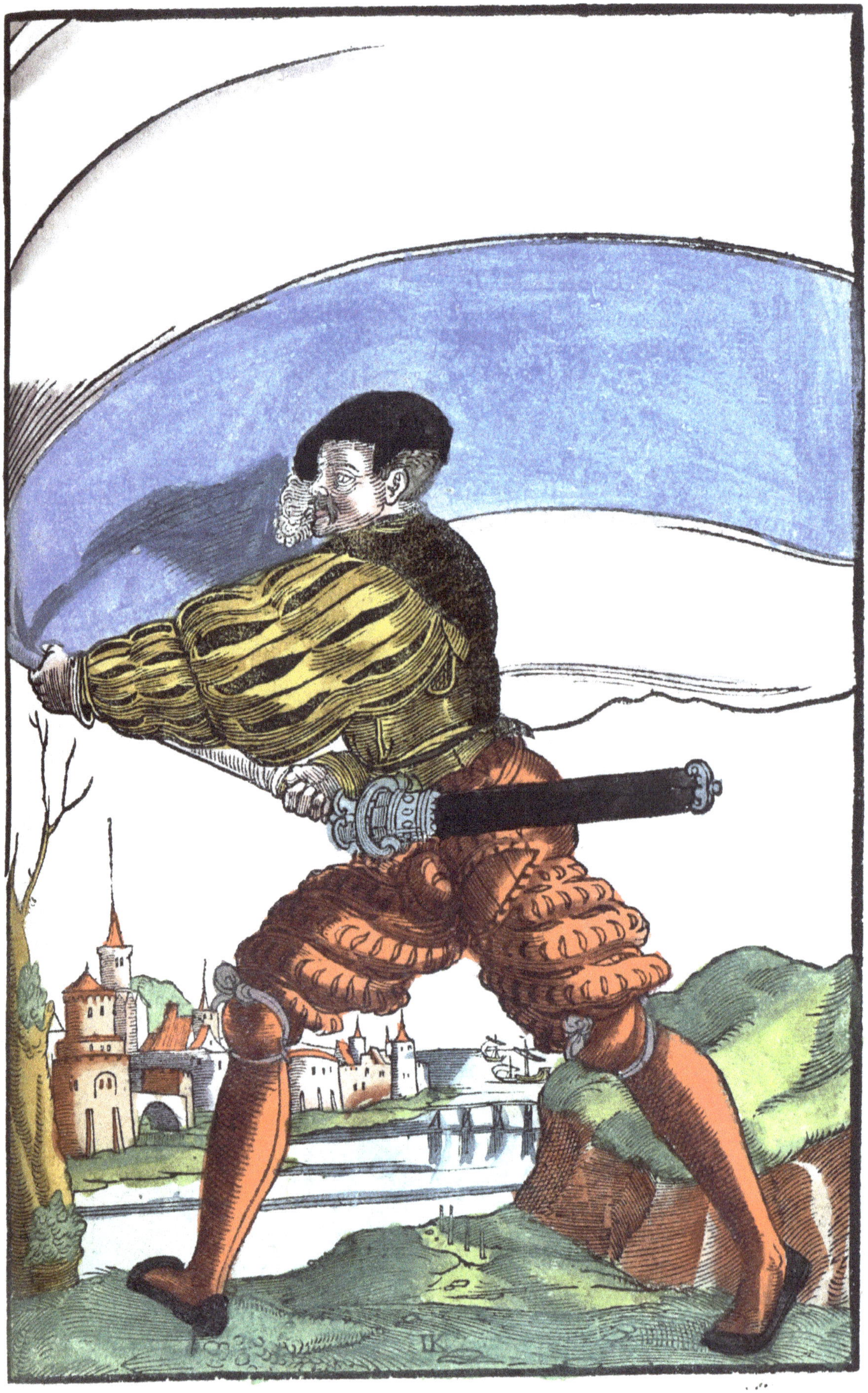

126. Unassigned Flag

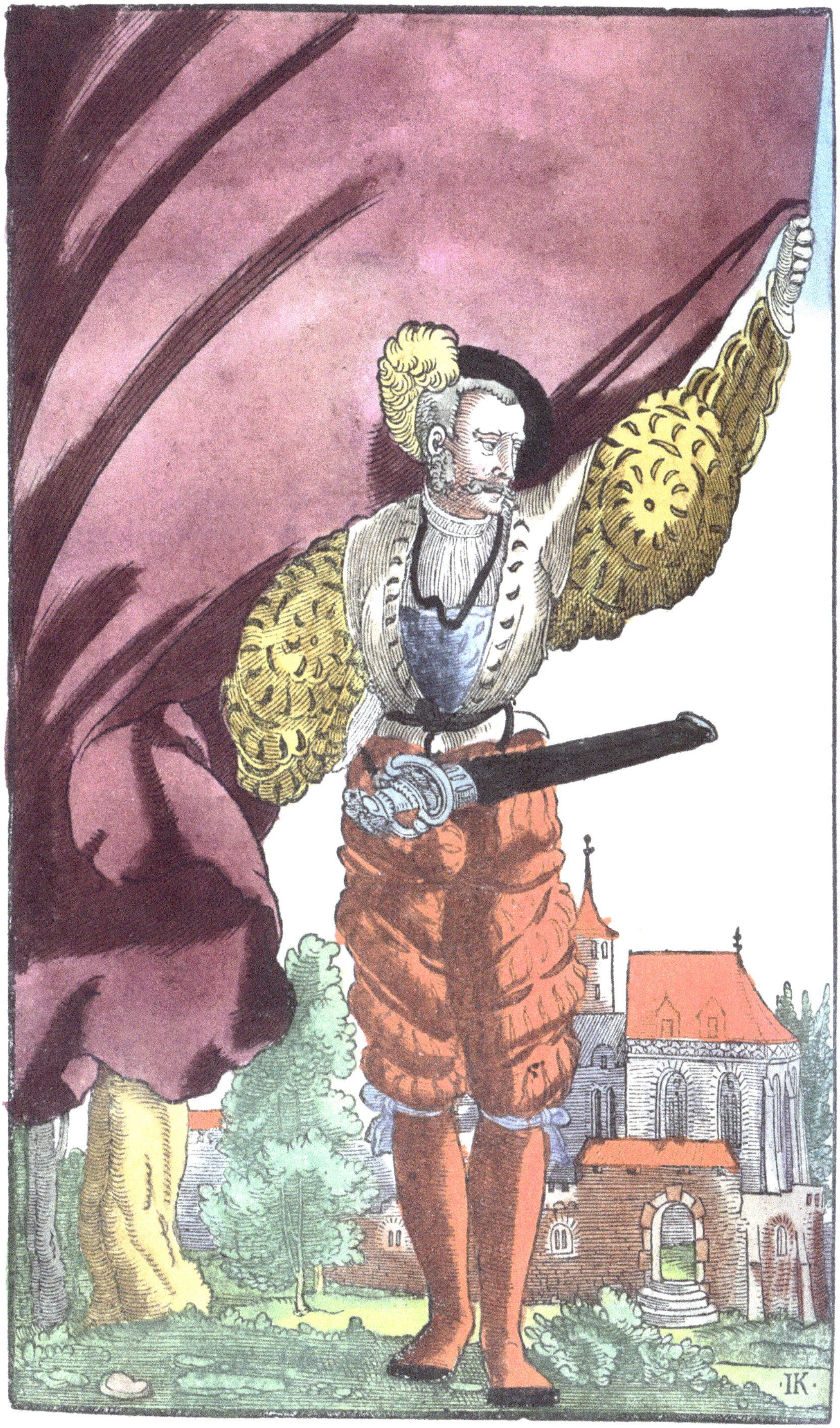

127. Unassigned Flag

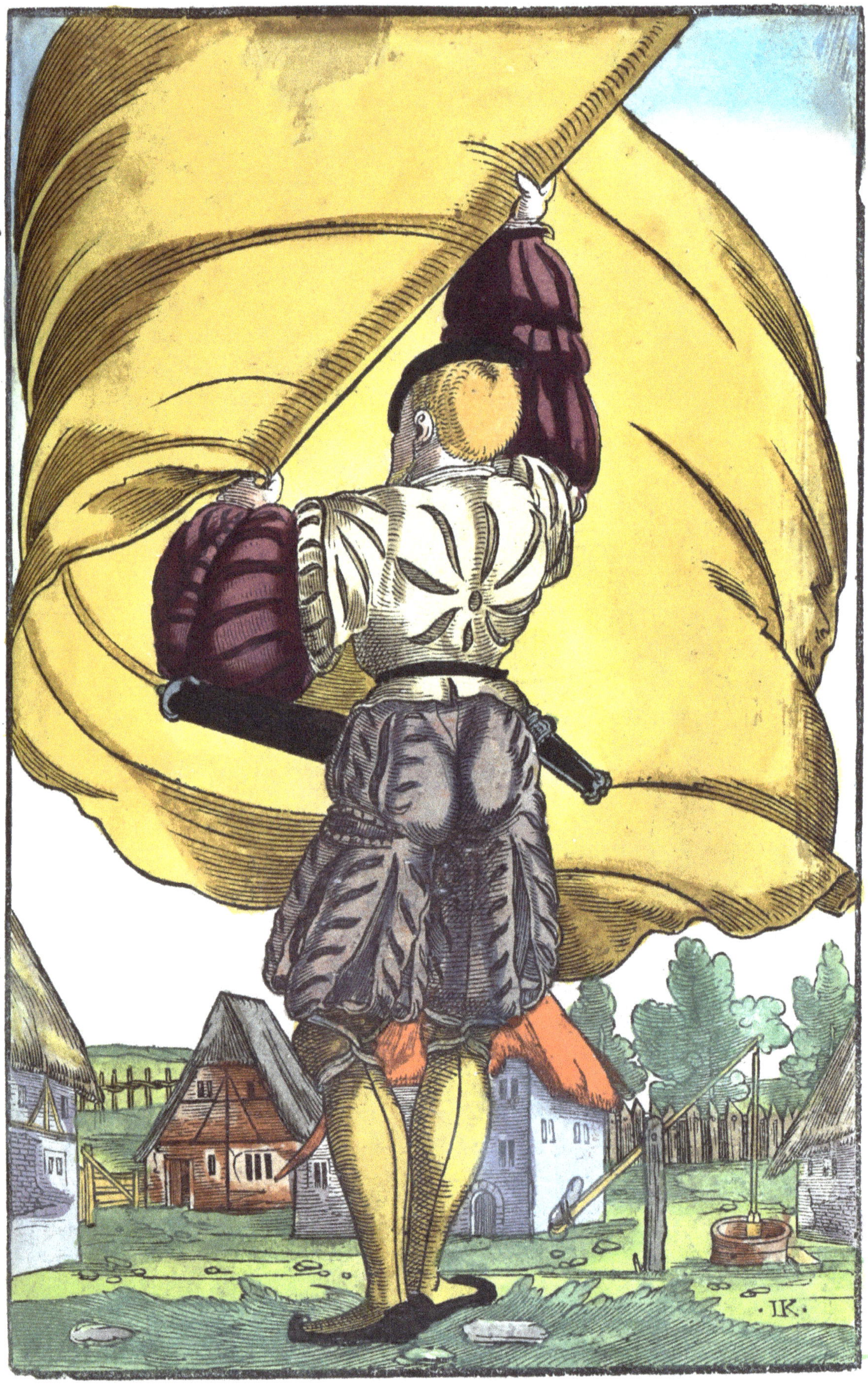

128. Unassigned Flag

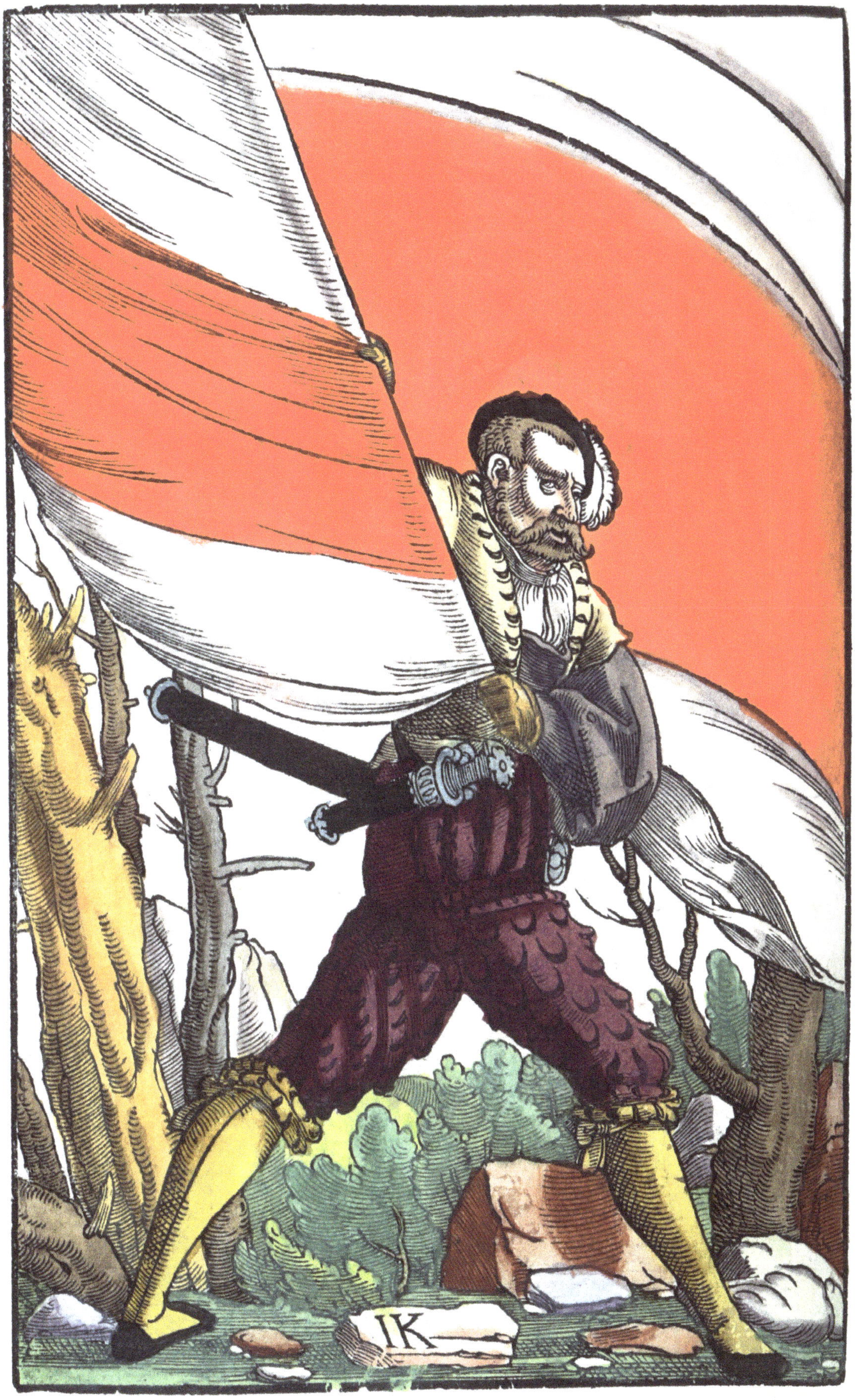

129. Unassigned Flag

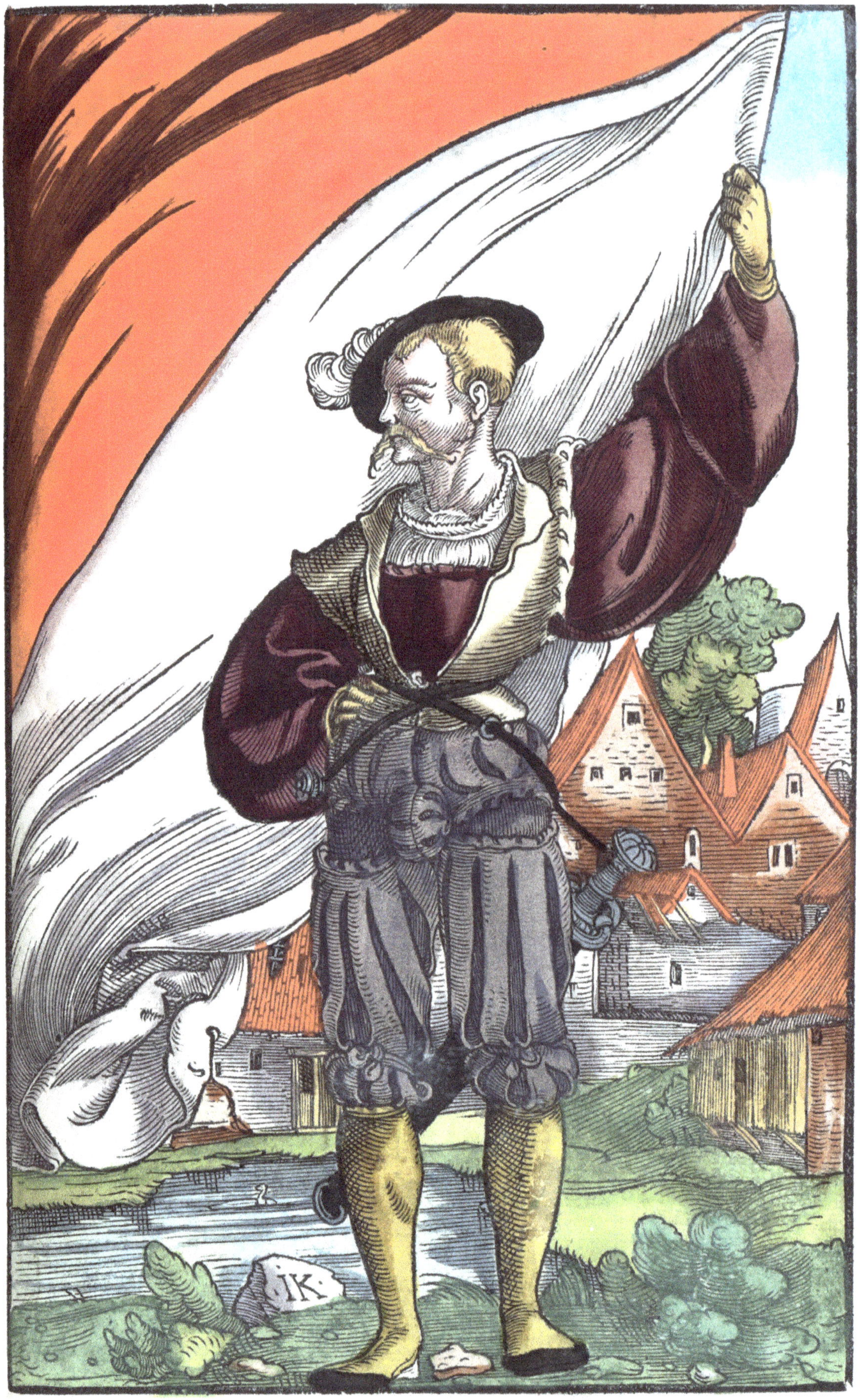

130. Unassigned Flag

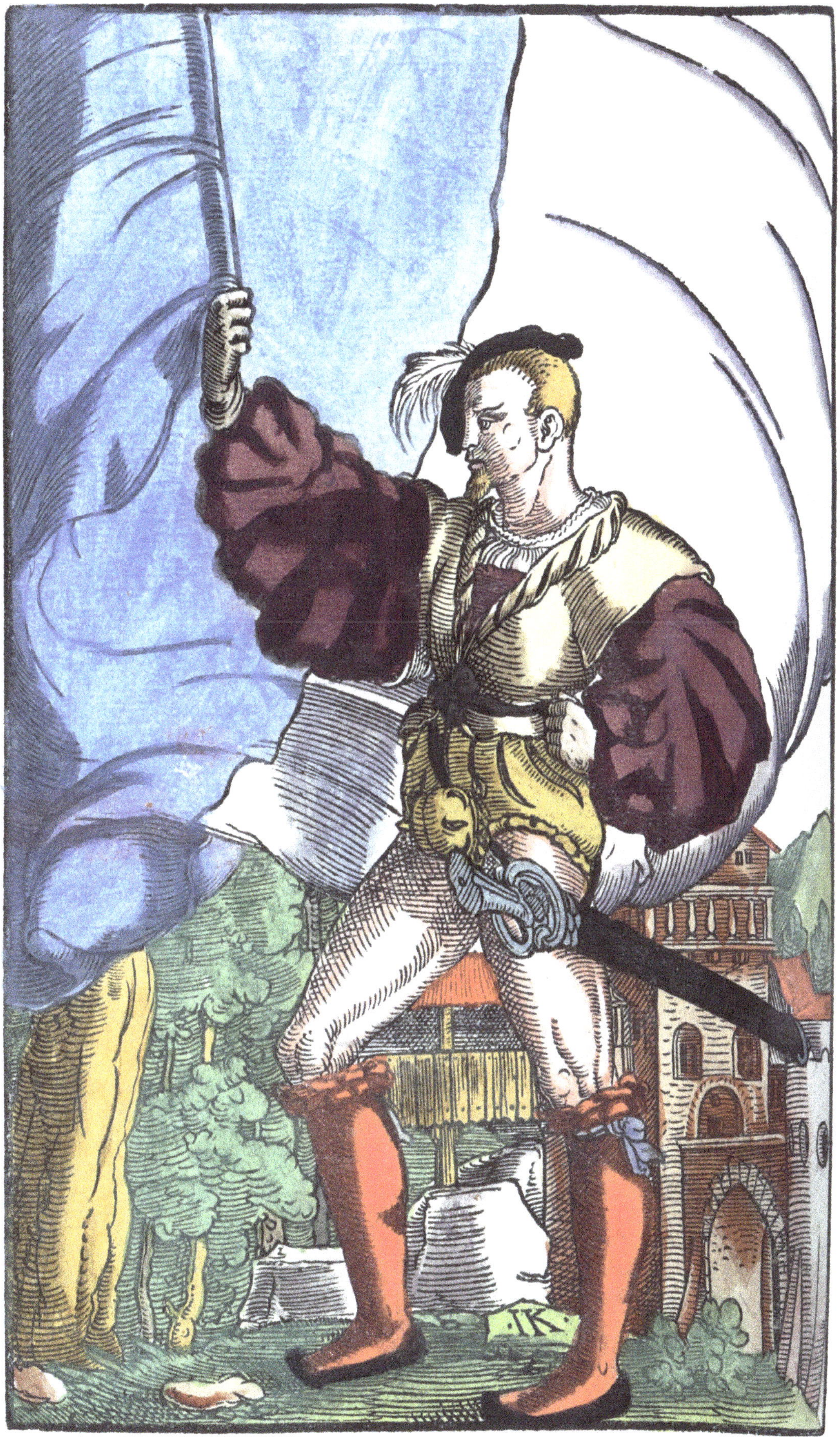

131. Unassigned Flag

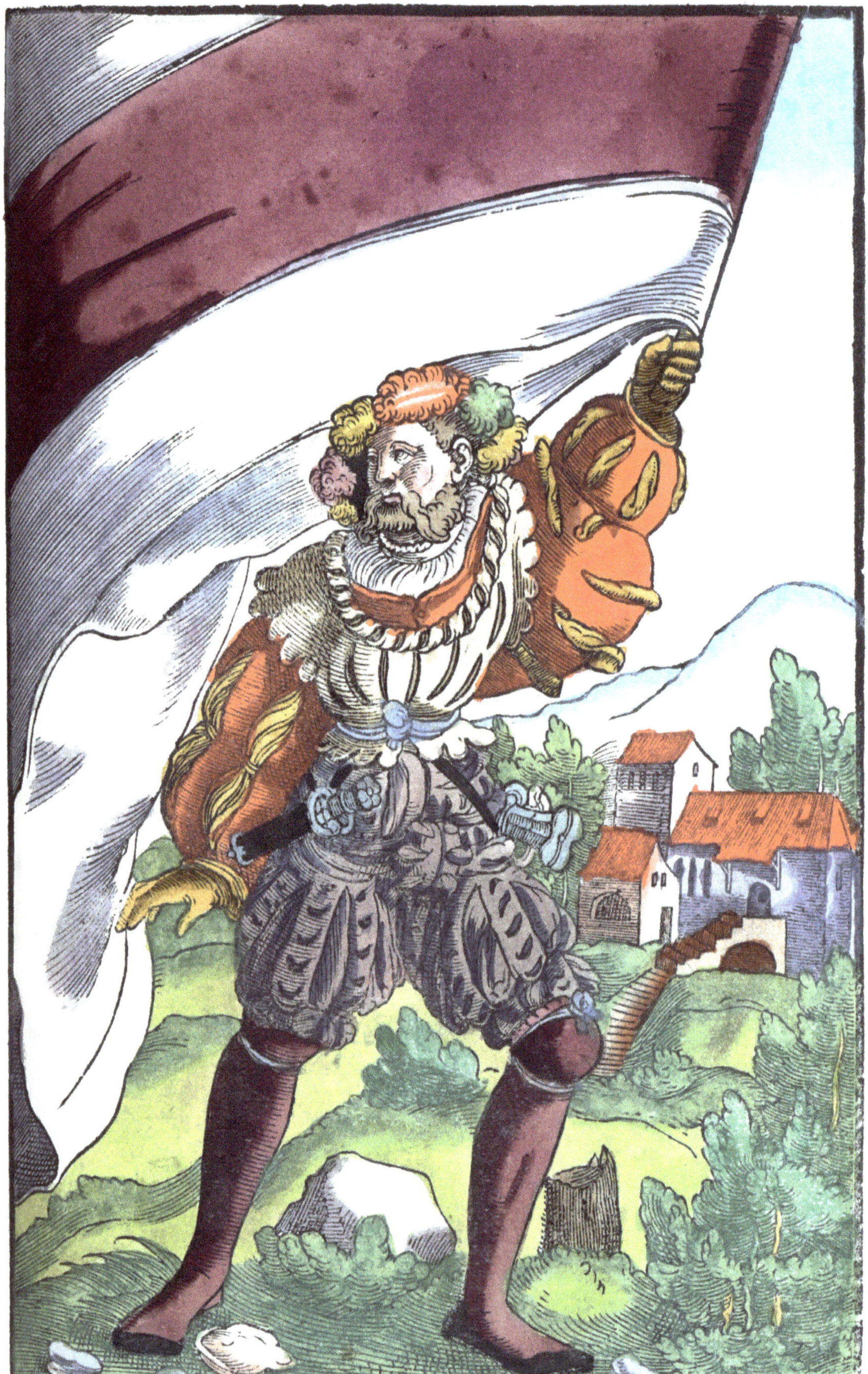

132. Unassigned Flag

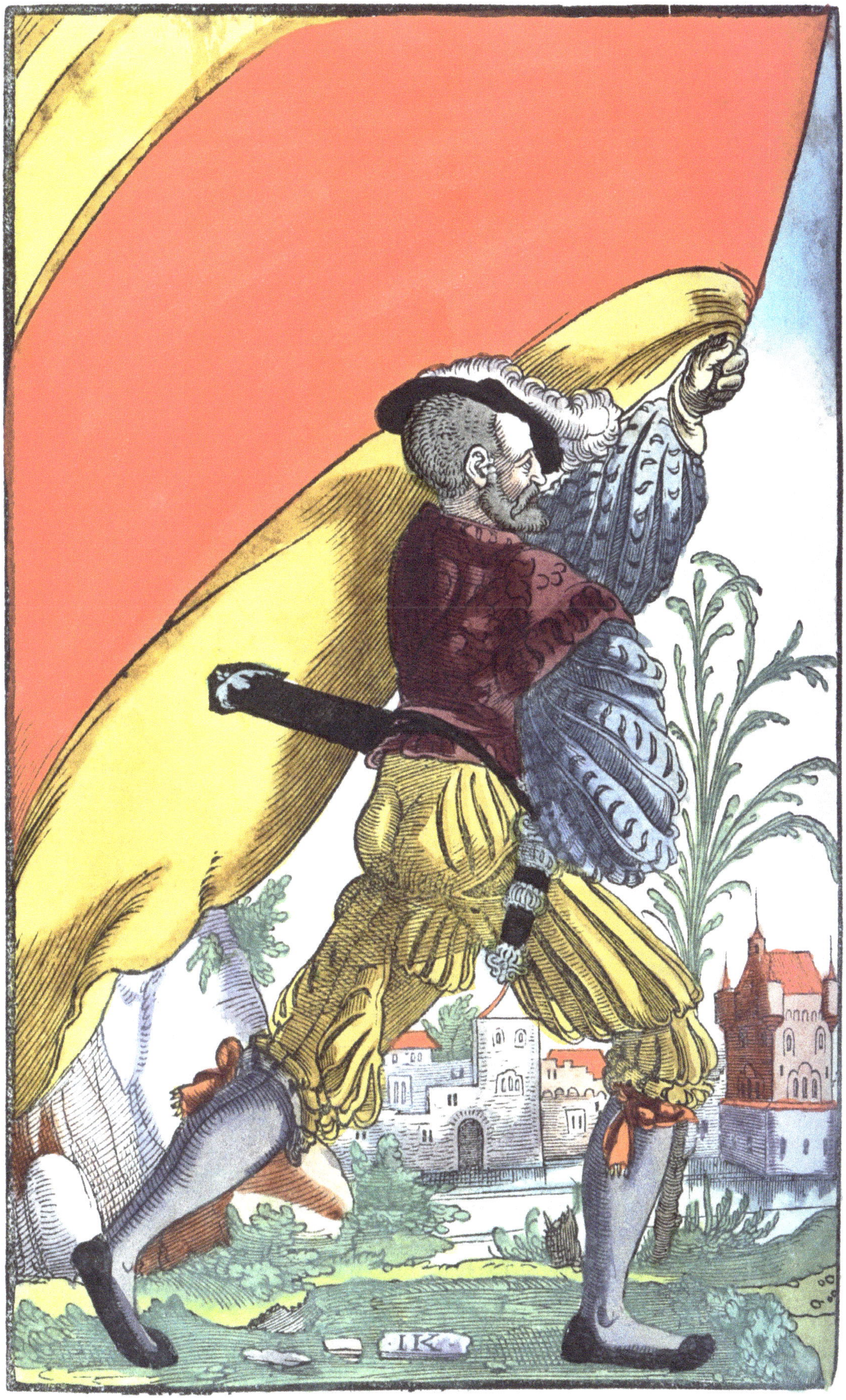

133. Unassigned Flag

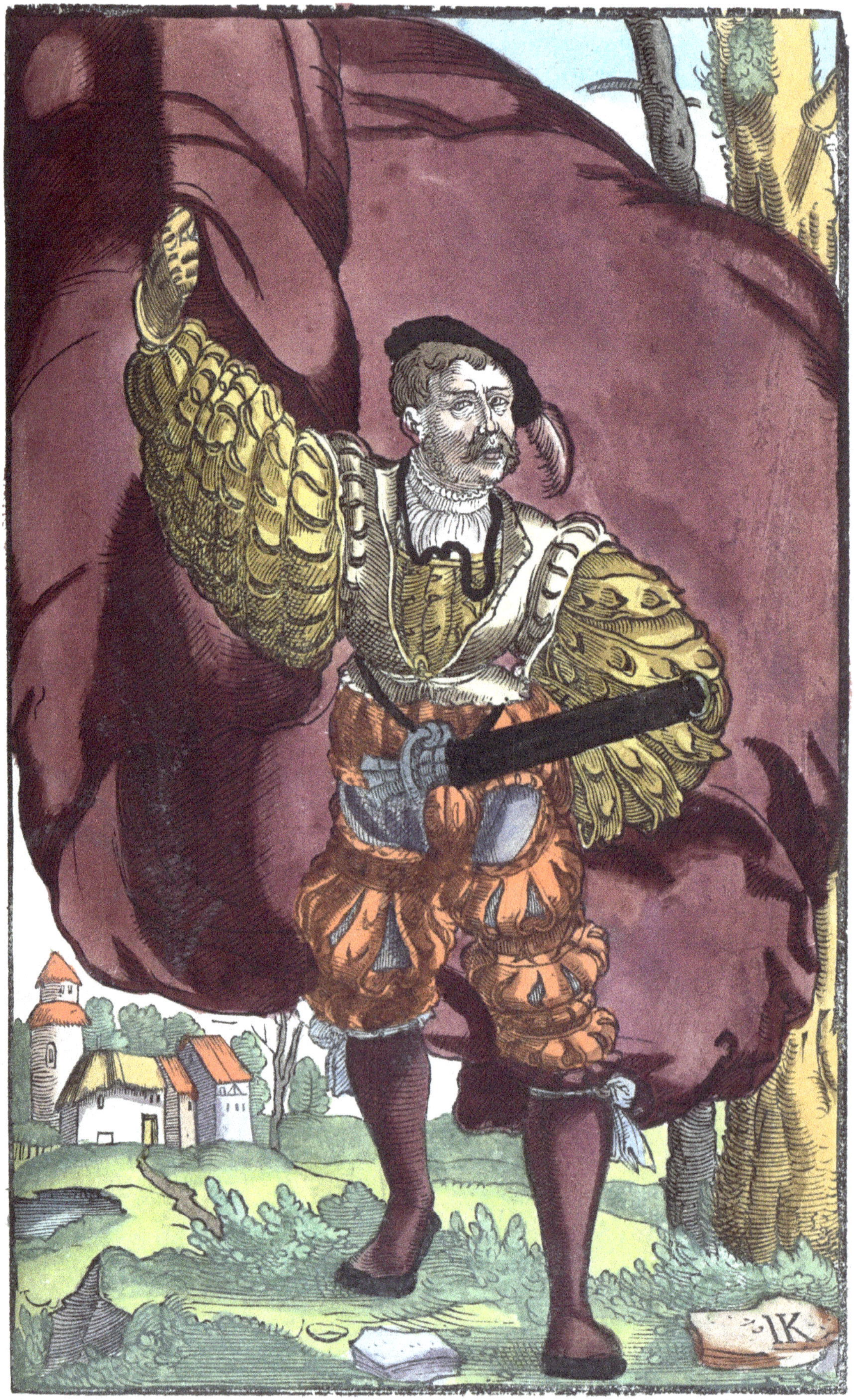

134. Unassigned Flag

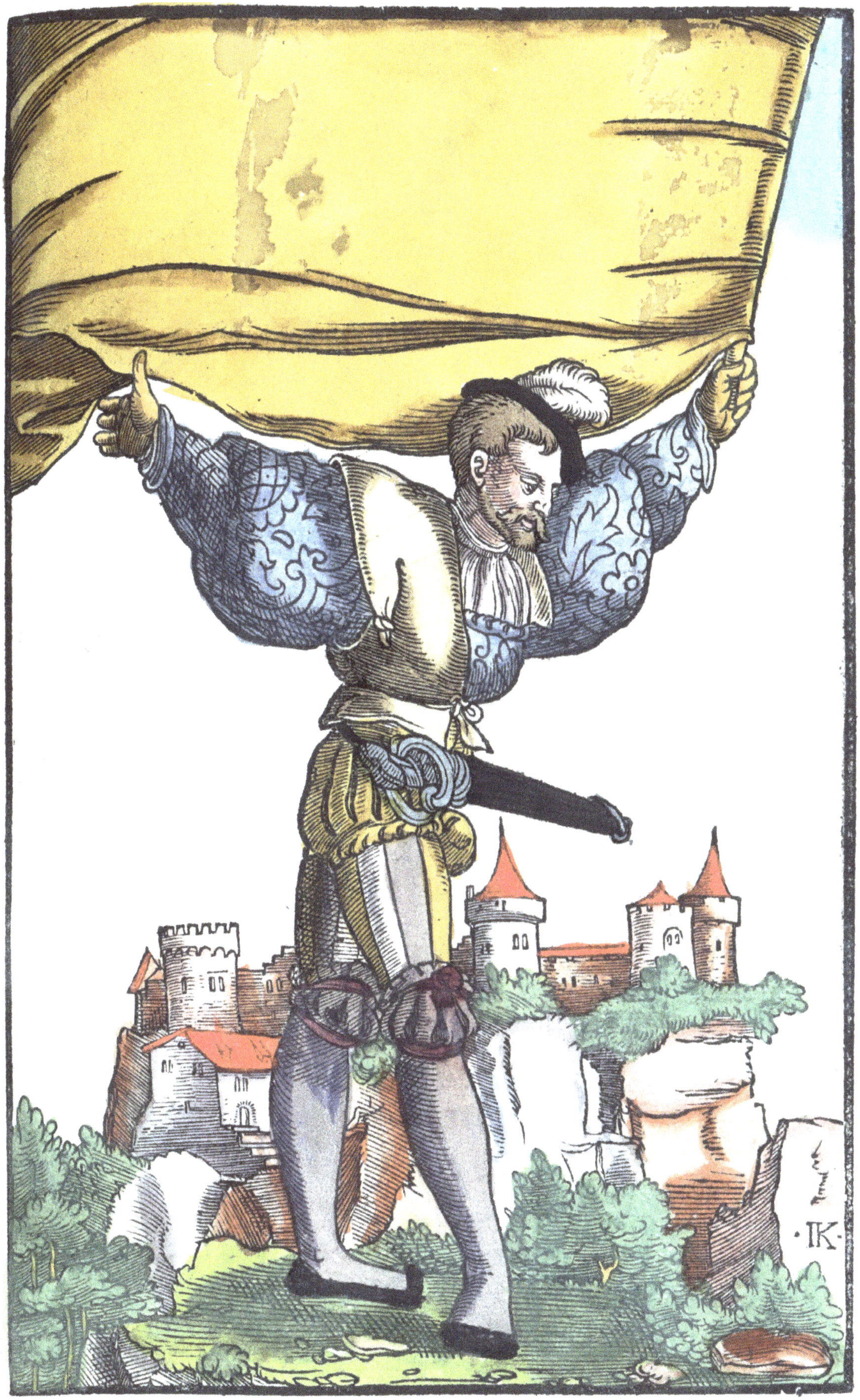

135. Unassigned Flag

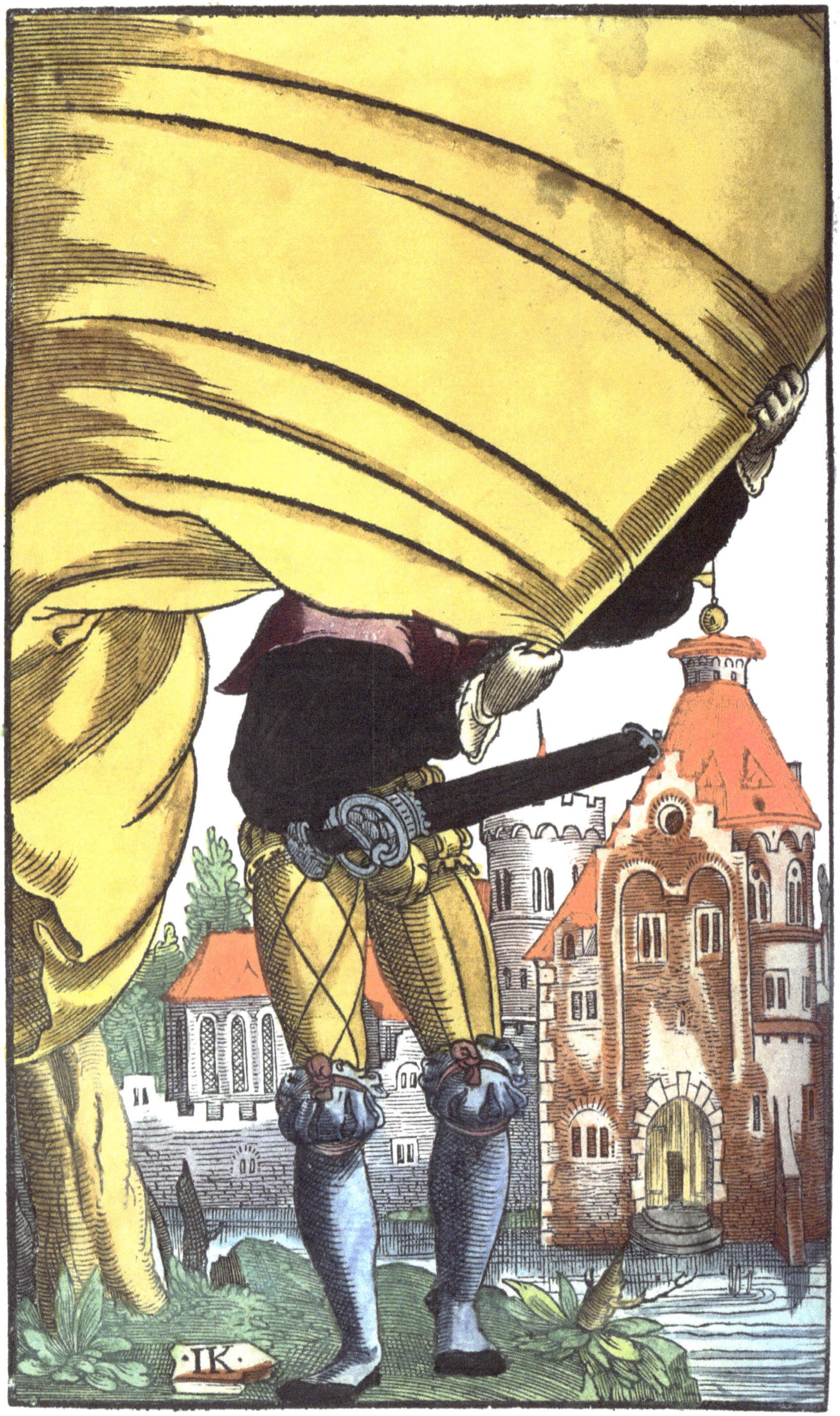

136. Unassigned Flag

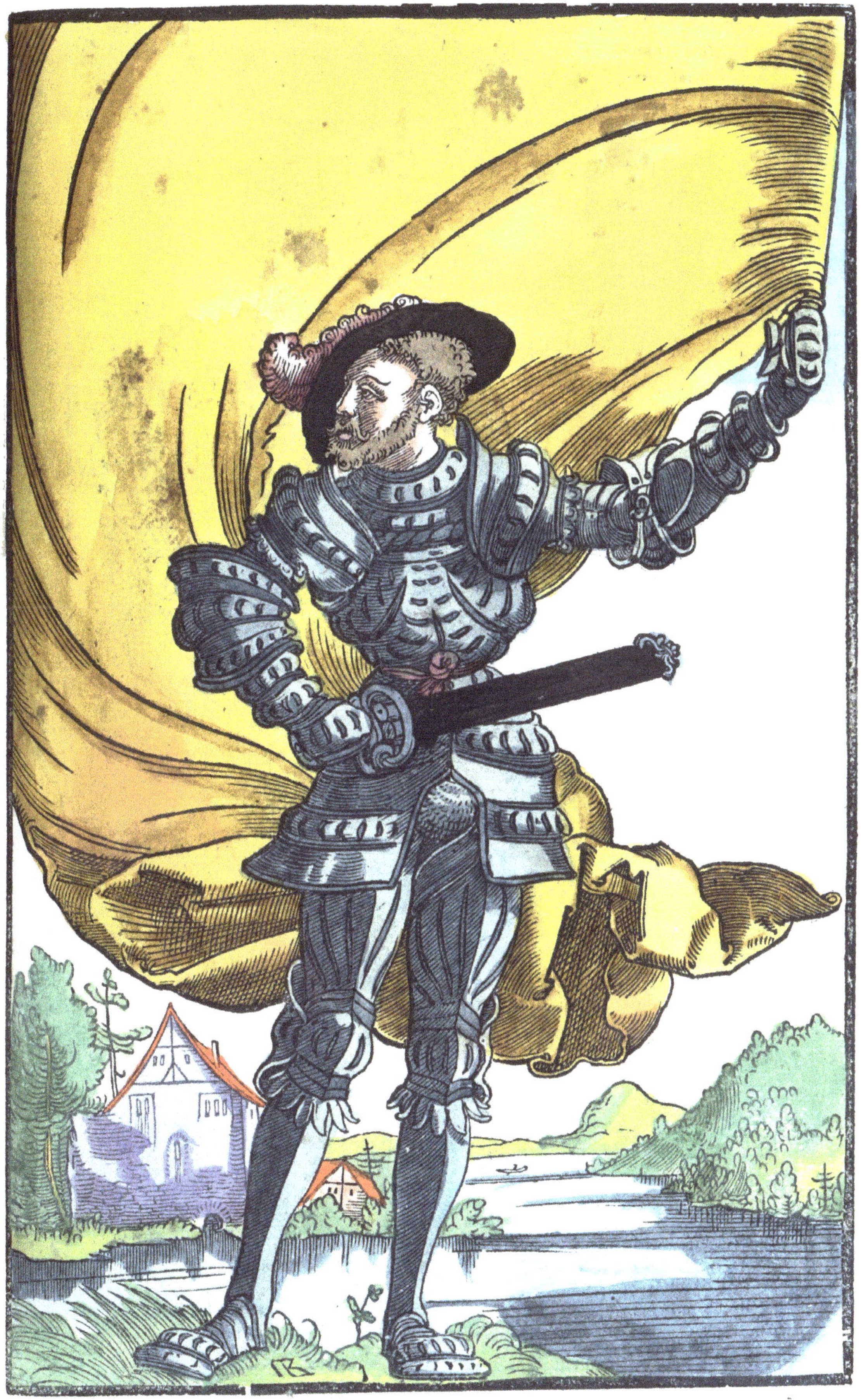

137. Unassigned Flag

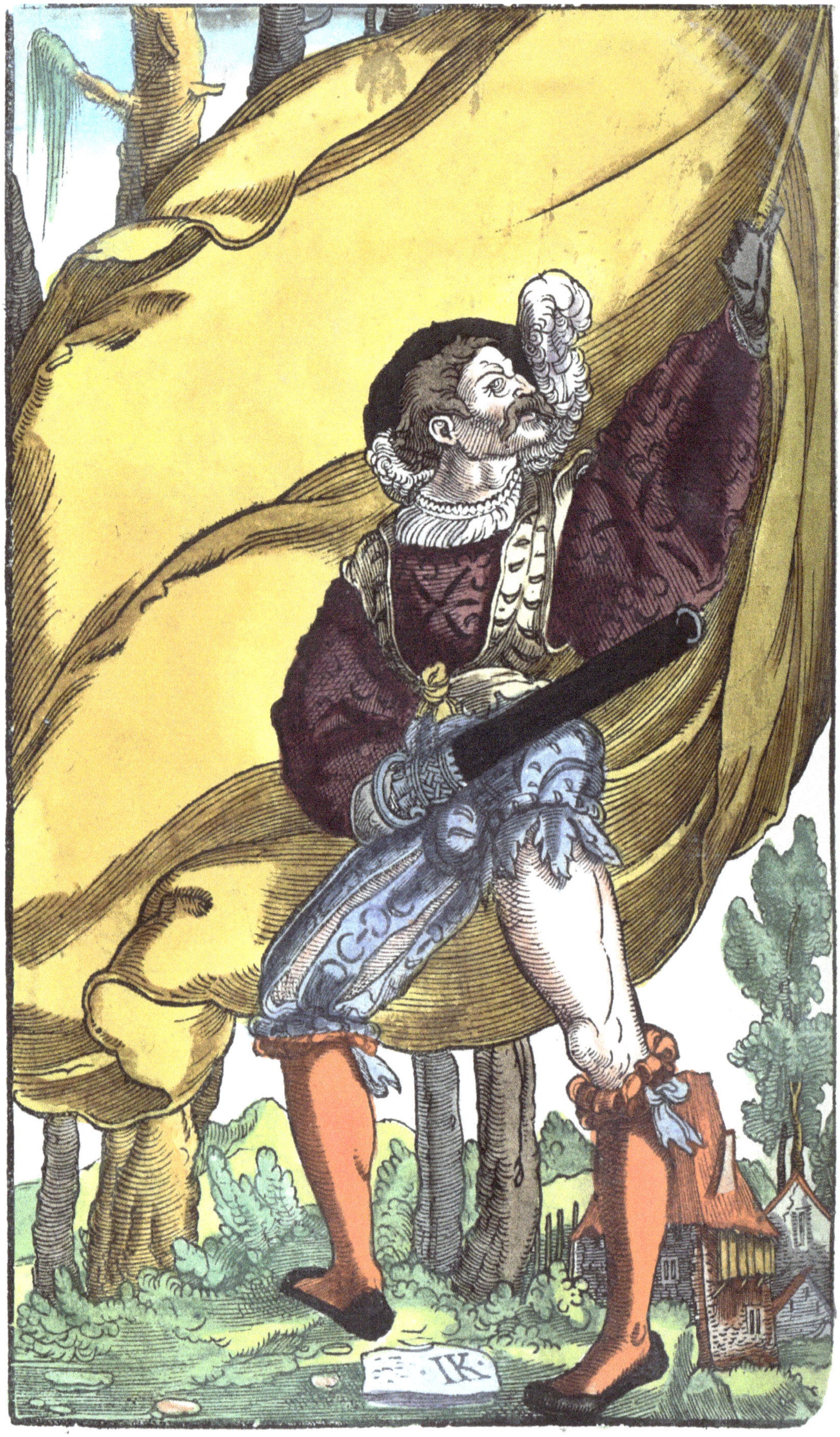

138. Unassigned Flag

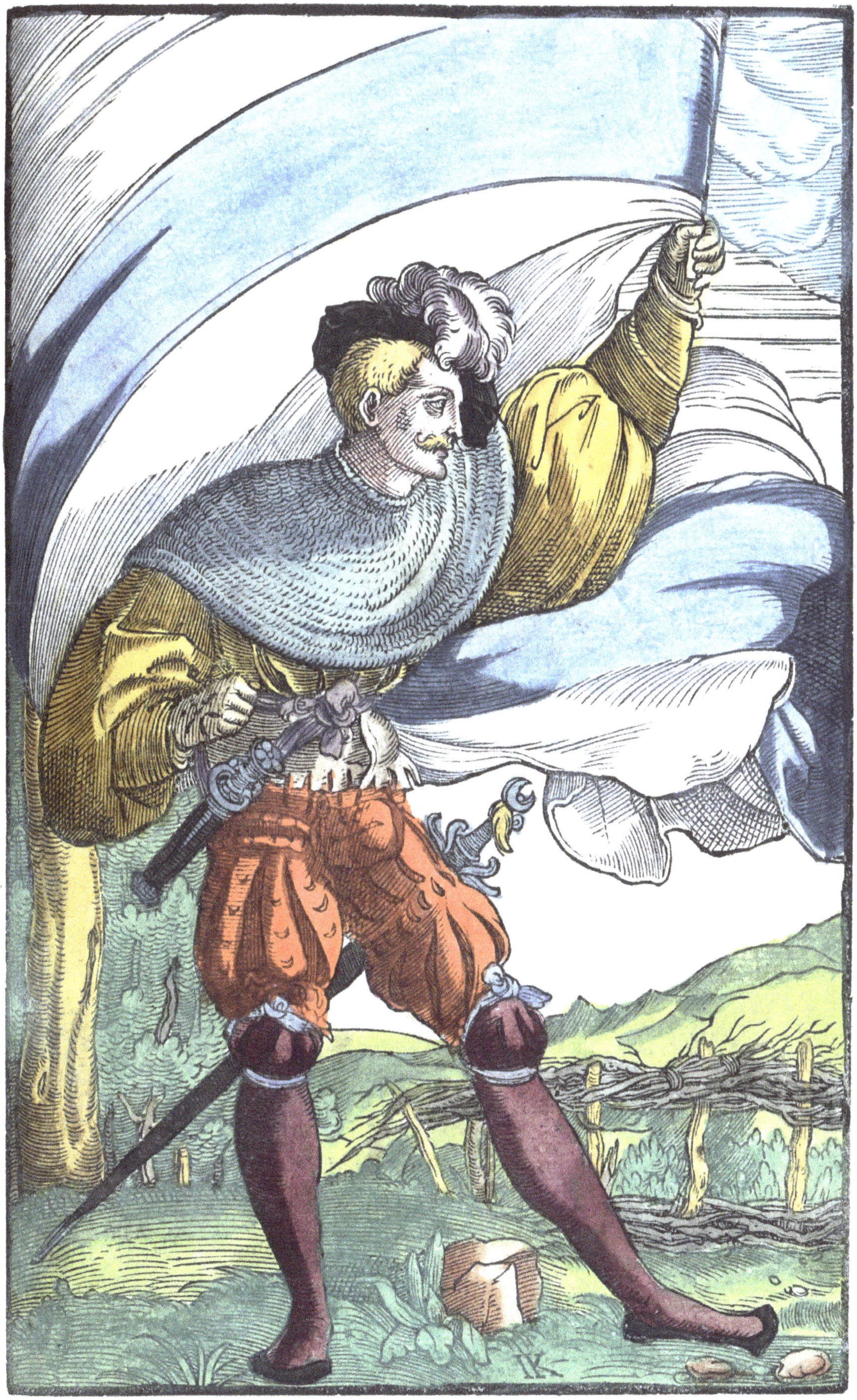

139. Unassigned Flag

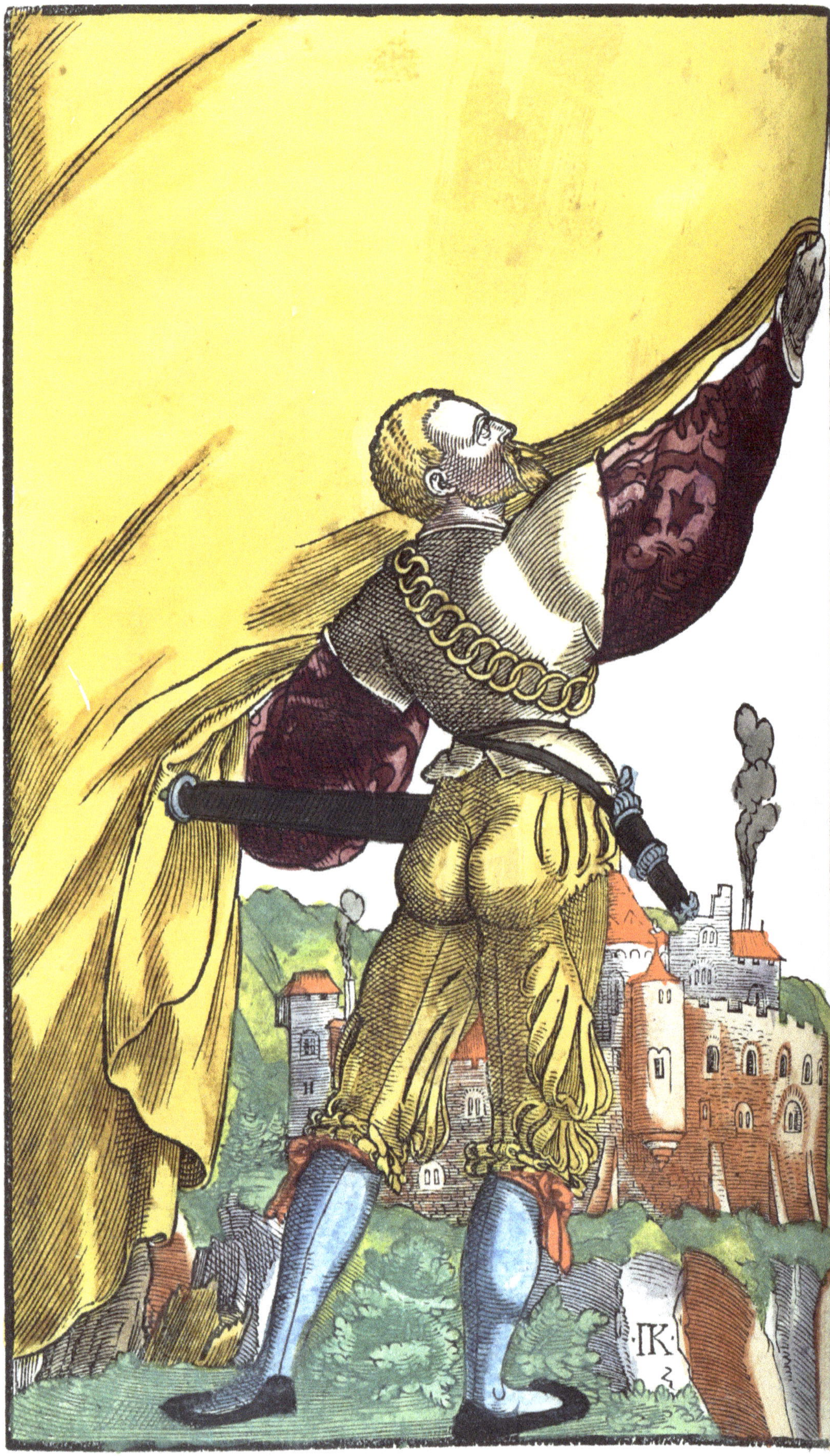

140. Unassigned Flag

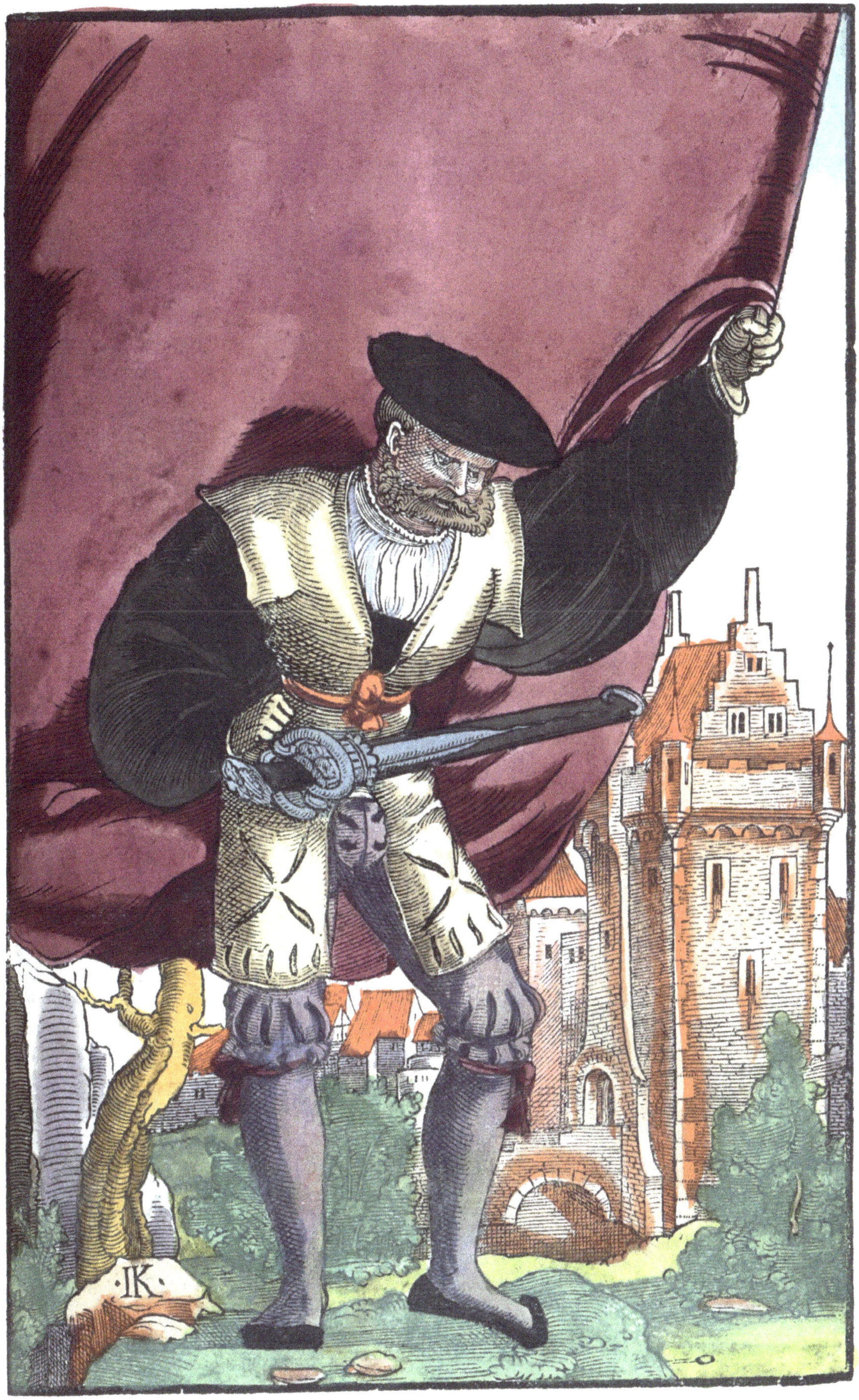

141. Unassigned Flag

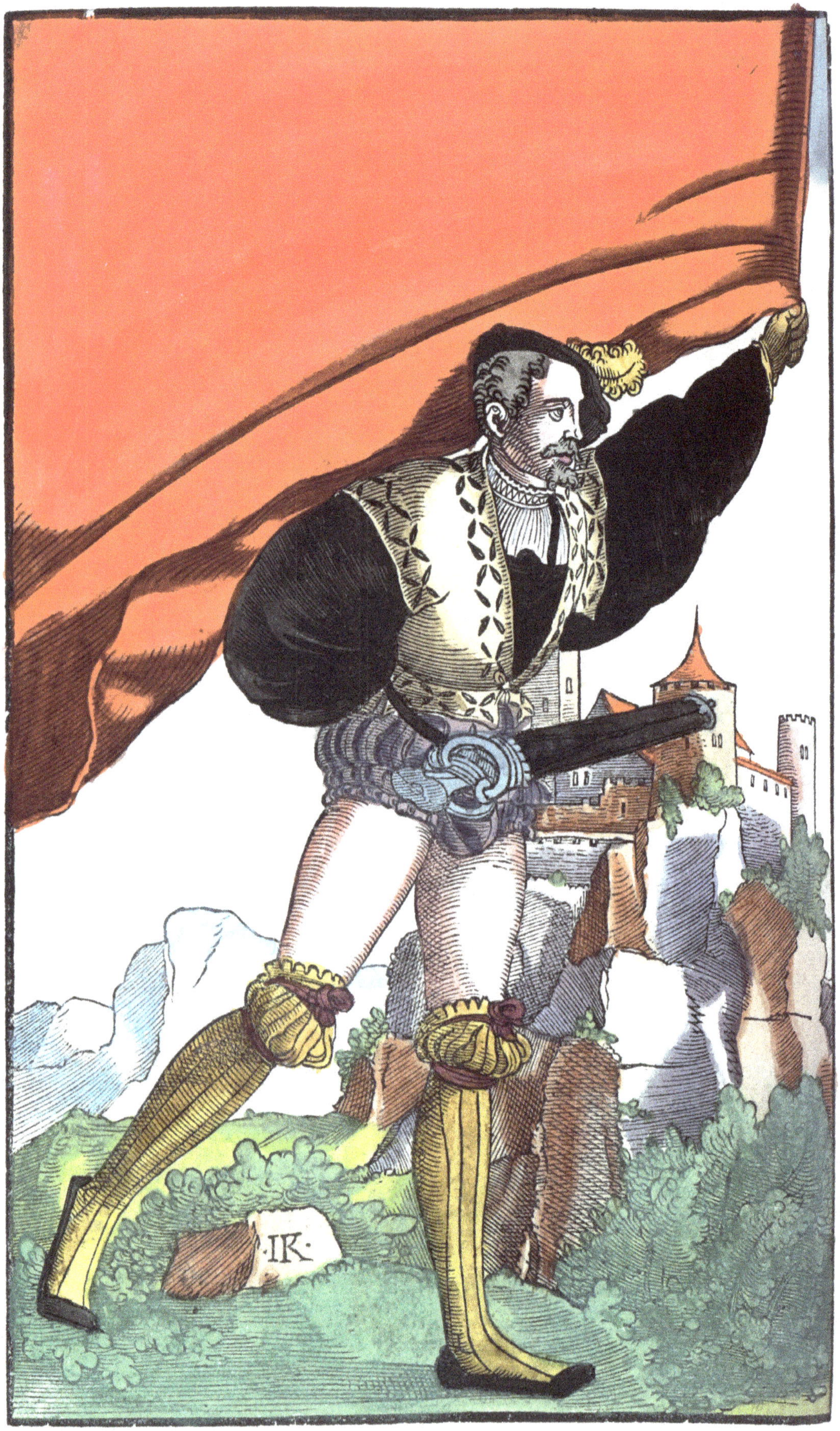

142. Unassigned Flag

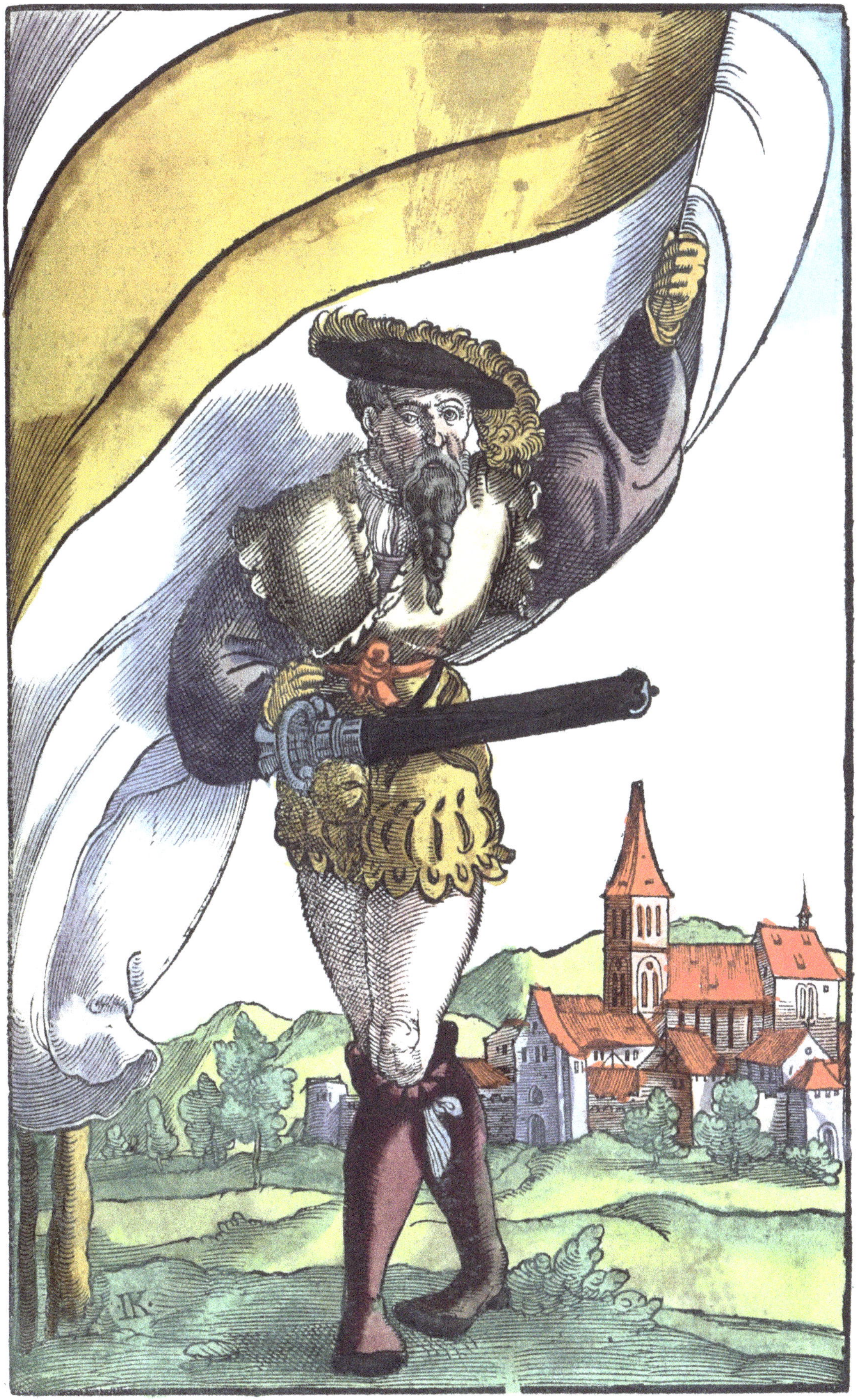

143. Unassigned Flag

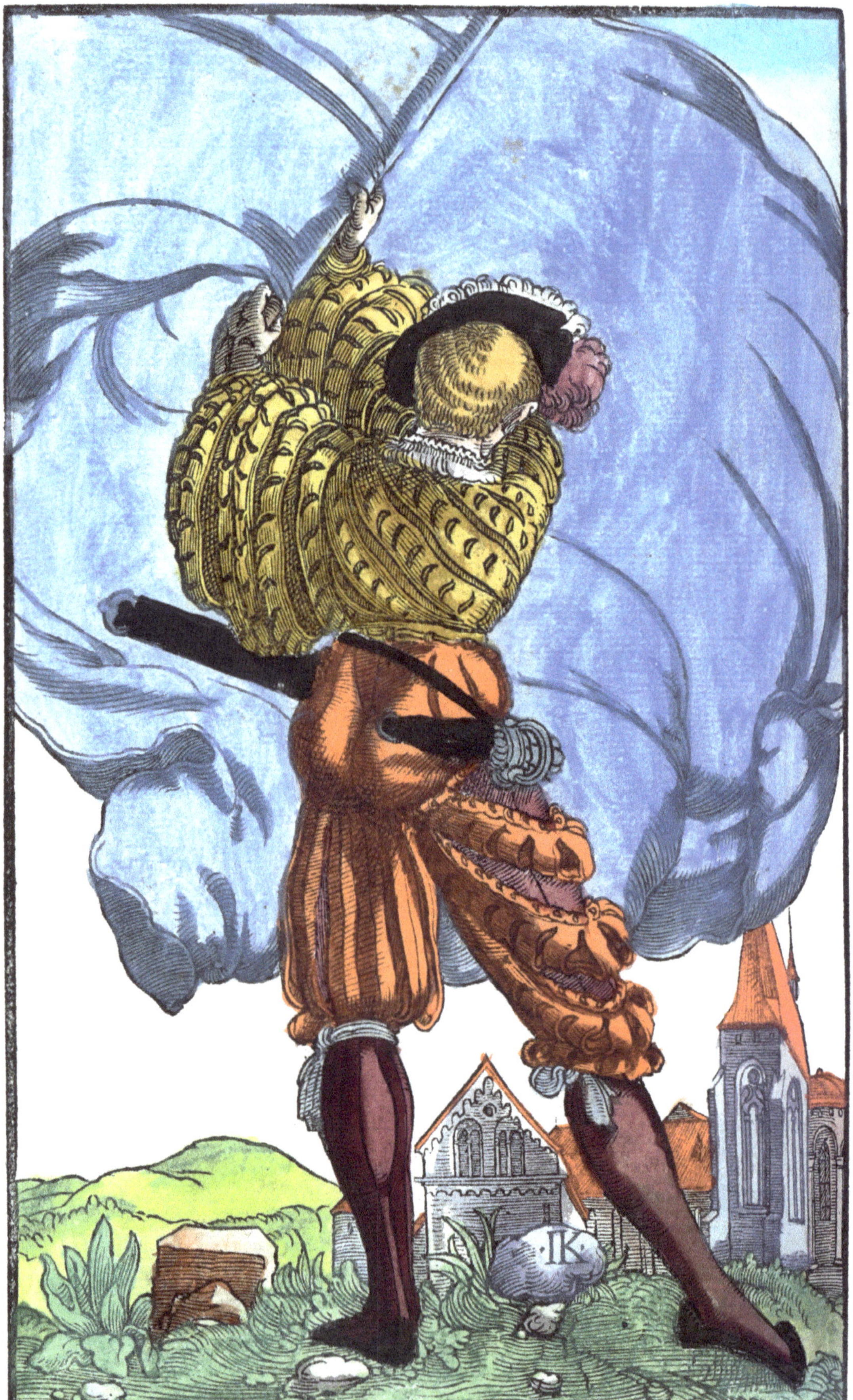

144. Unassigned Flag

Index

www.ingramcontent.com/pod-product-compliance
Lightning Source LLC
LaVergne TN
LVHW070854210926
848482LV00016B/929
* 9 7 8 0 9 9 8 5 9 7 7 7 5 *